*A Functional Biology
of Parasitism*

Functional Biology Series
Series Editor: Peter Calow, Department of Zoology,
University of Sheffield

A Functional Biology of Free-living Protozoa *
Johanna Laybourn-Parry
A Functional Biology of Sticklebacks *
R.J. Wootton
A Functional Biology of Marine Gastropods †
Roger N. Hughes
A Functional Biology of Nematodes †
David A. Wharton
A Functional Biology of Crop Plants #
Vincent P. Gutschick
A Functional Biology of Echinoderms †
John M. Lawrence
A Functional Biology of Clonal Animals
Roger N. Hughes
A Functional Biology of Sea Anemones
J. Malcolm Shick
A Functional Biology of Parasitism
Gerald W. Esch and Jacqueline C. Fernández

Available in the USA from
* University of California Press
† The Johns Hopkins University Press
# Timber Press

A Functional Biology of Parasitism

Ecological and evolutionary implications

Gerald W. Esch and
Jacqueline C. Fernández

Department of Biology
Wake Forest University
Winston-Salem
North Carolina, USA

 SPRINGER-SCIENCE+BUSINESS MEDIA, B.V.

First edition 1993

© 1993 Gerald W. Esch and Jacqueline C. Fernández
Originally published by Chapman & Hall in 1993

Typeset in M Plantin 10/12 pt by Expo Holdings Sdn Bhd, Malaysia

ISBN 978-94-010-5039-5   ISBN 978-94-011-2352-5 (eBook)

DOI 10.1007/978-94-011-2352-5

A catalogue record for this book is available from the British Library

Library of Congress Cataloging-in-Publication data available

Printed on permanent acid-free text paper, manufactured in accordance with the proposed
ANSI/NISO Z 39.48-199X and ANSI Z 39.48-1984

To Ann and Steve, and our parents.

Contents

Functional Biology Series: Foreword

Series Editor: Peter Calow, Department of Zoology,
University of Sheffield, England

The main aim of this series will be to illustrate and to explain the way organisms 'make a living' in nature. At the heart of this — their *functional biology* — is the way organisms acquire and then make use of resources in metabolism, movement, growth, reproduction, and so on. These processes will form the fundamental framework of all the books in the series. Each book will concentrate on a particular taxon (species, family, class or even phylum) and will bring together information on the form, physiology, ecology and evolutionary biology of the group. The aim will be not only to describe *how* organisms work, but also to consider *why* they have come to work in that way. By concentration on taxa which are well known, it is hoped that the series will not only illustrate the success of selection, but also show the constraints imposed upon it by the physiological, morphological and developmental limitations of the groups.

Another important feature of the series will be its *organismic orientation*. Each book will emphasize the importance of functional *integration* in the day-to-day lives and the evolution of organisms. This is crucial since, though it may be true that organisms can be considered as collections of gene-determined traits, they nevertheless interact with their environment as integrated wholes and it is in this context that individual traits have been subjected to natural selection and have evolved.

The key features of the series are, therefore:

1. Its emphasis on whole organisms as integrated, resource-using systems.
2. Its interest in the way selection and constraints have moulded the evolution of adaptations in particular taxonomic groups.
3. Its bringing together of physiological, morphological, ecological and evolutionary information.

P. Calow

Preface

It has been interesting to watch the way in which parasitology has evolved as a discipline. When the senior author began his professional career, the 'hot' areas were biochemistry and physiology. Then, it shifted to fine structure. After electron microscopy reached its pinnacle, attention was re-focussed on immunology. Seemingly, the emphasis has now moved to molecular biology at the one extreme and modern epidemiology at the other. Over a period of time, it appears that each one of these areas has 'pulsed' for a while and then settled somewhat. As trendy as parasitology appears to have been through the years, there also has been a cadre of parasitologists who have worked on the ecological aspects of parasitism. Interest in this approach was small initially, but it has grown considerably over the past three decades.

Workers in ecological parasitology have adopted many of the 'tools' and approaches taken by those working on free-living populations and communities. Always, however, these investigators have had to contend with the one problem not faced by those using free-living systems – the habitats in which their study organisms reside during most of their lifetimes are alive and potentially capable of mounting a powerful defence.

There are several, very good, general parasitology texts currently available. In addition, a number of specialized volumes also have been published recently on topics ranging from the molecular biology of parasitism to the community ecology of parasites. Many of these volumes have been well received if their reviews are any indication. Sixteen years ago, C.R. Kennedy wrote a book entitled *Ecological Parasitology*. For students and faculty working on the ecology of host–parasite systems, it was required reading. Peter Price's volume *The Evolutionary Biology of*

Parasitism, was published 5 years after Kennedy's and it too was a significant contribution. Then, in 1982, Klaus Rohde's *The Ecology of Marine Parasites* was published and was well received. Unfortunately, nothing of a broad scope has been published on the ecological aspects of parasitism in more recent years. This area has emerged as an exciting field for a growing number of parasitologists. In part because so little of a broad sweep has been written recently, and because of the substantial increase in literature in ecological parasitology, we undertook the task of writing this book. The primary aim is to provide a functional view of parasitism within an ecological context. Most of the effort was directed at endoparasitic helminths, primarily because most of the literature pertains to this group of organisms.

The book was written with the assumption that an undergraduate or a graduate student could handle both the ecology and the parasitology without an extensive background in either discipline. This is why, for example, life cycles of most of the parasites mentioned are illustrated in brief, without extensive elaboration on the details of morphology, intrahost migration, etc. We also believe the information in the book will be useful to the practising ecologist or parasitologist, regardless of their background or training. The book was written so that the teacher of parasitology could perhaps use the information as the basis for a section on ecology in a course in general parasitology.

There are several people whom we would like to thank most sincerely for their help along the way. There is Tim Goater who began the project with us, but had to pull out because of the pressure of teaching in his new and exciting job at Malaspina College in British Columbia. We want to thank Scott and Liz Snyder, Eric Wetzel, and Mohamed Abdel-Meguid for reading the various chapters and offering us the benefit of their very constructive criticisms. Dr. John Aho provided several comments for the section dealing with the population genetics of host–parasite systems. Next, we thank Zella Johnson and Cindy Davis for their secretarial help, especially when it came to securing all those permissions to reproduce from authors and publishers. The information contained in Table 7.1 was assembled by Dale Edwards as part of an 'open book' question for his preliminary examination for the PhD degree here at Wake Forest and we thank him for this special effort. Kym C. Jacobson was most helpful in preparing the photographs and we are grateful. Finally, we want to thank Phillippa MacBain and Helena Watson at Chapman & Hall for their assistance and especially Peter Calow for agreeing that the idea for such an approach was appropriate for his Functional Biology series.

A number of journals and publishers gave us permission to reproduce figures and tables. These include Academic Press; *The American Midland Naturalist*; Birkhauser Verlag AG; Blackwell Scientific Publications, Inc.;

Cambridge University Press; *Canadian Journal of Zoology*; Clarendon Press; *Evolution*; Freshwater Biological Association; *International Journal for Parasitology*; *Journal of Parasitology*; Macmillan Magazines Limited; Oikos; Oxford University Press; *Proceedings of the Helminthological Society of Washington*; *Quarterly Review of Biology*; Sinauer Associates, Inc.; Springer-Verlag; *Transactions of the American Microscopical Society*; University of Idaho Press. Figure 9.4 was reprinted with permission from *Nature*, **332**, 259, Copyright 1988, Macmillan Magazines Ltd.

We also thank all the authors from whom we secured permission to reproduce various figures and tables for our book.

1 Introduction

1.1 THE EXTENT OF PARASITISM

'By definition the utility of function is fitness, or genetic survival and reproduction' (Fisher, 1991).

The initial question regarding parasitism is, then, how successful is it as a life style? The extent of parasitism in nature is debatable, depending in large part on how one defines the term. In the broadest sense, some (e.g. Price, 1980) estimate that more than 50% of all plant and animal species are parasitic at some point in their life cycle. Rather than raising the question of how many parasite species there may be, however, it seems just as important to know how many plants and animals are parasitized. The number unquestionably approaches 100%. In other words, if any given animal is necropsied, the chances of finding a parasite would be excellent. Of course, it is important to know where to look within a host because parasites are found in virtually every cavity, canal and conduit in the vertebrate animal, and most invertebrates as well; all of these are endoparasitic in character. There is also a large class of ectoparasitic organisms that infect individuals in virtually all of the vertebrate and invertebrate taxa. The ectoparasites include fleas, ticks, fungi, protozoans, monogenetic trematodes, copepods, and even the glochidia of certain mussels and clams. Parasites come in almost every form and size, from those of but a few microns in diameter to the beef tapeworm of man that ranges up to 10 m in length, or the didymozoid flukes of some fishes that rival the beef tapeworm in total length.

The number of parasites present in a given host also will vary considerably. The range in number can be related to a variety of interacting factors, some associated with the host, some external to the host, and some inherent within the biology of the parasite itself. Ultimately, this array of interacting variables will affect both the numbers and kinds of parasites present in an individual host, and in a host population or community. It is the nature of these variables upon which attention will be focussed in order to gain a fundamental understanding of the functional biology of parasitism.

1.2 DEFINITIONS

The term **symbiosis** (Gr. *syn* = *sym*, together; Gr. *bios*, life) is an old one, having been coined by DeBary in 1876. The word was employed then, and continues to be used in most of the modern literature, as an umbrella to describe organisms that live together. When approached in this manner, there is no implication regarding the length or outcome of the association. Symbiosis, then, covers several forms of intimate interactions between organisms, the more common ones being **mutualism** (L. *mutuus*, reciprocal), **commensalism** (L. *com*, together; L. *mensa*, table), and **parasitism** (Gr. *para*, beside; Gr. *sitos*, food).

In mutualistic relationships the organisms are highly interdependent, to the extent that the two associates cannot survive without one another. The classic examples of mutualistic interaction include those between the flagellate protozoans, *Trichonympha* spp., and the common termite, *Zootermopsis augusticollis*, and between cattle and their rumen dwelling, ciliated protozoans. In each case, the enteric micro-organisms provide enzymes that convert cellulose into monosaccharides which they can then use, as well as share with the host. In return, the host provides their mutuals with a stable environment as well as the consistent access to the appropriate nutrient resource. In summary, there are two-way benefits and no harm to either partner.

In commensalistic relationships, on the other hand, there is only one-way benefit and, as in the case of mutualism, no harm is exerted in either direction. In many cases, such as with the clownfish, *Amphriprion clarkii*, and the stinging sea anemone, *Stichodactyla haddoni*, food is shared. However, in this particular interaction the sea anemone also provides protection from potential predators for the clownfish. The fish protection mechanism against the coelenterate stings seems to be highly evolved and genetically determined. The external mucus layer of the clownfish is comparatively thicker than that of closely related species and its chemical composition is different. It consists to a large extent of glycoprotein that contains a neutral polysaccharide; it lacks excitatory substances which are present in related, non-symbiotic fishes (Lubbock, 1981). Finally, while there is usually a feeding relationship between commensalistic organisms, it generally does not involve metabolic dependence.

Of all the intimate relationships involving heterospecific organisms, parasitism has received the greatest attention, for a variety of reasons. Before considering the nature and significance of these reasons, the term itself must be first defined. The classical definition of parasitism holds that it is an intimate relationship between two organisms in which one lives on, off, and at the expense of the other. The clearest implication of this type of lifestyle resides in the last part of the definition because it implies harm to one of the partners and benefit to the other. While this is a key element in the definition, it is also a major problem because as will become apparent subsequently, harm is a relative term and, as such, certainly is not quantifiable. Other definitions of parasitism rely on the idea of genetic complementation, or metabolic interdependence, or both. In each of these situations, the definitions attempt to codify an essential attribute of all parasitic associations, something that has become a basic paradigm for the parasitologist, and that is the concept of the host–parasite relationship. This concept has become an essential tenet in the study of parasitism because it provides the basis for understanding the manner in which the partners are tied to each other, both evolutionarily and ecologically.

Another definition, or perhaps more appropriately a description of parasitism, is that of Crofton (1971a, b). He suggested that parasitism is an ecological relationship between two organisms, one designated as the parasite and the other as a host. He indicated that the essential features of the relationship are:

1. physiological dependence of the parasite on its host;
2. heavily infected hosts will be killed by their parasites;
3. the reproductive potential of parasites exceeds that of their hosts; and
4. an overdispersed frequency distribution of parasites within the host population. That is, the variance (S^2) of the parasite population is significantly greater than the mean ($\bar{x}$) of the parasite population.

(The concepts of population mean, variance, and overdispersion, will be discussed more fully in a later chapter, but a point of clarification is required here. In his definition, Crofton (1971a) synonymized overdispersion and contagion (or clumping). As noted by Simberloff (1990), however, most ecologists who study free-living systems mean just the opposite when they refer to overdispersion; they are actually referring to underdispersion or a regular frequency distribution. Throughout this book, overdispersion will be considered in the manner of Crofton (1971a) since that is also the way in which it is regarded in most of the parasitological literature).

There are several features of Crofton's definition that deserve extended comment. For example, physiological dependence by the parasite would be automatically assumed by most parasitologists in any definition of parasitism; in this way, his approach is not diffferent from that of most workers in the field. Although many parasitologists ignore his last three ideas, it is the notion of overdispersion that makes Crofton's scheme a quantifiable one and sets it apart from other definitions. Moreover, while Crofton does not refer to expense directly, he clearly indicates what it means in his definition. He thus states that death of heavily infected hosts will be the outcome of the relationship (certainly the most expensive effect at the individual host level). By combining the notions of overdispersion and death, he created a new paradigm; in effect, if parasite-induced host mortality is accepted as a functional part of the host–parasite relationship, then there will also be significant implications regarding the ecology and evolution of both hosts and parasites. This concept will be pursued in greater detail in Chapters 8 and 9. Finally, he formalized the view that a parasite's reproductive capacity exceeds that of its host. While this certainly was not a new idea, it was the first time that it was incorporated into a definition of parasitism.

Competition is the interaction between two species of organisms in which one species may directly exclude the other from a given resource (**interference competition**) or in which one species may be using a resource necessary to the survival of other species (**exploitation competition**). **Predation** is the actual killing and consumption of individuals of one species by individuals of another species.

Many of the older and traditional ecology textbooks equate predation and parasitism (but see Begon, Harper and Townsend, 1990). This perception of the relationship, however, is totally unacceptable. A comparison of predation and parasitism within the context of different biological characteristics is shown in Table 1.1. While many of these differences are subtle and of little consequence, several are worth emphasizing. First, the feeding strategy of a predator is, quite simply, to kill and consume, while the parasite is best described as an insidious consumer. Second, the predator is usually not selective with respect to tissues consumed while the parasite is highly selective in terms of the body tissues or fluids upon which it feeds. Indeed, the notion of selective consumption by the parasite is directly related to the concept of site fidelity or preference by the parasite within a host. Third, as has been already stated, the predator inevitably kills its prey. The individual parasite, on the other hand, is generally not lethal since the effect of killing its host would be disastrous for the individual parasite. There are exceptions, obviously. Despite the proposed definition of Crofton (1971a), the extent of parasite-induced host mortality in nature has been debated widely since it relates to the matter of potential regulation of host populations by parasites. This issue is still not completely resolved and will be considered more thoroughly in Chapter 3. On the other hand, there are also situations in which a parasite

Table 1.1 A comparison of the biological features of parasites and predators. (Modified from Kuris, 1974)

Biological feature	*Parasites*	*Predators*
Feeding strategy	Insidious	Total consumption
Size	Smaller than host	Larger than prey
Reproductive potential	Greater than host	Less than prey
Population size relative to host	More numerous	Less numerous
Specificity	Variable, but usually specific	Usually not specific in prey choice
Outcome of single act	No reduction in host viability	Single predator consumes many prey
Superinfection by a single species	Frequently important; may reduce host viability	Does not apply
Density-dependent effects on host populations	Variable, but not that common	Variable, but apparently common
Lethality	Seldom	Always

may alter the behaviour of its host, making it more vulnerable to predation and thus increasing the likelihood of transmission to a second host. Finally, parasites are always far more numerous than their hosts while the reverse is almost always true for predators and their prey.

Before proceeding further with the more conceptual ideas regarding parasitism, it is necessary to consider several other terms. A **definitive host** can be defined as that host in the life cycle in which the parasite reaches sexual maturity. Considered another way, a host in which a parasite produces progeny by sexual means is the definitive host. An **intermediate host** is one that is required by the parasite in order to complete its life cycle and, usually, the parasite will undergo morphological or physiological change in it. The parasite also may reproduce within the intermediate host but, if it does, it will be by **agametic** methods. Agametic reproduction does not involve gametes; it is synonymous with asexual or vegetative reproduction, but excludes **apomictic parthenogenesis** which is reproduction in the absence of meiosis (Hughes, 1989; Chapter 5 in this volume).

The number of hosts required to complete a life cycle will vary from one species to another, but the number is usually fixed for each species. The life-cycle patterns among parasitic organisms can range from complicated, four-host cycles among hemiurid and strigeid trematodes, to direct, one-host cycles associated with many ascaridoid, strongyloid, and oxyuroid nematodes. It should be noted here that throughout this volume, life cycles will be displayed in figures without reference to descriptions of specific life-cycle forms; some representative larval stages of digenetic trematodes and cestodes are reproduced in Figs. 5.7 and 5.8 (readers should refer to standard parasitology textbooks such as Noble *et al.* (1989) for detailed accounts of parasite morphology). Some protozoans and filaroid nematodes require an insect **vector** in order to complete their cycles. Noble *et al.* (1989) defined a vector as 'an agent of transmission', but this is too simplistic. Cheng (1986) defined a vector in terms of 'arthropods and other invertebrates that serve as intermediate hosts, as well as carriers for protozoan and other smaller parasites', but this is also inadequate. Thus, a vector may be either a definitive or an intermediate host depending on whether the sexual phase of the parasite's life cycle occurs in it or not; but the term also implies a more active role in transmission rather than a passive one. For example, when a mosquito vector for the malarial parasite, *Plasmodium vivax*, transmits the parasite to an appropriate vertebrate host, the infective agent is actively inoculated into the next host by the mosquito. Moreover, the mosquito is capable of moving over a relatively wide geographic range. It should be noted that the insect vector in the case of malaria is also its definitive host since the parasite matures sexually in the mosquito.

In some parasite species, a **transport**, or **paratenic**, host may be employed. The paratenic host may or may not be required to complete the cycle but, if it is used, there will not be a detectable morphological change in the parasite. The paratenic host generally is one which acts to bridge an ecological barrier. It is also another way of moving the parasite through a trophic web. An excellent example of a parasite which may employ a paratenic host is the trematode, *Alaria canis*. The life cycle is shown in Fig. 1.1. The definitive hosts for this parasite are canids where the adults are found in the small intestine. Eggs of the parasite are released from the parasite and shed in the faeces. After a period of incubation, eggs hatch and release motile miracidia that seek out and penetrate individuals of any of several species of the pulmonate snail, *Helisoma*. After two generations of sporocysts, the daughter sporocysts begin production of furcocercous, or fork-tailed, cercariae. These free-swimming cercariae, when released from their snail hosts, are particularly sensitive to water currents. When tadpoles of any of several species of ranid frogs swim in the vicinity of the cercariae, the larval trematodes will attach themselves to the skin of the tadpole, drop their tails and penetrate the tadpole's surface. Once inside, the cercaria develops into an unencysted mesocercaria, a larval stage that is intermediate between a cercaria and a metacercaria. The life cycle may then go in one of two directions. If an infected tadpole or a metamorphosed adult frog is eaten by a canid, the mesocercaria will be freed from the ranid's flesh by digestion and penetrate the gut wall of the new

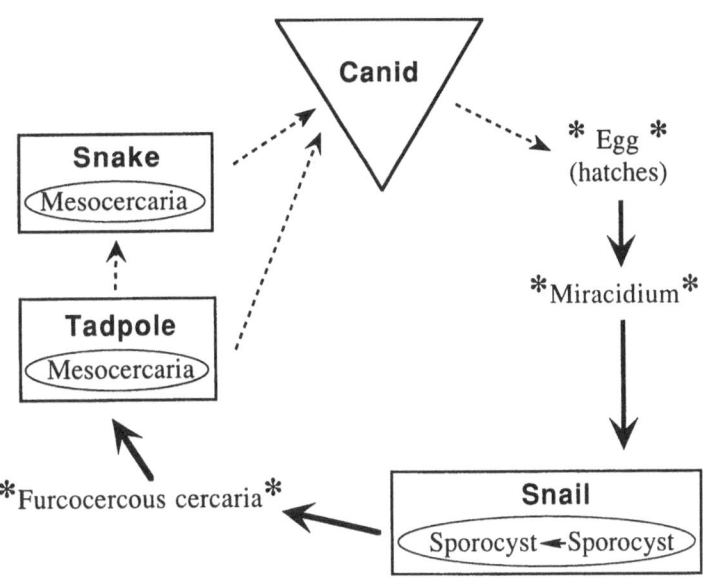

Figure 1.1 Schematic life cycle for the trematode, *Alaria canis*. Note that for this species, the canid is both an intermediate and the definitive host. *Symbols*: the triangles represent the definitive host and the rectangles are intermediate hosts. Asterisks represent 'free-living' stages; the bold characters inside the geometric figures are the hosts and the characters inside the ovals are the parasite larval stages in that host. Dashed arrows represent passive transmission; solid arrows indicate active transmission.

host. It will migrate through the diaphragm and into the lung where the mesocercaria transforms into a metacercaria. Eventually, the parasite then migrates up the trachea and is swallowed. In the small intestine of the canid definitive host, it matures sexually (note that the presence of meta-cercariae and sexually mature adults in the dog make it both an intermediate and definitive host, a most unusual arrangement). This unusual migratory odyssey is made even more remarkable by the fact that canids normally do not prey on either tadpoles or adult ranid frogs. In order to bridge this ecological barrier, a paratenic host is employed. In this particular case, water snakes eat infected tadpoles and adult frogs. Meso-cercariae are freed from the flesh of the second intermediate host and penetrate the gut wall of the snakes where they enter the coelom and remain as mesocercariae. Over a period of time, the numbers of meso-cercariae increase rather dramatically in the snake. When canids eat water snakes, they acquire mesocercariae which migrate as if the parasite had been directly transmitted to the canid by an infected tadpole or an adult frog.

A number of insects may be preyed upon by other insects in a rather unusual manner; in these special cases, one insect will lay its egg(s) inside the body of another insect and then, when the egg hatches, the larval stage that emerges will devour its host from within. These 'predatory' insects are referred to as **parasitoids**. Many members of the insect order, Hymenoptera (wasps and ants), are parasitoids at some stage during their life cycles.

1.3 THE CONCEPT OF HARM

When defining parasitism earlier, the idea of expense or harm was introduced as a functional characteristic of the host–parasite relationship. There is clear evidence that many parasites have the capacity to induce **morbidity** (= diseased state), and some even **mortalility** (= death). Indeed, when one considers the extent of human parasitic diseases in many Third World countries and the level of morbidity and mortality caused by parasites in these areas, the impact is truly staggering (Table 1.2). Moreover, the situation is exacerbated by the extensive malnutrition and other socioeconomic problems in many of these same Third World countries.

Thus, in some human host–parasite relationships, it can be argued that the parasite is having a significant negative impact at the level of the individual host. The same can be asserted for many other types of host–parasite combinations. For example, a well documented system involves the strigeid trematode, *Uvulifer ambloplitis*, and its second intermediate host, juvenile centrarchid fishes, *Lepomis macrochirus* (Lemly and Esch, 1984c). When cercariae of the parasite penetrate the flesh of these fish,

Table 1.2 Number (in millions) of human parasite infections in 1947 (from Stoll, 1947) and WHO estimates in 1985.

	1947	*1985*
Ascaris lumbricoides	644	900[*]
Hookworm	457	800
Malaria	–	300
Trichuris trichiura	355	500
Amoebae	–	480
Filarias	236	280
Schistosomes	114	200
Giardia lamblia	–	200

[*]Crompton (1989) estimated 1 billion infections for 1986.

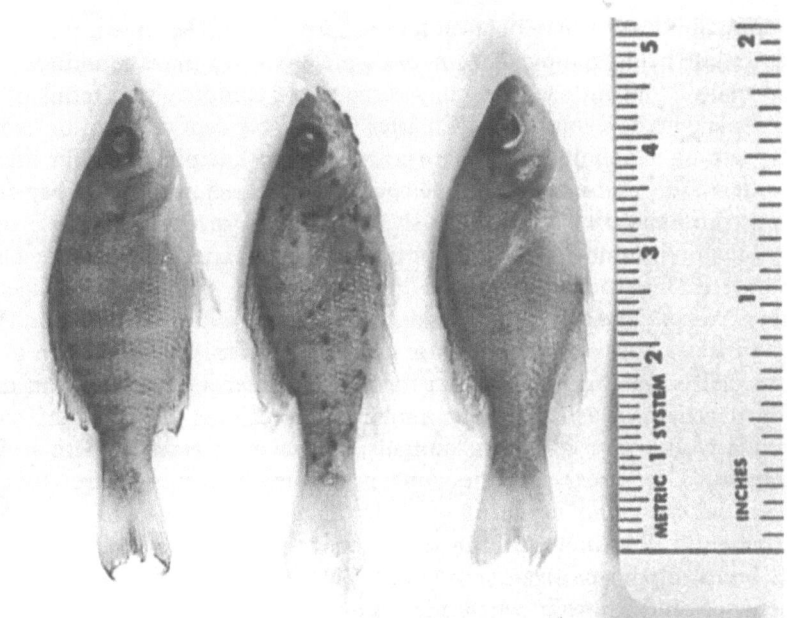

Figure 1.2 Black-spot disease in a fingerling bluegill, *Lepomis macrochirus*, caused by metacercariae of the trematode, *Uvulifer ambloplitis*, and its impact on the host's body condition. All the fishes are the same standard length. The fish in the middle is heavily infected (>50 cysts); the other two fishes are uninfected. (With permission, from Lemly, 1983.)

they encyst and become metacercariae. During this process, the host responds with a strong tissue reaction that is quite demanding metabolically for the fish host. The disease produced is known as black-spot (Fig. 1.2; only the middle fish is infected). If the host is heavily parasitized by *U. ambloplitis* (>50 metacercariae), the likelihood of the fish surviving the cold winter months is nil. The cause of the mortality is related to the reduction in body condition, or physiological fitness produced by the increased metabolism during the period when the host is responding to the invading cercariae. Simultaneously, there is a reduction in fat reserves required by the host to successfully overwinter. For the fish host that is heavily infected, the outcome is obviously disastrous. It was estimated (Lemly and Esch, 1984c) that the parasite intensities were high enough to eliminate 10–20% of the young-of-the-year fishes annually in one of their study sites. They concluded that the parasite, by inducing host mortality, was having a strong regulatory impact on the host population.

At the level of the individual host, a parasite clearly may be harmful, whether it be *Plasmodium falciparum* or *Uvulifer ambloplitis*, but at the population level it may be useful, especially if it can be shown that weaker hosts are being removed from the gene pool via parasite-induced host mortality. This idea will be considered more completely in terms of host and parasite co-evolution in Chapter 9. The concept of harm or expense as part of a definition for parasitism can be examined within another context and, that is, can a single species be viewed in terms of parasitism and commensalism simultaneously? In Africa, approximately 25% of the most agriculturally fertile parts of the continent cannot be used for raising domestic livestock because of the occurrence of the hemoflagellate, *Trypanosoma brucei*, and its tsetse fly vector, *Glossina morsitans*. Wild ungulates act as **reservoir hosts** (a reservoir host is defined as an organism that is a source of infection for man or other animals) and, in them, the flagellate is a harmless commensal, producing no apparent pathogenicity. In most domestic animals, however, *T.brucei* is lethal. With regard to *T. brucei*, then, the concept of harm becomes sharply focussed because the same organism can be highly pathogenic in what is an apparently abnormal host, but innocuous in what is its natural host. So, is *T. brucei* a parasite or a commensal? The obvious answer is, both, but it depends on the host in which it is found.

Unless the parasite kills, or it causes unmistakable morbidity, harm automatically becomes an even more relative term. Many workers would agree, for example, that the huge tapeworm of man, *Taenia saginata*, is absolutely harmless in its definitive host, producing pathology only in its bovine intermediate host. From a functional perspective, *T. saginata* is a parasite in one life-cycle stage and a commensal in another. In reality, however, this is the situation for a broad range of parasites, causing severe pathology on one host and no problem for the next host and, many times, it is the intermediate host rather than the definitive host that is adversely affected.

1.4 SOME ADAPTATIONS TO PARASITISM

1.4.1 Complex life cycles

To some, the idea that a complex life cycle could be advantageous to a parasitic lifestyle may seem almost paradoxical since the application of Occam's razor would tend to suggest otherwise. (William of Occam was a 14th century English Franciscan, who championed a nominalistic doctrine which argued that universals are but abstract terms, or predictables, and have no real existence. Occam's razor would hold that given the choice between two alternate hypotheses, the favoured one should be that which

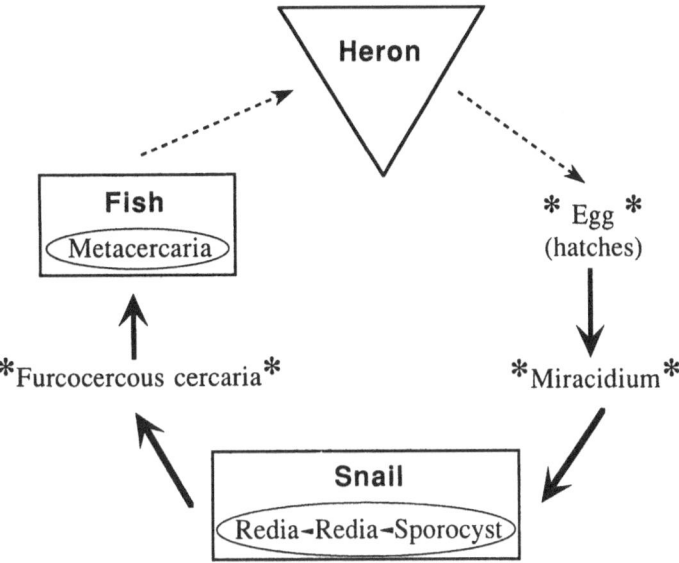

Figure 1.3 Schematic life cycle for the trematode, *Clinostomum campanulatum.* The symbols are explained in the caption for Fig. 1.1 (page 7).

makes the least number of assumptions, or that complex life cycles should be disadvantageous).

Consider the digenetic trematode, *Clinostomum campanulatum* as an example of a parasite with a complex life cycle (Fig. 1.3). The adult forms of the parasite live in the oesophageal–pharyngeal region of their definitive host, the Great Blue Heron, *Ardea herodias.* Eggs shed from adult parasites are swallowed by the heron and are passed to the outside via the faeces of the bird. Some eggs are viable when shed and hatch immediately while others require a period of time in order to embryonate before hatching. On hatching, a free-swimming, ciliated miracidium emerges. Within a matter of a few hours, the miracidium must locate an appropriate snail intermediate host or it will die. The first intermediate host for *C. campanulatum* is the planorbid snail, *Helisoma anceps.* On entering the snail, the larval parasite migrates to the hepatopancreas where it trans-forms into a sporocyst that gives rise, by the asexual process of poly-embryony (section 5.2.1. has further discussion of this process), to rediae. Again by polyembryony, a second generation of rediae is produced. In turn, daughter rediae give rise to extraordinary numbers of another larval stage, the cercaria. Cercariae are released from the daughter rediae, escape from the snail, and swim freely in the water column. Within a matter of hours, the cercaria must locate and penetrate the surface of a fish or, like the miracidium, it also will die. Over a period of several weeks following penetration into the flesh of a fish, a cyst is formed around   the parasite, now called metacercaria. The cyst wall is largely of host origin and becomes yellow in colour, hence the name 'yellow grub' for the meta- cercaria. When a heron eats a fish possessing metacercariae, the parasites undergo excystment in the stomach, migrate back up into the oesophageal–pharyngeal area, and become sexually mature.

Given the abbreviated life-spans of both free-living stages, the potential for a strong immune response (certainly by the fish intermediate host), and the vagaries of ephemeral site visits by the heron to a pond containing infected fish, it would appear that the potential for completing the complex life cycle of *C. campanulatum* would be reduced. On the contrary, the complexity of the life cycle contributes greatly to the probability of its completion; several factors are involved. First, specificity at the fish intermediate host level is low since metacercariae may be found in as many as 30 species of fishes (Olsen, 1986). Second, vagility of the defin- itive host means that the eggs of the parasite will be widely disseminated as the heron moves from pond to pond in search of prey. Third, as for many snail–trematode systems, once the snail acquires the parasite, it typically retains the infection throughout its life span that, in some areas of North America, could be two to three years. Once cercariae production has begun in the snail, it usually continues for as long as the snail lives. It also means that cercariae transmission to the second intermediate host is continuous as long as appropriate second intermediate hosts are present. Finally, visitation to a given feeding site is not continous by the definitive host and there is certainly no guarantee that the heron will consume an infected fish if one is present. However, once a fish is infected, the meta- cercariae may remain viable for several years. Thus, a heron visiting a feeding site for the first time does not need to be infected to become exposed to metacercariae of the parasite. Because the snails in a given pond need to be infected by the parasite only once, they can be involved in transmitting the parasite to second intermediate hosts in the absence of an infected definitive host. The only way in which this could occur is if the life cycle is complex and if it has a temporal, residual component asso- ciated with one or more life-cycle stages. This residual feature for certain life-cycle stages is a characteristic of many species of helminth and proto- zoan parasites, not just digenetic trematodes. The complex life cycle in many cases is also a clear illustration of the manner in which parasites employ predator–prey interactions in order to effect transmission from one host to another (see also Goater,1990; Goater, Browne and Esch 1990). Without question, this is an excellent strategy and certainly advant- ageous to a parasitic way of life. In short, complex life cycles represent ideal adaptations to exploit transient or ephemeral opportunities and to reduce intraspecific competition. The spectacular life cycles of many parasites also help to ensure dispersal in time and space, analogous to the reasons postulated for the evolution and maintenance of complex life cycles in free-living taxa such as some amphibians and holometabolous insects (Wilbur, 1980).

1.4.2 Physiological/biochemical adaptations

The range of physiologica/biochemical adaptations for a parasitic lifestyle is formidable. Consider, for example, that a large number of helminth and protozoan parasites are enteric, existing throughout their entire adult lives in a soup of highly effective enzymes, secreted by the host, that digest proteins, lipids and carbohydrates. Yet the parasites are totally unaffected because many have evolved the capacity to neutralize at least some of the enzymes released by the host; both trypsin and chymotrypsin, for example, are known to be irreversibly inactivated by the cestode, *Hymenolepis diminuta* (Pappas and Read, 1972a, b).

The covering, or tegument, of most species of parasitic flatworms (cestodes and trematodes) has been modified so that it is both syncitial and covered with microvilli, or microtriches (Fig. 1.4). Within the syncitial tegument are mitochondria that are involved in the production of ATP. Cestodes are without an intestine so that all nutrient uptake is through the external surface. While there is an incomplete intestine in digenetic trematodes, much of the nutrient absorption in trematodes is still via the external surface of these parasites. Mitochondria in the outer part of tegument are well positioned for providing the ATP necessary for the active transport of nutrients into the body of these parasites. The microtriches on the surface of parasitic flatworms also increase the surface area available for absorption of nutrient in much the same way that brush borders function on the epithelial surfaces of both the intestine and the convoluted tubules of kidneys in vertebrate animals. The outer surface of the cestode is covered by a glycocalyx comprised of a complex mixture of polysaccharides, glycolipids and glycoproteins. According to Bryant and Behm (1989), 'At physiological pH, the glycocalyx is negatively charged and acts as a polyanion. Positively charged substances are strongly bound to it, and the whole of the cestode surface thus acts as a complex and selective ion exchanger', certainly a clear example of adaptation to that particular environment.

Several of the most unique physiological adaptions to a parasitic way of life are associated with the pathways involved with intermediary carbohydrate metabolism. Parasites undoubtedly have evolved from free-living ancestors, organisms that probably lived in environments that were rich in oxygen. It is also likely that these ancestral organisms had the ability to

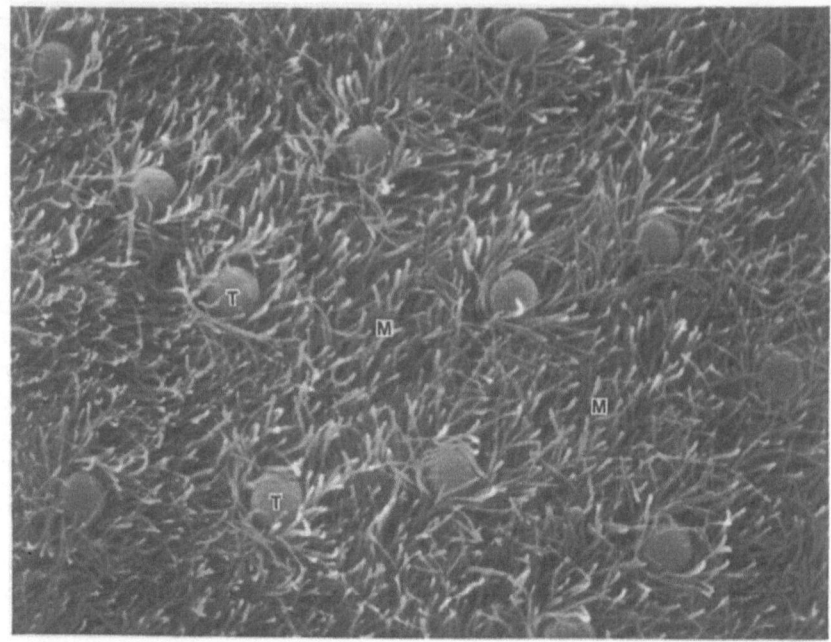

Figure 1.4 Electron micrograph of the scolex tegument of the cestode, *Bothriocephalus acheilognathi*, showing microtriches (MT) and tumuli (T) on the tegumental surface. (With permission, from Granath, Lewis and Esch, 1983.)

aerobically oxidize glucose to CO_2 and H_2O, freeing energy in the process and then capturing it in ATP via typical electron transport systems. During the long evolutionary transition, there has been the loss of certain metabolic capacities and the acquisition of new ones that enable parasites to function quite parsimoniously. Barrett (1981) refers to helminth metabolism as a kind of biochemical economy. In adult helminths, for example, it could be energetically more profitable to repress the genes that code for oxidative phosphorylation and keep them turned off or inactivated until a stage in the life cycle where oxygen was consistently available and typical aerobic metabolism possible (Kohler, 1985). A compensatory feature of the anaerobic environment of many adult helminths is the abundance of nutrients which means that parasites can squander (a not-so-parsimonious behaviour), without creating difficult circumstances for the host. Another interesting feature of helminth metabolism is that almost all of the end-products of intermediary carbohydrate metabolism still can be absorbed by the host and relatively large amounts of energy extracted for use by the host via aerobic pathways.

A significant functional adaptation of many parasitic helminths is associated with the glycolytic pathway. Normally, in the absence of oxygen, the end-product of glycolysis in most organisms will be lactic acid and the net production of 2 ATPs. However, many parasitic helminths have evolved modified pathways that actually involve the fixation of CO_2 and the increased production of succinate, a metabolite that is normally associated with the aerobic Krebs cycle. In these organisms, phosphoenolpyruvate is combined with CO_2 to produce oxaloacetate and guanosine triphosphate (GTP). The GTP is equivalent to ATP as an energy carrier in protein synthesis (Smyth, 1969), but the energy trapped in GTP can also be transferred to ATP in some parasites (Bryant and Behm, 1989). The enzyme mediator in the conversion of phosphoenolpyruvate to oxaloacetate is phosphoenolpyruvate carboxykinase. The oxaloacetate produced is then converted to succinate or proprionate and excreted from the parasite. A modification of this pathway involves the fixation of CO_2 to pyruvate and the formation of malate. In the process, $NADPH_2$ is oxidized; malic enzyme mediates the reaction and malate is then converted to succinate or proprionate.

One of the complexities faced by most parasitic organisms is the wide range of environmental changes with which they must cope in their life cycles. In some instances, there is evidence that parasites will adjust to these vagaries by shifting primary metabolic pathways (Bryant and Behm, 1989). For example, embryonated eggs of the large enteric nematode of man, *Ascaris lumbricoides*, possess an active cytochrome oxidase while in the adult, metabolism is largely anaerobic in character (Oya, Costello and Smith, 1963). Likewise, when young flukes of *Fasciola hepatica* first gain access to the liver of their sheep definitive host, the Krebs cycle is fully functional. Within 17 weeks and entry into the bile ducts and gall bladder, the metabolic emphasis has become almost totally anaerobic (Tielens, Van den Huevel and Van den Bergh, 1984). Bryant and Behm (1989) indicate that glucose-based metabolic pathways of certain bloodstream trypanosomes switch to an amino acid-based oxidative metabolism in the gut of the insect intermediate host, presumably an adaptive change in response to a different set of environmental conditions in the two sites of infection.

Physiologically and biochemically, parasitic organisms have developed complex methods for adjusting to sometimes harsh or hostile environments. In the process, they have also lost complex physiological/ biochemical abilities many times. For that reason, parasites are sometimes mistakenly considered as degenerate. However, the capacity to cope with digestive enzymes, to survive under both aerobic and anaerobic conditions, and to escape (or resist) a highly evolved immune response (see below) is not the mark of degeneracy. On the contrary, it is indicative of the high degree of evolutionary 'tracking' of a host by the parasite.

1.4.3 Behavioural adaptations

Parasites have evolved behavioural adaptations to find hosts, to locate sites of infection within definitive and intermediate hosts, to escape immune responses, to entice another host to consume the host in which they are residing, etc. Many of these mechanisms are highly complex, but they are sometimes also quite subtle. While the mechanisms or processes involved in a few parasites are understood, most are not.

One of the most spectacular is associated with the digenetic trematode, *Dicrocoelium dendriticum*. The life cycle is shown in Fig. 1.5. This parasite uses the bile ducts of sheep in which to mature sexually. Eggs produced by the adult trematodes are shed in the faeces and are accidentally ingested by certain species of land snails. Following two sporocyst generations, tailless cercariae are released, leaving within slime balls produced as a salivary secretion by their snail hosts (an adaptation to terrestrial transmission that provides a moist environment for the cercariae). Slime balls are then ingested by ants as they forage. The cercariae migrate into the haemocoele of the ant where they become metacercariae. Interestingly, one or two of the first cercariae to be acquired by the ant will always migrate to a region near the brain where they become metacercariae. These parasites induce a behavioural change that makes the ant more vulnerable to ingestion by the definitive host. Ambient air temperature begins

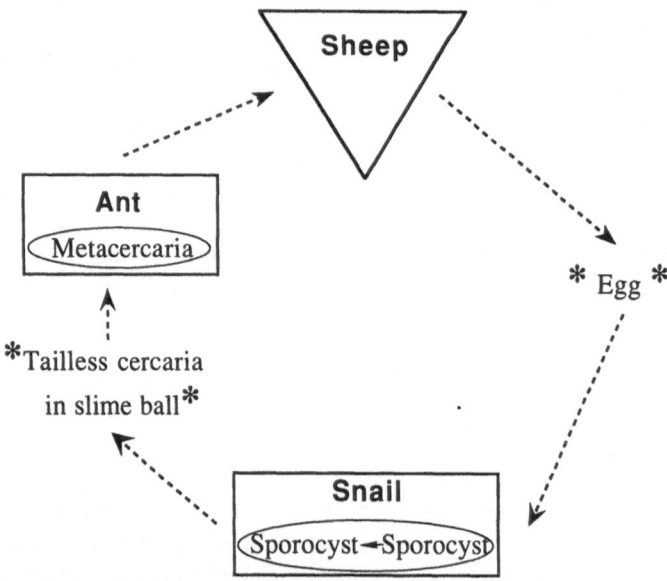

Figure 1.5 Schematic life cycle for the trematode, *Dicrocoelium dendriticum*. The symbols are explained in the caption for Fig. 1.1 (page 7).

to drop in the late afternoon and early evening; at some point, the infected ant is unable to open its jaws to disengage itself from a blade of grass upon which it might have been grazing. In this way, the infected ant is fixed in place for the accidental ingestion by sheep that tend to graze in late afternoon and evening, and again during the early morning hours. Then, when the air temperature rises in the morning, the parasite-induced form of 'lockjaw' disappears and the infected ant is free to move about again.

The same species of parasite can be used to illustrate another kind of behavioural modification, one that is common to virtually all parasites in one or more stages of their life cycle, and is associated with site selection and site fidelity within an intermediate or definitive host. In its definitive host, *D. dendriticum* is always found in the bile duct and gall bladder. When metacercariae are digested from the flesh of the ant, they inevitably migrate from the lumen of the sheep's intestine into the bile duct where they mature sexually. It is clear that they are responding to a chemical stimulus of some sort, the nature of which is not known. The route taken by *D. dendriticum* to reach its final site of infection is not complex, but the physiology that must be involved in sensing a chemical stimulus and then in responding through migration is certainly complicated. *Fasciola hepatica* is another digenetic trematode that uses the bile duct and gall bladder of sheep in which to become sexually mature. However, following their ingestion and excystment, metacercariae of this species do not move directly into the bile duct. Instead, they penetrate the gut wall and migrate into the body cavity. They then locate and penetrate the liver from the outside. The metacercariae wander for 7–14 days through the liver parenchyma, growing rapidly in size as they feed on the liver tissue and causing considerable damage in the process. After an appropriate period of time, they move into the gall bladder and bile duct where they become permanent residents as sexually mature adults. In the case of *F. hepatica*, excystment of the metacercariae occurs in the small intestine, where bile stimulates their active escape from the cyst. Extracts from the duodenum and liver (bile salts, supernatants from liver homogenates) significantly affect the rate of locomotion and orientation of excysted parasites; however, the mechanism(s) by which they orient is unclear (Sukhdeo and Mettrick, 1986). Recently, however, Sukhdeo and Sukhdeo (1989) were able to show that adult *F. hepatica* were capable of responding to certain gastrointestinal hormones and that there may be adaptive value in these behavioural responses.

Here are examples of two parasite species capable of occupying the same site of infection in the same definitive host, each using a completely different migratory route and, probably, a completely different set of chemical cues in order to get there. An important feature of this behaviour is that as long as the parasite reaches the proper host, it will almost always end up in the correct location. This strong site fidelity is, however, usually lost when the parasite is ingested by the wrong host. If the parasite survives in the abnormal host and if some sort of tissue migration is a normal feature of its life cycle, then it is likely that it will migrate to the wrong place and use the wrong route. When this happens, tissue damage, pathology, and even host death, are a real possibility. A disease known as sparganosis will occur in humans, for example, if they accidentally ingest copepods carrying procercoids (the second larval stage in its life cycle) of the pseudophyllidean cestode, *Spirometra mansonoides*. Upon ingestion, the procercoid will be freed in the small intestine and migrate to subcutaneous sites. There, it will transform into a plerocercoid (also knwon as a sparganum, hence the name sparganosis) that will ultimately grow to a length of several centimeters. It constantly moves under the skin causing severe pain and inflammation as it migrates. Man is the wrong host for the parasite at this stage in its life cycle; the parasite should have been in a frog. In the Far East, frog meat is often used as a poultice and direct transfer of plerocercoids from infected frogs to man has also been reported to occur when infected frog flesh is used as a poultice for an open wound on the surface of human skin.

There are many host–parasite interactions in which behaviour is modified in such a way that the probability of the host being preyed upon is enhanced. The case of *D. dendriticum* and its ant host already has been described. A large number of other parasite-induced, altered behaviours also could be described for both invertebrate and vertebrate hosts (Holmes and Bethel, 1972; Moore, 1983b, 1984a, 1984b; Barnard and Behnke, 1990). For example, *Ligula intestinalis* is a pseudophyllidean tapeworm that uses many wading and diving birds as a definitive host. The second intermediate hosts for *L. intestinalis* include fishes of a number of species (Fig. 1.6) that are infected when they consume copepods carrying procercoids of the parasite. In 1952, van Dobben reported that about 7% of roach, *Rutilus rutilus*, were infected with plerocercoids of *L. intestinalis*, while 30% of the roach in stomachs of cormorant definitive hosts were infected. Clearly, this implies selective predation on infected fishes.

Arme and Owen (1967) reported that the gonadal–hypophyseal hormonal relationship was interrupted in sticklebacks, *Gasterosteus aculeatus*, by some sort of secretory product released from plerocercoids of the pseudophyllidean tapeworm, *Schistocephalus solidus*. The result was reproductively sterile hosts. Infections were exceedingly heavy in some sticklebacks. They suggested that the distension of the body in these fishes could have caused an increase in vulnerability to predation by fish-eating birds, as well as the increased chance of transmission to the definitive host of the parasite. Note that in each of these cases, and most of the others that can

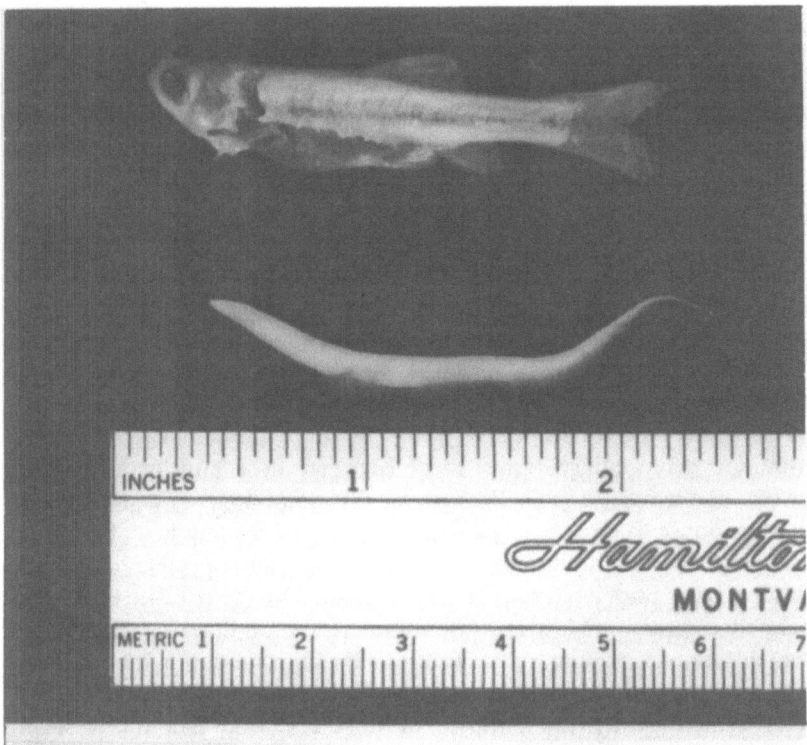

Figure 1.6 A plerocercoid of the cestode, *Ligula intestinalis*, removed from the body cavity of a fathead minnow, *Pimephales promelas*, from Belews Lake, North Carolina, USA.

be documented, the parasite does not kill the host. Instead, it alters the host's behaviour, causing the host to be killed and consumed by a predator and ensuring successful parasite transmission in the process.

Another highly unusual behavioural adaptation is associated with the filarial nematode, *Wuchereria bancrofti*. This parasite's definitive host is man and, when constantly infected by mosquito vectors over a long period of time, the nematode can cause blockages within the lymphatic system of the host, producing the horribly disfiguring disease known as elephantiasis. The insect becomes infected with microfilariae when it takes a blood meal from an infected human. It infects another human with microfilariae when it takes its next blood meal and the microfilariae develop into adults within the lymphatic system. The interesting feature in this process is that microfilariae in the peripheral circulatory system of humans are periodic in their appearance, being present between the hours of 2200 and 0200 (Larsh, 1964). This diurnal periodicity matches almost exactly the time during which the insect vectors are actively feeding. The periodicity

also ensures that other blood-sucking insects will not accidentally consume microfilariae and thus could also be considered as a rather clever
conservation mechanism on the part of the parasite. The periodicity of
microfilariae in the peripheral circulation in the human and the feeding
behaviour of the mosquito vector represents a marvellous example of
synchrony by the parasite, the insect vector, and the definitive host.

1.4.4 Evasion of the host's immune response

During the long period of co-evolution between many parasites and their
hosts, it is evident that the latter's immune response has become more and
more sophisticated. Likewise, however, the parasites have evolved a number
of mechanisms for evading, even suppressing, the ability of a host to
respond to the parasite immunologically. In effect, in many cases, there is a
continuous evolutionary 'arms race' between host and parasite (*sensu*
Dawkins and Krebs, 1979). Studies on the immunogenic capability of a
host in response to a parasite are obviously within the purview of immuno-
parasitology. While immunology is clearly in the realm of cell and molecular
biology, a basic understanding of host resistance by the parasite ecologist is
absolutely essential. This is because there are several host–parasite systems
in which there are regulatory implications for parasite population and even
community dynamics, topics discussed in more detail in Chapter 3.

The avoidance by the parasite of host rejection has taken several
different routes over the evolutionary history of various host–parasite
relationships. For example, *Leishmania donovani* is an intracellular
hemoflagellate that infects humans throughout many countries in tropical,
sub-tropical, and even desert, areas of the world. Interestingly, the cells,
primarily macrophages which the parasite attacks, are an important component of the host's immune system. On penetration, the macrophages
sequester the parasites in phagosomes that subsequently fuse with
enzyme-carrying lysosomes. Most organisms under these conditions
would be killed and lysed quickly, but not *L. donovani*. Indeed, within the
macrophage, promastigotes difffferentiate into amastigotes; these multiply
and lead to the destruction of the macrophage whereupon the parasites
invade new macrophages and the process is repeated. Leishmanias seem
to avoid the capacity of the macrophage to lyse by one, or a combination,
of three different mechanisms. The first involves the production of
dismutase- and peroxidase-like enzymes that inactivate or dismutate
hydrogen peroxide and superoxide anions contributed by the lysosome. A
second mechanism is by the production of a surface layer of molecules
that is resistant to acid hydrolases, enabling them to survive in the acidic
environment of the macrophage's phagolysosome. The last mechanism is
suggested to operate through the process of **molecular mimicry** (*sensu*

Damian, 1964), in which the surface molecules of the parasite resemble or mimic those of the macrophage. As a consequence, the lysosomal enzymes of the macrophage are unable to recognize the parasite as foreign and will not digest it (Chang, 1990).

Concomitant immunity is a special kind of immune response that is generally associated with species of the digenetic trematode, *Schistosoma*, several of which infect humans. These digenetic trematodes are highly pathogenic; they are also quite common (see Table 1.2) and are even increasing in terms of **prevalence** (the percentage of hosts infected in a given population) in many parts of the world. Concomitant immunity refers to the condition in which adult parasites in an initial infection are unaffected immunologically by a host, but in which challenge infections are. The effect is to keep the host from becoming overwhelmed by the parasite in subsequent invasions of infective agent that, in this case, are water-borne cercariae. As stated by Butterworth (1990), it is a 'subtle adaptation toward the establishment of a stable host–parasite relationship'. The mechanisms by which the adult parasites escape the host's immune response are as yet unknown despite intensive study over the years. However, some investigators are convinced that the parasite incorporates host proteins in a process known as **antigen sharing**, and thus masks the parasite from the host. Others suggest that the parasite becomes immunologically 'inert' with the loss of capacity by the parasite to express antigens in its tegumental surface that would be perceived as foreign by the definitive host (Butterworth, 1990).

The African trypanosomes alluded to earlier, *Trypanosoma rhodesiense* and *T. gambiense*, are haemoflagellates that cause an insidious disease in man known as sleeping sickness (not to be confused with American sleeping sickness that is caused by a virus). In African trypanosomiasis, the host mounts what ostensibly should be an effective immune response via the production of antibodies to a specific antigen coat on the surface of the parasite. This coating is called the **variant surface glycoprotein**, or VSG. However, the response to the circulating haemoflagellates is ineffective even though antibody titres to a given VSG may be high. What makes the immune response impotent is that about the time the antibody titre gets high enough to become effective, a few trypanosomes have switched to a new VSG and the ones with a new coat then increase rapidly. The process of VSG switching occurs sequentially, expressing genes that encode different VSG sequences. This continues for as long as the host survives and, in certain forms of the disease, it can last for up to 20 years. The sequence of **antigenic variation** and host response continues inexorably, with the host never catching up. Even more fascinating is that the parasite never produces the same VSG and that the VSG production sequence is different from one host to the next (Turner and Donelson, 1990).

American trypanosomiasis, also called Chagas' disease, infects some 15 million people throughout Central and South America, and there is no cure. Several mechanisms have been suggested as ways by which this protozoan parasite (*Trypanosoma cruzi*) eludes the immune response, but none of them has been completely determined. Most evidence indicates that the parasite is capable of modulating the host immune response through a generalized **immunosuppression**. In this situation, the parasite shuts down antibody production by the host even though the memory responses by the appropriate cells in the host are still retained. Extracellular forms (epimastigotes) may also escape by being sequestered in host cells; once sequestered, they transform into amastigotes, divide repeatedly, and eventually destroy the cell by causing it to burst. Freed amastigotes invade new cells and the process continues. Among the blood forms (trypomastigotes), however, there is incorporation of some host plasma proteins, including immunoglobulins and complement components, on to their surface membrane, and this may also provide a way for them to escape recognition. Additional evidence indicates that the parasites may achieve protection by shedding antigen–antibody complexes (a form of **antigen-shedding**) from the parasite's surface after they are formed (Brener and Krettli, 1990).

1.5 LIFE CYCLE AND ECOLOGY

The ecology of parasitic organisms is not really different from that of free-living organisms except for one factor: many of the habitats in a parasite's life cycle are alive. This clearly presents parasites with a large number of problems that are not faced by free-living organisms. For example, the parasite in a vertebrate host, and in many invertebrate hosts as well, will be subjected to the host's immune system or, perhaps, its digestive enzymes. In contrast to free-living organisms, most parasites are not confronted with the possibility of predation (excepting intramolluscan stages of some digenetic trematodes).

In completing their life cycles, most parasitic organisms are subjected to more physical and biotic environmental challenges than virtually any other group of organisms. Consider the life cycle of the digenetic trematode, *Clonorchis sinesis*, as an example (Fig. 1.7). Adult parasites live in the bile ducts of mammals, including humans. The temperature in the bile duct is about 38°C, the pO_2 ranges from 0–30 mmHg, and the pH is about 6.5. The bile itself is isotonic with the blood (data from Smyth and Halton, 1983). In the secretion from the liver are bile salts, inorganic salts, proteins such as bilirubin and biliverdin, and glucose at a concentration of about 80 mg% (80 mg/100 ml of fluid). This is the habitat of the adult

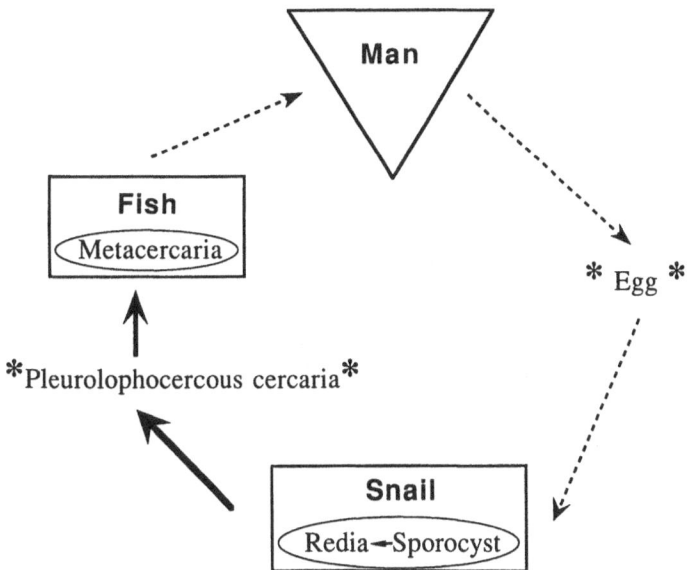

Figure 1.7 A schematic life cycle for the trematode, *Clonorchis sinensis*. The symbols are explained in the caption for Fig. 1.1 (Page 7).

parasite, or the place where it spends its life as a sexually mature individual. Adults produce eggs that are released to the outside via the host's faeces. In contrast to many species of digenetic trematodes, the eggs of *C. sinensis* do not hatch, and must be ingested accidentally by an appropriate snail intermediate host. Moreover, as will be discussed subsequently, the spatial distribution of the snail and the egg must overlap or the grazing snail will never come into contact with the egg. The idea of spatial scales in the transmission of parasites is not a new one, but its importance in the transmission of digenetic trematodes to snails has assumed a new dimension with several recent findings, discussed in Chapter 6.

Once eaten by the snail, the egg hatches and a miracidium emerges into the snail's gut where there is probably an abundance of CO_2 and little O_2, an acid pH, digestive enzymes and general reducing conditions. The parasite copes by using secretions from its penetration glands to move quickly through the gut wall. It then responds to appropriate stimuli provided by the host and migrates to the hepatopancreas. Note that it does not go to the gonad first or to the tissues of the foot or the kidney; it migrates directly to the hepatopancreas. The genome of the parasite thus includes the capacity to sort out a complex of potential chemical stimuli and respond appropriately. Once into the hepatopancreas, the parasite is surrounded by an entirely new set of environmental characteristics. Oxygen probably occurs in relatively high concentrations. The temperature within the snail is ambient with the surroundings of the snail itself and may fluctuate substantially on a diurnal and seasonal basis. There is probably little CO_2 and the freezing point depression is substantially lower than it was in the bile duct of the vertebrate host. There is an abundance of nutrients available. The miracidium immediately undergoes transformation into an amorphous sac, without a mouth or a gut. As a result, the new larval stage, now called the sporocyst, must obtain all nutrients from direct absorption through its tegumental surface. Within the sac, the parasite begins to produce new larval stages via the asexual process of polyembryony. The products of polyembryony are not amorphous forms like the sporocyst, but have a characteristic morphology. They come equipped with a mouth and a short, but functional, gut. These offspring, known as rediae, are released from the sporocyst and immediately begin the direct ingestion of the hepatopancreatic tissues. Within the rediae, polyembryony occurs again, but this time new larval forms called cercariae are released. Cercariae production is prodigious. In some species of trematodes, thousands will be released from the snail each day (Shostak and Esch, 1990a). Cercariae then move to the outside of the snail where they assume a free-living existence, being propelled by a powerful tail as they swim in yet another habitat.

This habitat is quite different from that inside the snail as the cercariae are now subject to all the biotic and abiotic vagaries to which any aquatic plant or animal might be subjected. That includes temperature, dissolved oxygen, light currents, predators, etc. Cercariae must then seek out and penetrate an appropriate piscine intermediate host. Moreover, they must do so within 18–36 h or they will die because they are non-feeding stages and their internal glycogen reserves are finite. Assuming the parasite locates the appropriate host, it will attach to the fish's surface using powerful suckers, or holdfast devices, evolved for that purpose. It will then employ secretions from penetration glands to move under the scales and enter the flesh of the fish host. Again the parasite is in another new habitat, one that is alive and potentially hostile. It sheds its tail, secretes a cyst wall around itself, and becomes sequestered; it is now known as a metacercaria. At this point, the metabolic activity of the metacercaria slows perceptibly as it rests within the cyst wall.

To successfully complete the life cycle, the infected host must be consumed by a human in a raw or undercooked state. A favourite dish in the Far East where the parasite is most common, is sushi, or raw fish, marinated in a concoction of aromatic herbs and spices. Into the new host's stomach goes the fish where it is subjected to powerful digestive enzymes, probably little or no O_2 and a pH of 1–3. The process of freeing the parasite (**excystment**) begins in the stomach and is continued after passage into yet another habitat, the small intestine, where once again the environmental conditions change. After completing excystment, the immature parasite must move back up the intestine, inching its way using two powerful holdfast organs, the oral sucker and the acetabulum. It locates the bile duct and enters it, finally completing the cycle by becoming sexually mature.

In completing this process, the parasite moved though eight distinct and different habitats. It contended with a tremendous array of abiotic and biotic conditions. As will be seen throughout this book, such a life cycle is certainly not atypical for that of many parasites. While some will have fewer stages, all will be subjected to sudden and dramatic changes in environmental conditions during their life cycles. It is the nature of these habitat shifts, in combination with variety of behaviour, morphological and physiological mechanisms used by parasites in coping with the dramatic habitat shifts, that makes the study of parasite ecology so exciting and challenging.

2 Population concepts

2.1 BACKGROUND

If the ecology of parasitic organisms is considered in its broadest form, then it has a long history, extending back to the middle of the 19th century when the connection between cysticerci and adult taeniid cestodes was identified by Dujardin in 1845, and when the life cycle of the canine tapeworm, *Taenia pisiformis*, was experimentally completed by Kuchenmeister in 1852. These discoveries were matched, if not exceeded, later in the century by Manson who described the life cycle of *Wuchereria bancrofti* in 1877 and by Ross who reported the life cycle of avian malaria in 1895; Sir Ronald Ross eventually was to receive one of the first Nobel prizes in physiology in recognition of this exceptional research achievement.

Modern parasite ecology can be traced to V.A. Dogiel and his colleagues (Dogiel, 1964). Many of these early studies had a decided natural history flavour, but they were nonetheless significant in their scope and approach. The research trends in parasite ecology continued to become more and more complex with time and the development of new technologies. However, it was not until Holmes (1961; 1962a, b) published his now classic studies on the interspecific competitive interactions between the rat tapeworm, *Hymenolepis diminuta*, and the acanthocephalan, *Moniliformis dubius*, that a quantitative perspective became solidly entrenched in the literature on parasites. These papers also became the standard for parasite community ecology. Ten years later, Crofton (1971a, b) established a similar standard for a quantitative approach in the study of parasite population dynamics. MacDonald (1965) was the first to extensively employ mathematical modelling in the study of host–parasite relationships. Over the last 30–35 years, investigations on the ecology of parasitic organisms and their hosts have expanded significantly.

2.2 GENERAL DEFINITIONS

Ecology can be considered within the context of two, quite different methodologies. The first, termed **synecology**, deals collectively with groups of organisms of different species that live together. For synecology, the approach is thus at the community level. A **community** of organisms is defined as a group of populations of different species occupying a similar habitat or **ecosystem**. An ecosystem encompasses the community of organisms, plus their physical surroundings. **Autecology** is concerned with the study of individual organisms or species. The approaches taken here can be varied but, plainly, population biology is autoecological in nature. Studies on the parasite population biology of humans form the discipline **epidemiology** while studies on the population biology of parasites in other animals are in the realm of **epizootiology**. A **population** can be defined as a group of organisms of the same species occupying a given space in time and comprising a single gene pool. Each population can be described by several parameters. Some of these include birth rate, death rate, age distribution, biotic potential (reproductive potential), dispersion, growth form, and density.

Most of these concepts were developed for free-living organisms, but there are certain problems when the concepts are applied to parasites. For example, do all members of a given parasite species within a single host constitute a population? Or, do all members of a given parasite species within all hosts in an ecosystem represent a population? Another difficulty with the concept as applied to parasites is that populations of free-living organisms increase in number through birth, or immigration, or both, while most helminth parasites (certainly adult forms) in a host increase only through immigration, or **recruitment**. In an effort to clarify these issues, Esch, Gibbons and Bourque (1975) proposed that parasites within a single host and those within all hosts in an ecosystem be separated and considered independently. They developed the concepts of the **infrapopulation** and **suprapopulation** to address these issues. An infrapopulation was described as one that includes all of the parasites of a single species in one host. A suprapopulation was defined to include all of the parasites of a given species, in all stages of development, within all hosts in an ecosystem. Subsequently, Riggs, Lemly and Esch (1987) coined the term **metapopulation** to describe all of the infrapopulations of a species of parasite within all hosts of a given species in an ecosystem.

The number of studies on infra- and metapopulations are substantial and far exceed those at the suprapopulation level, mainly because the logistics are far too complex in the latter case to cover all stages of a parasite's life cycle in a single ecosystem at the same time. Several investigations have been attempted on suprapopulations (Hairston, 1965; Dronen, 1978; Jarroll, 1979, 1980), or at least partially attempted (Riggs, Lemly and Esch, 1987; Marcogliese and Esch, 1989), and these will be discussed subsequently. Since the development of this schematic for parasite population ecology, a similar hierarchical approach has been adopted by those working in the area of parasite community ecology (Bush and Holmes, 1986a; Holmes and Price, 1986; Esch, Bush and Aho, 1990; Esch *et al.*, 1990).

The conceptualization of parasite population structure can be viewed in terms of a nested hierarchy (Fig. 2.1). Consider, for example, the three-host life cycle of a hypothetical digenetic trematode, one that involves a snail as the first intermediate host, a fish as the second intermediate host and a bird as the definitive host. For this parasite, active recruitment (solid lines in Fig. 2.1) is associated with free-swimming miracidia (asterisks) and cercariae (plus signs). The transmission of metacercariae to the definitive host is a passive process, almost always depending on predator–prey interactions (dashed lines). A spatial component in the transmission process is introduced with the positioning of the symbols for miracidia and cercariae. The large box that circumscribes the system represents the suprapopulation; observe that it includes all of the meta- and infrapopulations, as well as the free-living stages. The large triangle represents the metapopulation of parasites in snail hosts which, in turn, is comprised of a series of individual snails (smaller triangles), each with its own infrapopulation of parasites. The same kind of nesting is associated with the large rectangle and the large circle. The latter represent meta-populations of metacercariae in fish second intermediate hosts and the rectangles are definitive hosts. Infrapopulations within individual fish are smaller circles and individual birds are the smaller rectangles. In considering the suprapopulation as a nested hierarchy, the ideas of parasite flow and of recruitment and turnover can also be visualized. More importantly, even though the nature of the processes that impact on the dynamics of the system have not been illustrated, the idea of gene flow within the suprapopulation is also implicit.

2.3 FACTORS AFFECTING PARASITE POPULATIONS

The diagram in Figure 2.1 is a simplified way of illustrating the idea of parasite flow through a series of hosts within the framework of a supra-

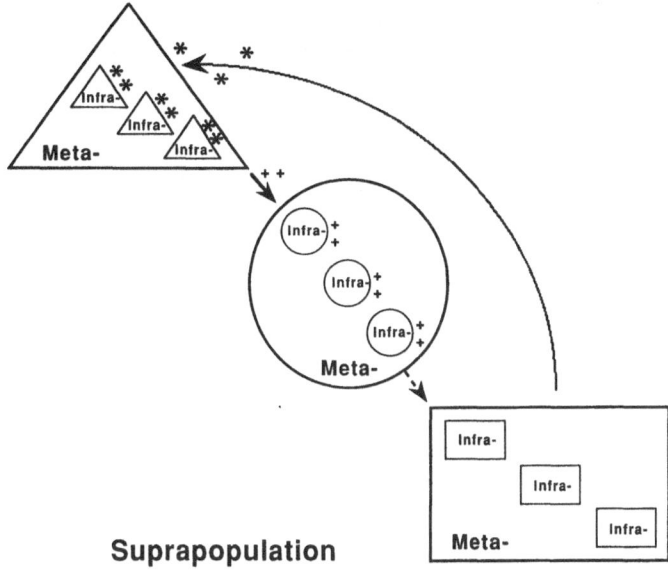

Figure 2.1 A schematic representation for the hierarchical organization of parasite infra-, meta-, and suprapopulations. While it illustrates a hypothetical digenetic trematode, the scheme is applicable to any host–parasite system. The small asterisks represent miracidia in the vicinity of snails; the small triangles are intramolluscan infrapopulations; the large triangle represents a metapopulation of intramolluscan larval stages. The small plus signs are cercariae in the vicinity of the second intermediate host; the small circles are infrapopulations of metacercariae in second intermediate hosts. The large circle is a metapopulation of metacercariae. The small rectangles are infrapopulations of the parasite in the definitive host; the large rectangle is the metapopulation of the adult parasite. Solid lines represent active transmission and dashed lines are passive transmission. The large rectangle surrounding all of the symbols in the diagram is the parasite suprapopulation.

population. Parasite recruitment and turnover are, however, influenced by a range of factors both external and internal to the host. Recruitment into a host can be passive, requiring no expenditure of energy by the parasite, or it can be active in which case an expenditure of energy is required. Table 2.1 provides an idea of the wide-ranging tactics that parasites have evolved in achieving recruitment success, by either active or passive means. The list is somewhat distorted, however, since the process of active recruitment is almost entirely restricted to parasites that require water to complete their life cycles. Exceptions here would include certain nematode species and, even then, their free-living rhabditiform and filariform larvae require special soil conditions (moisture, shade, etc.) in which to survive.

Table 2.1 Examples of parasite species using passive or active recruitment strategies for infecting the intermediate (I) or definitive (D) hosts.

Passive (stage involved-host type)	Active (stage involved-host type)
PROTOZOANS	
Entamoeba histolytica (cyst-D)	*Trypanosoma cruzi* (metacyclic trypomastigote-D)
Giardia lamblia (cyst-D)	*Ichthyopthirius multifilis* (theront-D)
Plasmodium spp. (gametocyte-D)	
CESTODES	
Taenia pisiformis (cysticercus-D)	*Proteocephalus ambloplitis* (coracidium-I)
Moniezia expansa (egg-I)	*Diphyllobothrium latum* (coracidium-I)
Hymenolepis diminuta (cysticercoid-D)	*Ligula intestinalis* (coracidium-I)
TREMATODES	
Clonorchis sinensis (egg-I)	*Schistosoma japonicum* (cercaria-D)
Uvulifer ambloplitis (metacercaria-D)	*Crepidostomum cooperi* (miracidium-I)
Dicrocoelium dendriticum (cercaria-I)	*Benedenia melleni* (oncomiracidium-D)
Alaria canis (mesocercaria-D)	*Paragonimus westermani* (miracidium-I)
NEMATODES	
Ascaris lumbricoides (egg-D)	*Necator americanus* (filariform larva-D)
Wuchereria bancrofti (microfilaria-I)	*Strongyloides stercoralis* (filariform larva-D)
Pneumostrongylus tenuis (L_1 larvae-I)	*Uncinaria lucasi* (filariform larva-D)

The turnover of parasite infrapopulations will result from natural mortality, inter- and intraspecific competition and from the effects of host immune responses. In some cases, the turnover processes will be directly affected by external as well as internal environmental factors.

2.3.1 External environmental factors

Free-living stages are directly affected by external environmental factors. For helminth parasites, such stages would include, among others, miracidia and cercariae of digenetic trematodes, coracidia of many pseudophyllidean cestodes and the filariform, or L_3 larvae, of some rhabditoid and strongyloid nematodes. The eggs of helminth parasites and the cysts of many parasitic protozoans are also affected by the external environment.

Consider the host-finding capability of free-swimming miracidia and the combination of physical and chemical factors that influence this behaviour (Christensen, Nansen and Fransen, 1978; Callinan, 1979; Kennedy, 1975; Kearn, 1980, Esch, 1982). Host-finding involves light and gravity stimuli, ensuring that the miracidia are spatially distributed in such a way as to enhance the probability of contact between the molluscan host and the parasite. Then, once the miracidia are swimming in the vicinity of their first intermediate hosts, they are further stimulated by appropriate chemical stimuli released from snails or clams, causing them to seek out and penetrate these hosts. The chemical stimuli are not complicated, consisting mainly of simple amino acids and fatty acids. The processes of host-finding by trematode miracidia are thus not random, but occur in a definite and predictable sequence.

Callinan (1979) has emphasized that 'an understanding of the ecology of the free-living stages of sheep nematodes is essential if maximum use is to be made of the natural processes controlling nematode populations'. Free-living, filariform larvae (L_3) of many nematode species employ host-finding mechanisms that are not unlike those of many miracidia. Evidence suggests that optimum soil conditions, including humidity and temperature, are absolutely necessary for the development and transmission of free-living larval nematodes to their normal definitive hosts (Callinan, 1979). Deviations from optimum microgeographic conditions will disrupt life cycles and, in turn, affect both the local and the global (zoogeographic) distributions of these parasites and the diseases they produce. On the other hand, a changing climate may also affect microgeographic conditions in such a way as to be conducive to colonization by new parasites.

The patterns of dispersal and transmission of digenetic trematodes, and of nematodes having free-living larval stages, are influenced by many microgeographic physical and chemical forces, some of which are very subtle in character. These same microgeographic factors also may affect the dispersal of parasites that have stages that are transmitted passively. Consider the digenetic trematode, *Halipegus occidualis*, as an example. This parasite has received considerable attention in recent years and much is known regarding its life cycle, population biology, and transmission dynamics. The life-cycle pattern (Fig. 2.2) of this hemiurid trematode is much more complicated than that of most other digeneans as it has four, rather than three, obligate hosts (Goater, Browne and Esch, 1990). Adults live under the tongue of the green frog, *Rana clamitans*, throughout temperate areas of North America. Eggs are swallowed and then shed by the frog when it defecates. Eggs must then be eaten by the snail intermediate host, *Helisoma anceps*. Once inside the snail gut, the eggs hatch and a motile larval stage penetrates the gut wall and migrates to the hepatopancreas where sporocysts develop. After a sporocyst generation, rediae are released from the sporocysts and eventually spill over into the gonads. These voracious larvae consume the gonads and cause host castration. Cercariae are then produced by the rediae and released from the snail. One of the unique characteristics of this parasite is the unusual

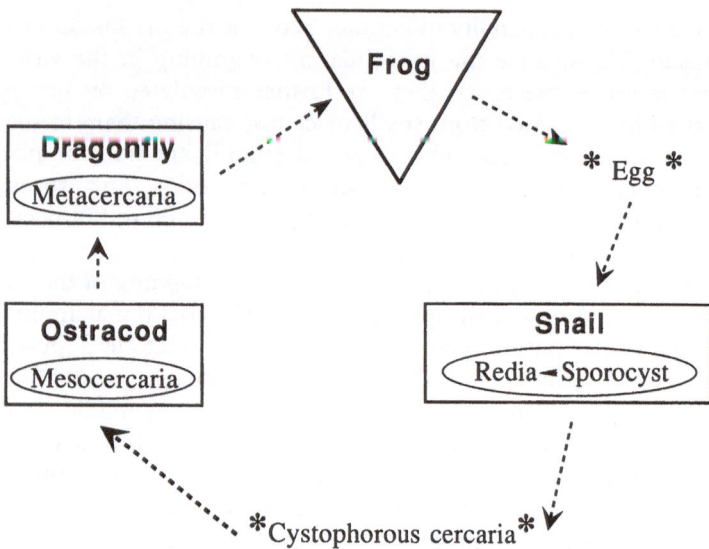

Figure 2.2 A schematic life cycle for the trematode, *Halipegus occidualis*. The symbols are explained in the caption for Fig. 1.1 (page 7).

cystophorous cercaria, also called a cercariocyst (Fig. 2.3a, b). These cercariocysts are non-motile, but they have a relatively long life span (Fig. 2.4) that is inversely related to water temperatures in the pond (Shostak and Esch, 1990b). There is a strong indication that cercariocysts are in an anhydrobiotic state when they are released from the snail (Goater, Browne and Esch, 1990) and that this accounts for their extended life span. The latter authors speculated that long-term survival by the cystophorous cercariae has evolved as a temporal mechanism for dispersal in time as opposed to the spatial dispersal strategies employed by free-swimming, but short-lived, cercariae of most other species.

The cystophorous cercariae are accidentally ingested by benthic copepods and ostracods in which the parasites penetrate the gut wall and develop into mesocercariae within the haemocoele. An unusual feature of transmission to the benthic crustaceans is associated with the presence of what appears to be a handle (Fig. 2.3a, b) on the surface of the cercariocyst. When stimulated by the crustacean, a long, barbed delivery tube (Fig. 2.3c) explodes from the cercariocyst and penetrates the gut wall of the crustacean. The body of the cercaria (Fig. 2.3d) is then simul-taneously, and explosively, 'shot' through the tube directly into the crustacean's haemocoele. Within the haemocoele, a mesocercaria devel-ops; this is an intermediate stage between a cercaria and a metacercaria.

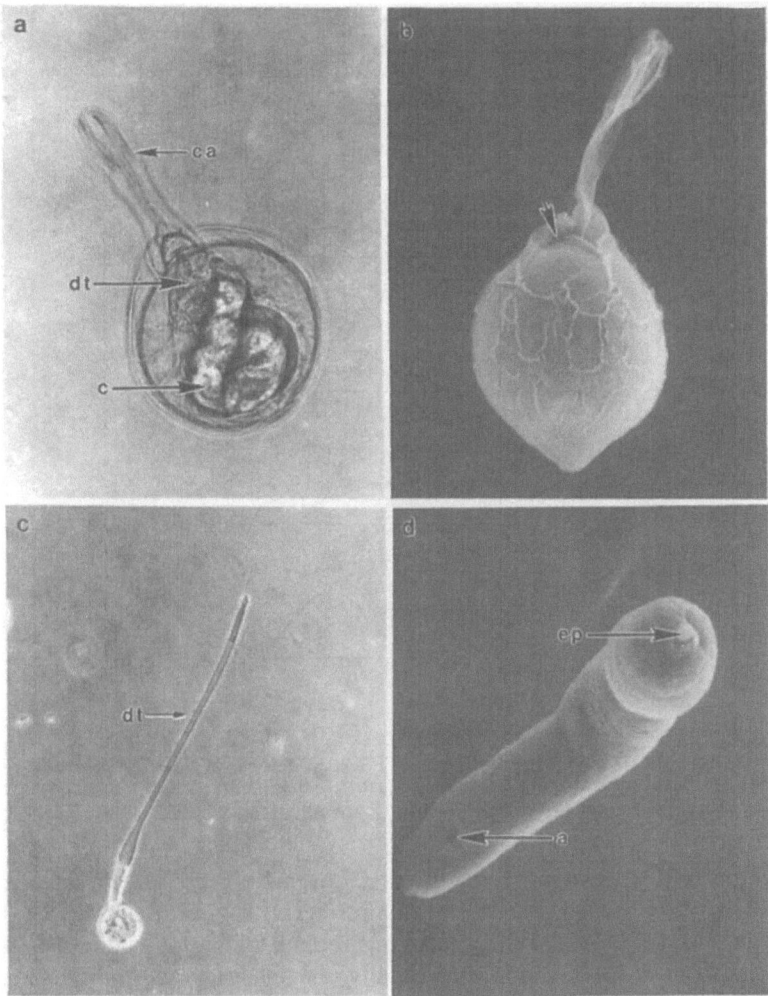

Figure 2.3 Developmental stages of the trematode, *Halipegus occidualis*: (a) Phase contrast photograph of a live, fully developed cystophorous cercariae of *H. occidualis*. Protruding from the transparent cyst is the caudal appendage (ca). At its base and within the double-walled cyst is the cercarial body (c) and the coiled delivery tube (dt); (b) Scanning electron micrograph of an intact cystophorous cercaria of *H. occidualis*. The outer cyst membrane is wrinkled. The arrow indicates the opening through which the delivery tube is everted upon manipulation of the caudal appendage by the feeding microcrustacean; (c) Phase contrast photograph of the everted cyst of *H. occidualis*. The cercarial body has been ejected through by the delivery tube (dt) in response to a gentle cover-glass pressure; (d) Scanning electron micrograph of the cercarial body of *H. occidualis*. The tegument is covered with microvilli and the excretory pore (ep) and small acetabulum (a) are clearly visible. (With permission, from Goater, Browne and Esch, 1990.)

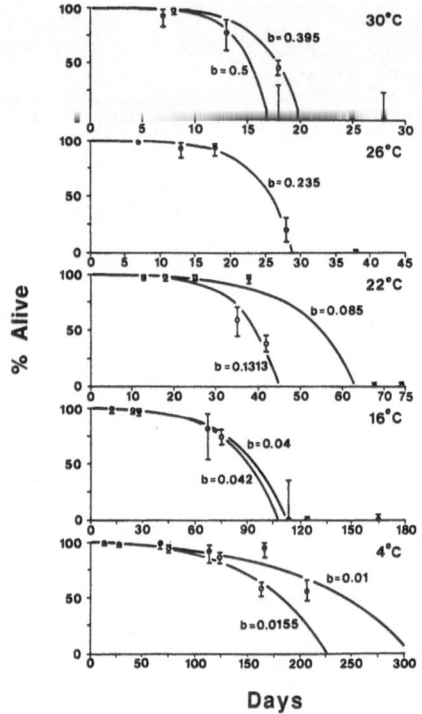

Figure 2.4 Survival of cercariae of *Halipegus occidualis* at constant temperature. Values are percent and 95% confidence limits. Solid and open circles show results from two experiments. (With permission, from Shostak and Esch, 1990b.)

The infected crustaceans are, in turn, eaten by dragonfly nymphs of a number of species representing at least seven different families. Meta-cercariae develop in the dragonfly, but remain within the midgut and do not move into the haemocoele. Finally, as the dragonfly nymphs meta-morphose and begin to emerge from the pond, they become vulnerable to predation by the ranid definitive host. On being digested from the dragon-fly in the frog's stomach, the immature parasites crawl back into the frog's mouth and mature sexually.

The distribution of *Halipegus*-infected snails was examined from a microgeographic perspective in a small 0.02 km² (2 ha) pond in the Piedmont area of North Carolina, USA (Williams and Esch, 1991). They found that infected snails were not randomly distributed in the littoral zone of the pond, but were concentrated in very shallow water, less than 5.0 cm, immediately adjacent to the shoreline (Fig. 2.5) and were found significantly more often on leaf substrata in the shallow waters (Fig. 2.6). These observations on the microgeographic distribution of *H. occidualis*

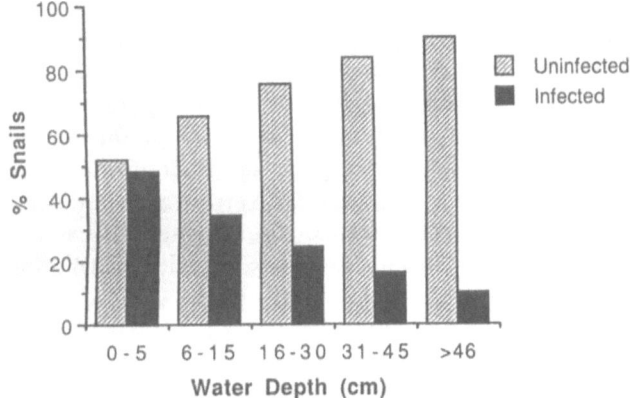

Figure 2.5 The prevalence of *Halipegus occidualis* in the pulmonate snail, *Helisoma anceps*, as a function of water depth. (With permission, from Williams and Esch, 1991.)

serve to emphasize the concept of scales in considering the overall population dynamics of different species of trematodes and probably other parasites as well (the notion of scale will be examined more thoroughly in later Chapters). They explained the distribution of the infected snails by correlating it with the behaviour of the definitive host. Green frogs are ambush predators, spending most of their time submerged in very shallow water along the pond's edge. It is here that they defecate and shed eggs in

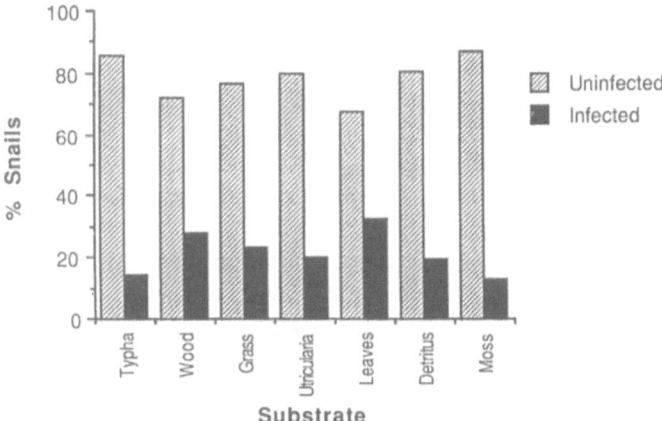

Figure 2.6 The prevalence of *Halipegus occidualis* in the snail, *Helisoma anceps*, with respect to the substratum from which the snails were captured. (With permission, from Williams and Esch, 1991.)

high concentrations. Snails such as *H. anceps* graze the leaf litter and other substrata in the littoral zone. As they do, they encounter aggregations of parasite eggs, primarily in shallow water and on leaf litter that, in effect, function as enhanced transmission foci for the parasite. The pond in which they worked is a small, lentic system, but the microgeographic qualities associated with this habitat are no different than they would be for a lotic habitat such as a stream or river, or even the ocean. Indeed, very few homogeneous environments exist anywhere. Because of the great amount of heterogeneity in most habitats, parasite distributions in space will also be patchy. For example, Gleason (1984, 1987) postulated that the distribution of the acanthocephalan, *Pomphorhynchus bulbocolli*, was restricted to riffle locations of a small Kentucky (USA) stream because the amphipod intermediate host, *Gammarus pseudolimnaeus*, was also restricted to these habitats.

Of the external factors affecting parasite infrapopulation biology, diet is probably the most basic. Thus, almost all of the enteric parasites within a definitive host are present because the host ingested an infective stage of some parasite. Dogiel (1964) made a strong case for the importance of diet in influencing the parasite fauna within hosts when he said, 'the parasitological indicators of diet are among those clues that allow us to make deductions from the type of parasite fauna about various aspects of the ecology of the host'. He illustrated this point by noting that herbivorous cyprinids within an aquatic system were virtually devoid of enteric helminths while carnivorous cyprinids within the same system possessed rich helminth faunas. The implication is that a carnivorous lifestyle is more likely to result in recruitments of parasites than an herbivorous one, but there are many exceptions to this pattern. For example, slider turtles, *Trachemys scripta*, are typically omnivorous, but also feed extensively on aquatic vegetation. It is their herbivorous diet that results in the recruitment of an extensive acanthocephalan fauna of the genus *Neoechinorhynchus* (Esch, Gibbons and Bourque, 1979a, b; Jacobson, 1987). These parasites use various species of ostracods as first intermediate host. The ostracods are a part of the epiphytic fauna associated with aquatic vegetation and are accidentally consumed as the turtles feed.

Horses are herbivores and they also often harbour rich faunas of strongylid nematodes in their gut. These parasites are accidentally acquired as filariform larvae while the horses graze. Some detritivores such as mullet (Mugilidae) also have rich trematode faunas that are acquired as the fish feed on the benthos of estuarine and marine coastal areas, accidentally ingesting metacercariae of the parasites present in the detritus (Paperna and Overstreet, 1981; Fernandez, 1987).

Only a few of the external environmental factors known to affect the infrapopulation biology of parasites have been discussed here. It is hoped, however, that these examples will serve as an introduction; other factors will be considered more fully in subsequent chapters. These will include seasonal changes in photoperiod and light intensity, ecological succession, the many ramifications of nutrient loading and pollution, habitat stability, dispersion and other spatial distribution patterns such as those discussed above for *H. occidualis*. The zoogeographical factors influencing the global distribution of both parasites and their hosts also will be considered.

2.3.2 Internal environmental factors

The range of internal environmental factors affecting the infrapopulation biology of parasites is probably not as great as the external ones but, in many ways, internal factors are more complex than external ones. This is true primarily because phenomena such as behaviour, host and parasite genetics, natural and acquired resistance, ontogenetic factors associated with the aging process, and sex are involved; moreover, several of these frequently act in concert. Interactions among infrapopulations within and between species may also affect parasite densities and fecundity. It is well known, for example, that egg production by some enteric helminths will rise with increasing parasite density, but will reach a threshold or even decline if parasite densities increase above a certain point (see section 3.4.3 for a full discussion of competition).

The two most complex internal factors affecting parasite infra-population biology are associated with the host's immune system and with the genetics of both the parasite and the host. This complexity is compounded when the genetics of the host impacts directly on its immune capabilities. One of the best studied cases of such an interaction involves the deer mouse, *Peromyscus maniculatus*, and the tapeworm, *Hymenolepis citelli*. Wassom, Guss and Grundmann, (1973), Wassom, DeWitt and Grundmann, (1974) and Wassom *et al.* (1986) reported that the tapeworms were contagiously distributed within the definitive host and that the incidence was low, less than 5%. They attributed these observations to the heterogeneous distribution of infected intermediate hosts (camel crickets, *Ceuthophilus utahensis*) within the habitat of the deer mice. Moreover, they presented evidence that a relatively high percentage (75%) of the deer mice are endowed with natural resistance that is related to a single, autosomal dominant gene. This resistance results in rejection of the tapeworms before they can become sexually mature and thus contributes to both the low incidence and the contagious frequency distribution of the parasite.

Host behaviour, age and sex are factors that may influence the infra-population biology of parasitic organisms. In some cases, all three factors can be inextricably linked. A good example of this linkage is associated

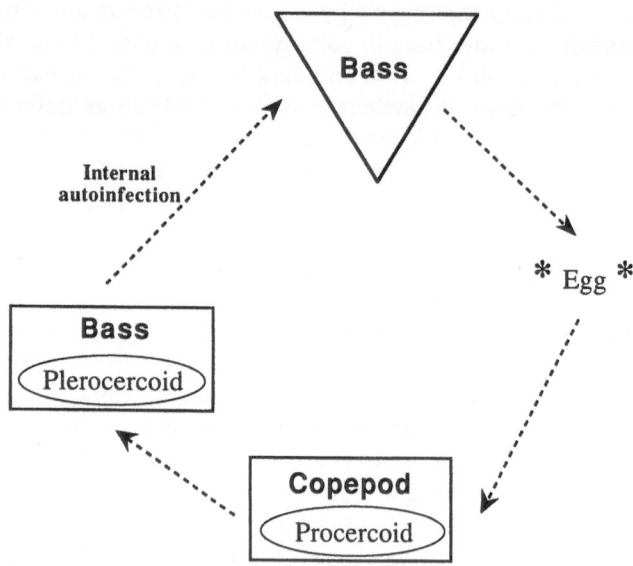

Figure 2.7 A schematic life cycle for the cestode, *Proteocephalus ambloplitis*. The symbols are explained in the caption for Fig. 1.1 (page 7).

with aspects of the infrapopulation biology of the cestode, *Proteocephalus ambloplitis*. Definitive hosts for this parasite are smallmouth bass, *Micropterus dolomieui*, and largemouth bass, *Micropterus salmoides*. The smallmouth bass is a cold-water species found commonly in the northern USA and southern Canada, and at higher elevations in the southern USA. After eggs are released from the adult tapeworm, they are shed in the faeces. The life cycle is shown in Fig. 2.7. Eggs are freed from the faecal material and float in the water column, using a peculiar set of dumb-bell shaped membranes that surround the egg for buoyancy. Eggs are consumed by several species of planktonic copepods. An oncosphere emerges from the egg and penetrates the gut wall of the copepod; it enters the haemocoele and develops into a procercoid. The copepods are then eaten by planktivorous fishes. It should be emphasized that, for the most part, only species of smaller fishes and fingerlings that will grow to much larger sizes will normally feed on copepods in freshwater systems. We thus have an example of an age-related or, perhaps more precisely, a size-related transmission tactic. Within the gut of the small fish, the procercoid is freed from the copepod by digestion and the liberated parasite penetrates the gut wall where it develops into a parenteric, i.e. outside the intestine, plerocercoid.

Another age-related component of the parasite's life cycle should also be considered. In order to complete the cycle, the plerocercoid must first reach a parenteric site within the definitive host, almost always a sexually mature smallmouth or largemouth bass. This can be accomplished in one of two ways. First, the plerocercoid-infected small fish or fingerling can be eaten by a larger bass. If this happens, the plerocercoid will be digested from the flesh of the intermediate host and then migrate through the gut wall of the bass into the body cavity. Once inside, it will migrate into various organs such as the gonads, spleen, and liver. Apparently, it continues to wander in these organs, growing in size until an appropriate stimulus is received.

In the second route, the copepod containing the procercoid is consumed by a fingerling smallmouth bass, in which case the parasite migrates into a parenteric site within the fingerling and changes into a plerocercoid. It wanders through various abdominal organs, growing in size until the bass becomes sexually mature. The infrapopulation of plerocercoids within a bass that becomes sexually mature for the first time will thus consists of a mix of parasites acquired directly from procercoids in copepods and those acquired through predation on small fishes and fingerlings infected with plerocercoids that were obtained from copepods infected with procercoids.

The final step in the cycle involves several phenomena about which not much is understood. In the spring of each year, when water temperatures rise, adult tapeworms begin to accumulate in the pyloric caeca of sexually mature bass (all adult worms are lost each autumn and are replaced the following spring by a new infrapopulation). Under laboratory conditions, Fischer and Freeman (1969) were able to demonstrate that rising temperature was apparently the stimulus for migration of parenteric plerocercoids from sites within the abdominal cavity into the lumen of the pyloric caeca. However, this migration also coincides with increases in the level of circulating spawning hormones that are released from the gonads and pituitary gland in the spring. The spawning act in smallmouth bass is precisely determined by water temperature. An interesting feature in this process is that when sexually mature bass begin to acquire new plerocercoids from plerocercoid-infected fishes as the summer progresses, the new plerocercoids migrate into parenteric sites and do not develop into adult tapeworms until the following spring. In other words, all plerocercoids must spend at least part of one annual cycle in parenteric sites within a sexually mature bass before migrating back into the intestine and developing into adults within that bass. This could be interpreted as a mechanism to minimize competition between an already established adult infrapopulation and new, potential adults. Plerocercoids occurring in other centrarchid species such as the bluegill, *Lepomis macrochirus*, will not migrate out of parenteric sites after the fishes become sexually mature even though subjected to rising spring temperatures. This clearly implies that the stimulus for migration from parenteric sites in bass is not rising temperature alone, but is something specifically associated with the appropriate definitive host.

The precise nature of the stimulus in smallmouth bass is not known. Fischer and Freeman (1969) and Esch, Johnson and Coggins (1975) suggested that it was a combination of rising temperatures and increasing levels of hormones since migration coincided with the spawning act and since juvenile bass are not infected with adults even though they usually possess plerocercoids. The problem with this hypothesis is that in South Carolina (southern USA), adults of *P. ambloplitis* are present in large-mouth bass during winter and disappear in the spring at about the time of spawning (Eure, 1976). This is opposite to the pattern that occurs in the north. Whatever the migration cue, this system is an elegant one to highlight the complex relationships between parasite infrapopulation biology and host behaviour, age, and sexual maturation.

Another excellent example of an hormonal effect is associated with the synchronization of the life cycles of the monogenean, *Polystoma integerrimum*, and its frog host, *Rana temporaria*. *Polystoma integerrimum* enters the frog bladder before the tadpoles metamorphose and the juvenile frogs become terrestrial. Accordingly, transmission of the parasite to new hosts can occur only when frogs invade temporary bodies of water to reproduce. When frogs are preparing to enter the water prior to copulation (after 3 years on land), the reproductive system of the monogenean develops. As the frog spawns for the first time, maturation of *P. integerrimum* occurs and eggs are released. In this way, the synchronized mechanism ensures that, when the monogenean eggs hatch, tadpoles will be available for infection. The high correlation between the host's and the parasite's reproductive cycles suggests that reproductive development in *Polystoma* is controlled by the hormonal activity of the frog. Indeed, experiments have demonstrated that maturation and stimulation of gamete production in *P. integerrimum* can be elicited by injecting the frog host with a pituitary extract. However, it is not known yet whether the effect is a direct one through the injected pituitary hormones, or via the host gonadal hormones that are produced in response to those from the pituitary. A similar synchrony in life cycles has been demonstrated for *Polystomum stellai* and the tree frog, *Hyla septemtrionalis* (Stunkard, 1955). In contrast, however, polystomatids parasitic in more aquatic amphibians, e.g., *Protopolystoma xenoposis* in *Xenopus* spp., exhibit development that is not influenced by season of the year, reproductive condition of the host, or by experimental treatment with sex hormones. In these cases, the uninterrupted reproductive activity of the parasite is in some manner related to the continuous availability of its aquatic host (Tinsley and Owen, 1975).

2.4 THE DISPERSION CONCEPT

Before further considering any basic ideas regarding parasite population biology and regulatory interactions, it is first necessary to identify and discuss several additional population terms and concepts. The sample **mean**, $\bar{X}$, is defined as 'the sum of all measurements in the sample divided by the number of measurements in the sample' (Zar, 1984). It is one of the most important measures of what statisticians refer to as central values or central tendencies. The **variance**, (S^2), of a population mean is a measure of mathematical variability within the population. The $\bar{X}$ and S^2 are among the most important parameters for any population, but they are only estimates since everything cannot be known about any population (for a complete review of these terms, refer to Zar, 1984).

Since the mean and the variance for a given population are only estimates, it becomes critical that sampling be conducted both accurately and randomly. In the sampling of free-living populations, it is a matter of going into the field and counting individuals within the population in a precise, reproducible, and statistically sound manner. Some species are clearly more difficult than others to enumerate and, accordingly, a variety of sampling protocols have been developed for various plants and animals (Tanner, 1987).

Parasite infrapopulations present special problems. For all but a few species, hosts must be killed before parasites can be counted. The problem with this approach is that by killing the hosts, both the hosts and the parasites they carry are removed from both their overall populations and gene pools. In some cases, this does not create a serious problem because the population size of the host is large enough so that the removal of a relatively few individuals will not significantly affect the reproductive capabilities of the remaining hosts or their parasites. It is, therefore, necessary to obtain certain basic information on the population biology of both the host and the parasite and evaluate it carefully before undertaking a kill-and-count procedure.

The variance of the parasite metapopulation is an important estimate when assessing dispersion of parasites among the hosts. Consider the six distinct metapopulation frequency distributions in Figure 2.8. In each of these cases, the mean is identical, but note how the metapopulations differ in their dispersion characteristics. If an assessment is made regarding how much the dispersion patterns of various metapopulations differ from each other, then substantially more information about the metapopulations could be generated. Such a measure would provide an idea of the spread, or variability, within the metapopulation.

Frequency distributions of metapopulations in space typically come in three distinct patterns (Fig. 2.9). These distributions are termed **random**

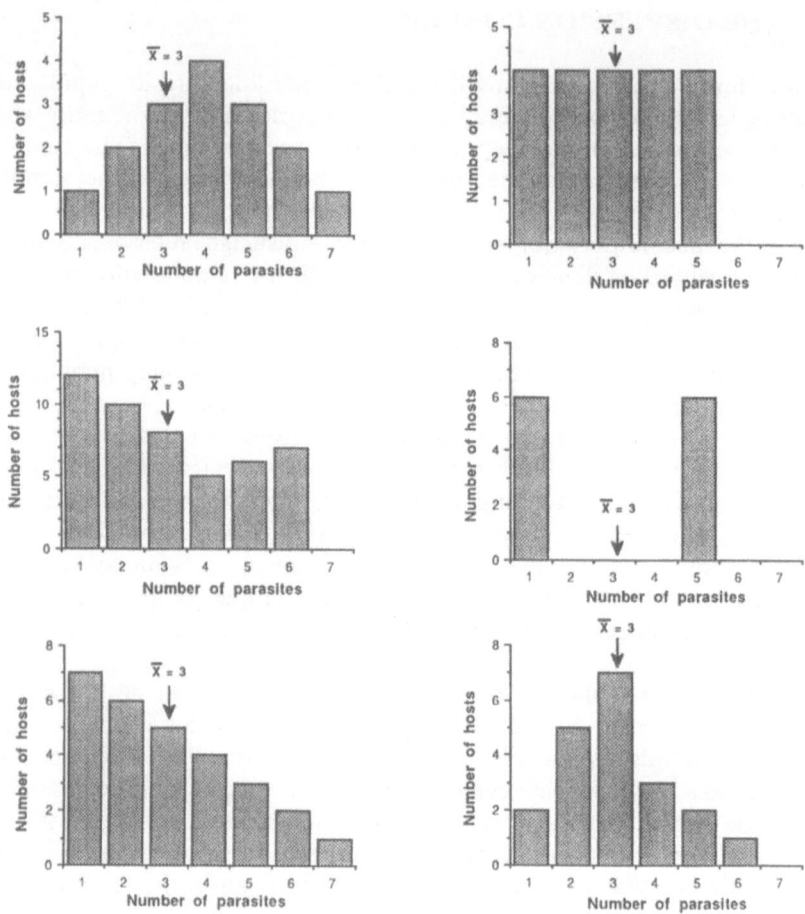

Figure 2.8 An example of the variability with which different populations can be dispersed around the same mean. In each case the mean is the same ($\bar{x}=3$), but the distribution of the observed numbers is different.

or Poisson, **underdispersed** or regular, and **overdispersed**, contagious, or clumped. Random distributions occur when the position of one individual is completely independent of any other and when each segment of the habitat has the same probability of being colonized. Random frequency distributions are most often associated with newly colonized habitats, while regular distributions are usually identified with competition or **allelopathy** (antibiosis). A classic example of the latter phenomenon is with creosote plants in the southwestern desert of the USA; these shrubs produce volatile terpenes and phenols that prevent the germination and

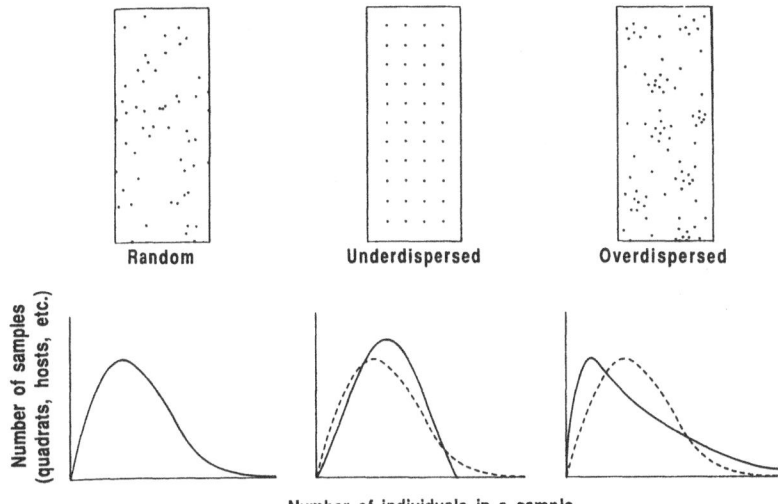

Figure 2.9 Distributional patterns in space (above) for infrapopulations that are random, underdispersed, and overdispersed. Below are the frequency curves which describe these distributions. In a regular distribution, most hosts would have approximately the same number of parasites; in an overdispersed distribution, there would be a relatively large number of hosts with a few parasites and a relatively small number with most of the parasites. The dashed line is the same random distribution for comparison.

establishment of other creosote plants in their vicinity, thereby creating dispersion patterns that are regular.

The importance of knowing the variances and means of parasite metapopulations is apparent when these two estimates are computed for the three hypothetical metapopulations given in Table 2.2. For each, the mean $\bar{X}$, the variance (S^2), and the variance/mean ($S^2/\bar{X}$) ratio, or **coefficient of dispersion** have been calculated. Note that for each metapopulation, the mean ($\bar{X}$) is the same ($\bar{X} \approx 6.0$). The statistical significance for goodness of fit can be assessed using a simple chi-squared test. In metapopulation I, the variance (S^2) is not significantly different from the mean ($\bar{X}$); thus, $S^2 \approx \bar{X}$. In this case, the metapopulation is randomly distributed. Whenever the variance/mean ratio is at unity, or close to it, a metapopulation is randomly distributed. In metapopulation II, the variance/mean ratio ($S^2/\bar{X}$) is greater than unity ($S^2 >> \bar{X}$) and it is overdispersed. In metapopulation III, the variance/mean ratio ($S^2/\bar{X}$) is less than unity and, mathematically, this metapopulation is regularly distributed, or underdispersed ($S^2 << \bar{X}$). In other words, the distribution of values around the mean in each metapopulation is quite distinct and each distribution can be described mathematically; they will be either

Table 2.2 Mean (X), variance (S^2) and variance/mean ratios for three, hypothetical populations. (X_i = number of parasites in individual hosts, for example; N = number of hosts sampled in each population; $\overline{X}$=mean number of parasites per individual host sampled; S^2 = variance of the parasites in the N hosts sampled).

Population I	Population II	Population III
X_i = 3, 7, 10, 5, 8, 7, 2	X_i = 10, 11, 2, 3, 11, 3, 2	X_i = 6, 5, 8, 7, 6, 5, 5
N = 7	N = 7	N = 7
$\overline{X}$ = 6.0	$\overline{X}$ = 6.0	$\overline{X}$ = 6.0
S^2 = 8.0	S^2 = 19.3	S^2 = 1.33
$S^2/\overline{X}$ = 1.33	$S^2/\overline{X}$ = 3.22	$S^2/\overline{X}$ = 0.22
$S^2 = \overline{X}$	$S^2 > \overline{X}$	$S^2 < \overline{X}$
Random distribution	Overdispersed distribution	Underdispersed distribution

random, underdispersed, or overdispersed. The significance of understanding the frequency distribution of parasite metapopulations will become more apparent when Crofton's (1971a, b) definition of parasitism is discussed in greater detail. Since his seminal work, the concept of overdispersion has become a central component for paradigms involving parasite population biology and of models employed in describing regulatory interactions between host and parasite populations (Anderson and May, 1979; May and Anderson, 1979).

Underdispersed frequency distributions are uncommon among parasites. Cestodarians of the genus *Gyrocotyle* are one of the few exceptions. Studies of different species of *Gyrocotyle* in holocephalan hosts such as *Chimaera monstrosa*, *Hydrolagus colliei*, *Callorhynchus millie* and *C. callorhynchus*, reveal some striking similarities. In most instances, prevalence is very high (90–100%) and parasite densities are low, with infrapopulations composed mostly of two individuals (Halvorsen and Williams, 1967/68; Dienske, 1968; Simmons and Laurie, 1972; Allison and Coakley, 1973; Fernández, Villalba and Albina, 1983). Four possible mechanism have been proposed to explain why underdispersed distributions occur among these species of parasites. The first suggests a reduction in the probability of infection caused by ontogenetic changes in host biology, e.g. in food selection or habitat. The second proposes that parasite recruitment rates are equal to death rates. A third explanation suggests that heavier infections are lethal. Finally, it has been suggested that further infections are prevented through intraspecific competition, or the host's immune response, or both. Although not much is known about the biology of either hosts or parasites, the evidence available favours explanations based on the last mechanism, with regulation mediated by antagonistic, parasite–parasite interactions (Williams, Colin and Halvorsen, 1987). Another example of an under-dispersed distribution is *Deretrema philippinensis*, a gall bladder trematode in the flashlight fish, *Anomalops katopron*. Burn (1980) and Beverley-Burton and Early (1982) consistently found two trematodes per gallbladder, with an incidence of almost 100%. Again, however, the mechanism responsible for this distribution pattern has not been determined.

Neither random nor regular frequency distributions are that common in nature. By far the most common form of frequency distribution is overdispersion, or contagion. This pattern is as common for parasite populations as it is for free-living organisms. Overdispersed frequency distributions can be described by several different theoretical models, including the log normal, log series, Neyman type A, and the negative binomial. The significance of these models and their application to parasite population dynamics will be discussed in Chapter 3.

2.5 DYNAMICS OF POPULATION GROWTH

Generally speaking, there are two patterns of growth exhibited by most populations. In an ideal environment that has unlimited resources, growth may be exponential; that is to say, numbers per unit of space will increase 1, 2, 4, 8, 16, 32, etc. (Fig. 2.10). Curve (a) in Fig. 2.10, which illustrates this growth, can be described by the following equation:

$$dN/dt = rN$$

where N equals the number of individuals at time t, and r equals the per capita rate of natural increase. The rate r, is obtained (in the absence of net immigration or emigration) by subtracting the individual birth rate, b_o, from the death rate, d_o. Thus, r is said to 'represent in a single number all of the physiological responses of all members of a population to a given set of environmental factors' (Hairston, Tinkle and Wilbur, 1970). Species with growth curves that are exponential in character are said to be r-selected (Dobzhansky, 1950; MacArthur and Wilson, 1967; Pianka, 1970; Esch, Hazen and Aho, 1977). Species that employ such a growth strategy are also frequently described as opportunistic in the sense that they tend to maximize their reproductive capacities and do not direct their energies toward enhancing the survivorship of their offspring.

Many species do not exhibit exponential growth. Instead, they follow a logistic pattern (curve (b) in Fig. 2.10); only when these species are colonizing a new habitat, or perhaps at the beginning of a new reproductive season, will they approximate an exponential pattern of growth. Even then, however, it is not truly exponential because of the resistance

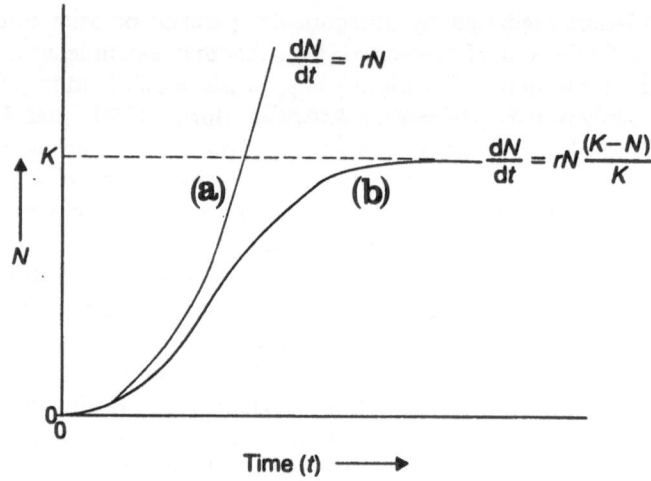

Figure 2.10 Population growth curves: (a) exponential growth; (b) logistic or sigmoidal growth. N=number of individuals; K=carrying capacity of the environment; r = rate of increase.

created by many environmental factors. The logistic growth curve can be described by the equation:

$$dN/dt = rN [(K - N) / K]$$

where K=the carrying capacity of the habitat, or that point in time when dN/dt=0. The term $[(K - N) / K]$ means that as N increases, dN/dt decreases. When N=K, the term will be 0 and dN/dt also will be 0. When N is close to 0, or when the population density is beginning to increase rapidly, dN/dt is nearly the same as $r - N$; at this point, the growth curve is as close to being exponential as it can be unless the carrying capacity is later increased in some manner. The term $[(K-N) / K]$ is the simplest method of expressing the manner in which a population can expand up to equilibrium K. If N exceeds K, the term will become negative and N will approach K from the other direction.

The shape of a logistic curve is characteristic ((b) in Fig. 2.10). When a population initially colonizes a habitat, growth is slow. This is a time for physiological adjustment to new surroundings. This period of time is followed by a near-exponential increase in population size. The difference between the shape of the growth curve at this point in time and a true exponential curve is said to represent the biotic resistance of the environment. Eventually, however, environmental pressures begin to slow population growth. Resources such as nutrients or space may become limiting, or toxic materials produced by members of the population may begin to accumulate. Finally, the population growth stabilizes; at this time, birth and death rates become equal. Once the carrying capacity is reached, population size will remain reasonably constant except for erratic or irregular fluctuations in response to short-term shifts in various resources. Competition, predation, and parasitism may also influence population fluctuations. There are some populations that oscillate after reaching their carrying capacities; in these cases, there are regular changes in density over fixed periods of time. The classic case is the 10-year lynx/ hare cycle recorded by the Hudson Bay Company since the mid-19th century. In this type of oscillation, population densities of the lynx track those of the hare with a high degree of fidelity. Some have speculated that the population density of the lynx is directly related to that of the hare, although some have also challenged this assumption.

If limiting factors are increased, then the initial phase of the growth curve may be repeated until some new carrying capacity is reached and then it will level off again. In some cases, the population will undergo senescence and become locally extinct. In these situations, perhaps the population has become too specialized for a changing environment, or perhaps a competitor enters the habitat and is more successful than the previously established population. Species that exhibit a logistic growth curve are said to be K-selected.

The concept of *r*- and K-selection was developed by Dobzhansky (1950). He suggested that in the tropics, selection favoured species with low fecundity and slow development whereas in temperate climates the opposite was true. Subsequently, MacArthur and Wilson (1967) coined the term *r*-selection to describe forces that would produce high fecundity and rapid development in a species and K-selection to describe selection forces that caused species to evolve growth and development strategies that were just the opposite. Pianka (1970) proposed the notion of an *r*–K continuum. At one end, the *r*-strategy was to place 100% of resources into reproduction. At the other end of the continuum, a K-strategy would be for production of highly fit progeny and for self-maintenance. Later, Esch, Hazen and Aho (1977) examined parasitism within the context of *r*- and K-selection, concluding that parasites exhibited many characteristics of an *r*-strategist but that, as with most generalizations, it was impossible to apply the concept consistently. During the past several years, the ideas regarding *r*- and K-selection have been highly debated (Sibly and Calow, 1985). The reasons for this turn of events and the current thinking regarding selection models will be considered more thoroughly in Chapter 5.

Generally speaking, populations that consistently grow in an exponential fashion throughout their existence will be influenced by density-independent factors. These factors are abiotic and are largely equivalent to the external environmental factors mentioned earlier in connection with parasite flow through a host. Such factors would include temperature, light, humidity, etc. The list of abiotic factors that may affect parasite population dynamics is not unlimited, but it is extensive. Within the context of r- and K-selection, organisms affected by abiotic factors would be considered as 'boom–bust' species, or toward the r- end of the r–K continuum.

Organisms whose populations exhibit logistic growth, in general, will be influenced by density-dependent factors such as competition for space or nutrient resources. Parasites that induce an immune response that is a function of parasite numbers may be considered as being regulated by density-dependent immune responses. Populations regulated by density-dependent forces are more likely to be at the K end of the r–K continuum.

In concluding this Chapter, a number of questions regarding parasite population dynamics should come to mind. For example, how do parasites fit into the exponential–logistic schemes of population growth? Are parasite populations regulated primarily by abiotic factors or by density-dependent forces? If the latter, are they consistent over time? If the former, are most parasite species of the 'boom–bust' variety? Are growth patterns of most parasite species influenced by the stability or instability of their physical environment? Most parasite metapopulations are over-dispersed within their host populations, but what is the significance of this sort of dispersion pattern within the framework of parasite population dynamics? May and Anderson (1979) claimed that parasitic organisms are highly effective in regulating host populations. Indeed, they stated that 'parasites (broadly defined) are probably as important as the more usually studied predators and insect parasitoids in regulating natural populations'. In general terms, there may be some validity to this statement if one includes viruses and bacteria as parasites (and they did). On the other hand, if one includes only that which the traditional parasitologists consider as parasites (mostly protozoans, helminths and arthropods), then the accuracy of their assertion would probably be challenged. Issues of this kind will be considered in the next two chapters.

3 Factors influencing parasite populations

3.1 DENSITY-INDEPENDENT FACTORS: INTRODUCTION

According to Scott and Dobson (1989), 'regulation occurs through the action of processes that tend to reduce the per capita survival or reproduction of the members of a population as the density of that population increases.' In other words, regulation implies the influence of density-dependent forces. Probably, however, most parasite infrapopulations are not regulated by density-dependent constraints. Instead, they are mostly affected by density-independent factors such as temperature or other general climatic conditions and by some biotic factors, primarily those associated with host behaviour.

There are several generalizations that have been made (Bradley, 1972) regarding density-independent factors and the role they play in the dynamics of parasite infra-, meta-, and suprapopulations. These factors are almost always extrinsic to both the hosts and parasites. Many of these species may be influenced by a large number of physical characteristics in the environment. Other species may be identified with an aspect of host behaviour that the parasite exploits in order to enhance the probability of transmission. In some of these cases, the parasite would capitalize on predator–prey interactions in such a way that the host is more likely to be eaten by the next host in the life cycle (Moore and Goniteli, 1990).

Some evidence suggests that the birth and mortality rates of parasites in most hosts under natural conditions are independent of infrapopulation density. Instead, reproduction by the parasite may be influenced by seasonal changes in air or water temperature, for example. Or, parasite density or prevalence could be a function of natural senescence and mortality within the host population. Thus, for example, many freshwater snail are annuals, with a new population cohort replacing an existing one each year. This means that as the snails die, the parasite community associated with them is also lost. Therefore, the replacement of an existing trematode fauna within a snail population having an annual cycle would also be an annual event, and occur independent of any density-related effects.

The densities of certain parasite species under the influence of density-independent factors may fluctuate erratically over the long-term. In reality, this statement is more of a prediction than a fact. The problem here is that there are far too few long-term studies to make an acceptable generalization. Kennedy and Rumpus (1977) contended that annual persistence in the prevalence of some parasites is not the result of density-dependent regulation, but rather is due to long-term stability within the ecosystem in which the parasite is found; the parasite metapopulations are therefore non-equilibrial even though that is not what they seem to be. In these situations, if the ecosystem is perturbed, then the population biology of the parasite will be affected in some unpredictable fashion. Certainly, Marcogliese, Goater and Esch (1990) in their study of the allocreadid trematode, *Crepidostomum cooperi*, in a eutrophic lake in Michigan (USA) would support such a contention. This parasite uses centrarchid fishes as definitive hosts and sphaeriid clams as the first intermediate host. Fig. 3.1 demonstrates the life cycle. Cercariae released from the clams encyst as metacercariae in nymphs of the burrowing mayfly, *Hexagenia limbata*. When nymphs emerge from the lake as sub-imagos, they are highly vulnerable to predation by many species of fish. After 48 hours the mayflies molt the last time, mate, and the females return to the lake to oviposit. They are again highly vulnerable to predation. If an infected mayfly is eaten by an appropriate fish, then the life cycle of *C. cooperi* will be completed.

Several hundred sub-imagos were collected each August over a 20 year period and the metacercariae counted (Marcogliese, Goater and Esch, 1990). For the first 8 years, parasite densities increased significantly (Fig. 3.2). They declined slightly for the next 6 years and then increased again, sharply. In 1986, the metapopulation density of metacercariae crashed and continued to decline over the next 3 years, reaching a 20-year low in the summer of 1989. The increasingly higher densities over the first 16 years corresponded with an environmental perturbation, namely progressive, cultural eutrophication. A sewer system for the lake's drainage basin was constructed, made operational in 1984, and eutrophication was reversed. Esch *et al.* (1986) earlier had speculated that mayfly nymphs were driven from deeper anoxic areas of the lake over the first 16 years (1969–1984) into parts of the lake where transmission of the parasites to

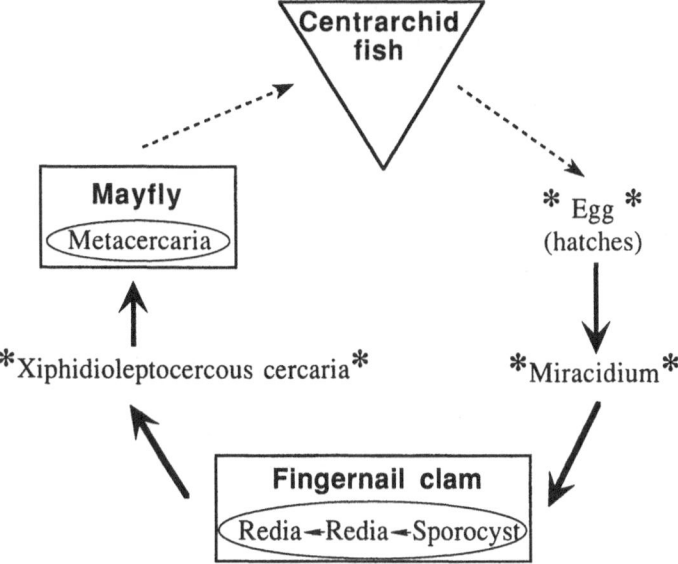

Figure 3.1 A schematic life cycle for the trematode, *Crepidostomum cooperi*. The symbols are explained in the caption for Fig. 1.1 (page 7).

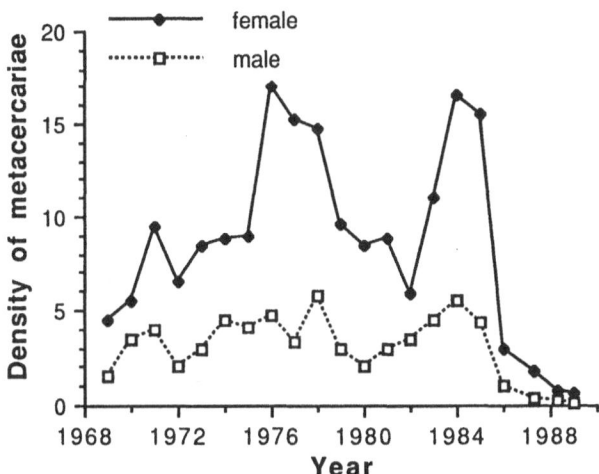

Figure 3.2 Densities of *Crepidostomum cooperi* metacercariae in female and male, *Hexagenia limbata*, sub-imagos from 1969 to 1989 in Gull Lake, Michigan, USA. (With permission, from Marcogliese, Goater and Esch, 1990).

the mayflies was enhanced. Then, as eutrophication was reversed, they predicted that nymphs would move back into deeper parts of the lake and away from higher densities of infected sphaeriid clams. The result would be a diminution in the transmission of cercariae to the mayfly nymphs and a decline in metacercariae prevalence and density. Based on the follow-up study by Marcogliese, Goater and Esch (1990), this appears to have happened in the way Esch *et al.* (1986) had predicted. However, one of the serious problems with such investigations, as pointed out by Marcogliese, Goater and Esch (1990), is that there are too few long-term studies with which their results could be compared. Moreover, without adequate controls or the ability to set up and conduct experimental protocols over the long-term, many of the conclusions, or speculations, or both, must necessarily be inferential in character. This makes the 'science' of long-term studies more difficult to assess and evaluate.

Another characteristic of parasite species that are influenced mainly by density-independent factors, is that they are frequently subject to local extinction. Again, however, in order to know if extinction has occurred, the system must be followed over a period of time and, as mentioned, such studies are virtually non-existent. In general, local extinction is most likely to occur in a perturbed ecosystem, or one in which colonization by infected definitive hosts is an annual, but erratic, event. An example of the latter situation is with the digenetic trematode, *Hysteromorpha triloba* in a lake in South Dakota, USA (Hugghins, 1956). The definitive hosts for this parasite are cormorants, *Phalacrocorax auritus*, that used the lake as a nesting site but, in 1955, following the opening of a public recreation area nearby, they abandoned the habitat. With the disappearance of the birds, the parasite also disappeared.

The sudden appearance of a new parasite in an ecosystem is also a characteristic of many species whose densities are primarily controlled by density-independent factors; it is, in effect, a colonization phenomenon. A case in point is in Slapton Ley in the southwest of England where C.R. Kennedy and several of his co-workers have been working for a number of years. For at least 12 years before 1973, the great-crested grebe, *Podiceps cristatus*, had not been present (Kennedy and Burrough, 1977). In 1973, the digenetic trematode, *Tylodelphys clavata* (the identification of the parasite was subsequently changed by Kennedy (1987) to *Tylodelphys podicipina*) suddenly appeared in perch, *Perca fluviatilis*, the second inter-mediate host of the parasite. The appearance of the parasite in perch coincided with the onset of nesting activity by grebes in the lake. Interestingly, this host–parasite system also became the focus of a long-term study (Kennedy, 1987). For 14 years, the prevalence and density of *T. podicipina* were followed in Slapton Ley. Densities of the parasite increased in perch until they peaked in 1981. There then followed an almost steady decline. The decrease was paralleled by a catastrophic decline in the perch population, making continued study of the parasite impossible. Kennedy (1987) referred to the *T. podicipina* metapopulation in perch as being non-equilibrial and unstable, even at the time when metacercariae densities were at their highest, >9/ host. He considered the possibility of density-dependent regulation (*sensu* Scott and Dobson, 1989), but rejected the idea. He concluded that the consistency of the parasite in the system prior to the decline of the perch population was more likely a function of the narrow transmission window for the parasite and of various density-independent factors affecting the parasite's life cycle.

Survival of parasites that are influenced by density-independent factors is uncertain. The uncertainty of survival can be considered at the infra-, meta-, and suprapopulation levels. Many parasites with complex life cycles require a special combination of physical conditions in the environment and specific sequences in host life-history events in order to survive at the suprapopulation level. The *T. podicipina* system described by Kennedy (1987) and the *C. cooperi*–mayfly described above (Marcogliese, Goater and Esch, 1990) are clear examples of how precise the physical conditions in a habitat must be in order for the parasite to be successful over the long term.

Many parasites that are influenced by density-independent phenomena are found in short-lived hosts. Most of these parasites are highly seasonal in their transmission patterns, especially since their short-lived hosts are usually seasonal with their own life-cycle patterns. Indeed, many of their hosts are also affected by density-independent factors. In these cases, it is absolutely necessary for the life cycles of the parasites to be highly synchronized with those of their hosts in order for them to be successful. In many situations, both hosts and parasites are responding to the same external stimuli, probably because they co-evolved, or co-adapted, and responded to some of the same selection pressures in evolutionary time. Parasites in short-lived hosts generally will not build up large numbers because the life spans of the hosts are not long enough to permit the generation of large numbers of parasites. There are, however, a number of exceptions to this pattern; these will be considered subsequently.

Finally, many parasites that are near the limit of their geographic ranges will be affected by density-independent factors, sometimes much more so than they would be otherwise. This is because their physiological tolerances are usually quite narrow and they are simply not highly adaptable to extreme environmental vagaries such as would occur at the edges of their range.

3.2 DENSITY-INDEPENDENT FACTORS: CASE HISTORIES

3.2.1 *Caryophyllaeus laticeps*

The order Caryophyllidea is considered by many as a minor taxon in the subclass Eucestoda. However, a number of species in this group have been extensively studied from a variety of perspectives and several have emerged as useful models for the study of parasite population biology. All species are monozoic, or unsegmented, and the scolex is quite simple morphologically. Adults of *C. laticeps* are found within the intestines of freshwater fishes (their life cycle is shown in Fig. 3.3). Unembryonated, operculate eggs of the parasite are shed from definitive hosts. After an appropriate incubation period, the eggs become infective for benthic, tubificid worms. On being ingested by the tubificid, the egg hatches in the intestine; an oncosphere emerges, penetrates the gut wall, and migrates into the haemocoele where it develops into a procercoid. When the infected annelid is eaten by a bottom-feeding fish, the parasites develop into what some refer to as neotenic plerocercoids (plerocercoids because they are monozoic and neotenic because, if they are plerocercoids, then they reproduce sexually while still in larval form (Olsen, 1986)). Whether neotenic plerocercoids or not, they mature sexually and produce eggs, completing the life cycle. The life cycle of *C. laticeps* is representative of most other caryophyllaeids.

Caryophyllaeus laticeps has a fairly cosmopolitan distribution in the northern hemisphere. It has a wide range of definitive hosts and probably intermediate hosts as well. The ease with which definitive hosts can be

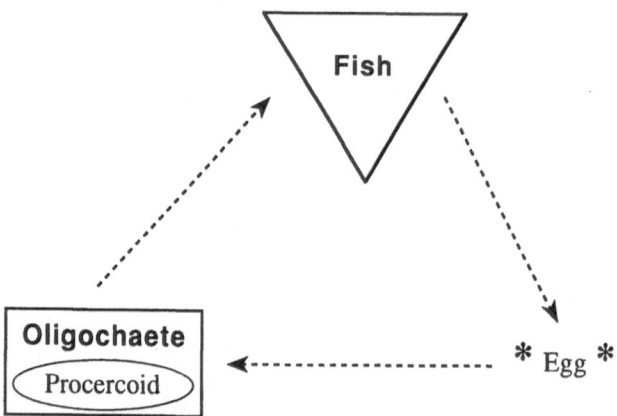

Figure 3.3 A schematic life cycle for the cestode, *Caryophyllaeus laticeps*. The symbols are explained in the caption for Fig. 1.1 (page 7).

Table 3.1 Seasonal variation in the degree of infection of dace, *Leuciscus leuciscus*, by the cestode, *Caryophyllaeus laticeps*. (From Kennedy, 1968).

Year and month		Number of dace		Percent infection	Number of worms found	Mean worm burden	Temperature (°C)	
		Examined	Infected				Mean	Range
1966	March	27	16	59.2	144	9	7.8	6.6–8.9
	April			No samples taken			9.25	5.0–11.1
	May	30	18	60.0	153	8.5	11.5	8.3–14.4
	June	27	6	22.2	7	1.1	14.6	13.3–15.5
	July			No samples taken			14.9	14.3–17.6
	August	30	0	0	0	0	15.5	13.3–17.7
	September	21	0	0	0	0	14.9	11.2–17.7
	October	28	1	3.6	1	1	11.1	8.3–13.3
	November	27	0	0	0	0	8.6	7.2–10.0
	December	23	2	8.7	3	1.5	6.8	4.4–8.9
1967	January	26	6	23.1	35	5.8	6.9	4.4–10.2
	February	25	14	56.0	262	18.9	7.3	4.4–10.2

collected and the relative simplicity of the parasite's life cycle make it an attractive model for study. It has, therefore, been the focus of fairly intense research efforts over the years. The choice of *C. laticeps* for illustrating the effects of temperature on the population biology of a given parasite species was made with a clear purpose. A review of the literature shows a striking correlation between water temperature and parasite population biology, especially in aquatic poikilotherms; it also demonstrates how differently the parasite behaves from one host species to another, even within a relatively narrow geographic range.

Consider first the study by Kennedy (1968) on *C. laticeps* in dace, *Leuciscus leuciscus*, in the River Avon, Hampshire, England. His investigation was conducted for 12 consecutive months. Two sets of data from his paper are instructive with respect to the influence of temperature on the metapopulation biology of the parasite. The first, in Table 3.1, shows the clear seasonal component of both the prevalence and density, i.e. mean worm burden, of *C. laticeps* in dace. The second, in Fig. 3.4, shows the changing pattern of maturation by the parasites after recruitment from infected tubificids. Several conclusions can be drawn from these data. Thus, recruitment of juvenile *C. laticeps* began in December and was continuous through June when it stopped completely. Mature, egg-shedding adults were present from January to June when they disappeared completely. The highest prevalence of all worms, disregarding the state of maturity, occurred from January to June. The highest densities were from January to May, with a few present in December and June. An examination of the data in Table 3.1 shows an almost perfect relationship between rising temperature and both the recruitment and maturation of the parasite. Despite these strong correlations, Kennedy (1968) proposed

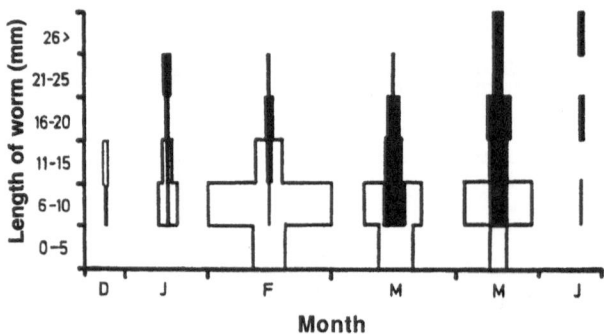

Figure 3.4 Monthly pattern of changes in length and state of maturity of *Caryophyllaeus laticeps* in dace, *Leuciscus leuciscus*. The data are expressed as actual numbers of mature (unshaded) and gravid (shaded) worms. (With permission, from Kennedy, 1968.)

that a direct relationship did not exist, arguing that it was spawning hormones that were probably triggering maturation. He also suggested that elimination of the parasite from dace in early summer was due to some sort of rejection response that was stimulated by the rising temperature. Kennedy and Walker (1969) subsequently presented evidence that purported to support their contention.

One facet of the transmission process not examined by the 1968 Kennedy study was infection in the tubificid intermediate host, *Psammoryctes barbatus*. Subsequently, however, Kennedy (1969) found that tubificids were infected throughout the year and that the feeding behaviour of dace was unaltered from one season to the next. The implication was that dace were constantly being exposed to infected tubificids but, that during certain times of the year, the parasites were incapable of remaining established even though they were still being actively recruited. Based on this observation and the one made by Kennedy and Walker (1969), he asserted that the overriding controlling factor was the temperature-dependent rejection response in the fish host. Subsequently, Kennedy (1971) conducted a series of experimental studies on *C. laticeps* using the orfe, *Leuciscus idus*, a species of fish that is closely related to dace. Based on these results, he concluded that circulating antibodies were not involved in rejection, but he still could not characterize the exact nature of the response.

Anderson (1974, 1976) also studied the metapopulation biology of *C. laticeps*, but the definitive host for his investigation was the bream, *Abramis brama*. The results he obtained were similar in some ways to those of Kennedy (1969, 1971), but different in others. For example, Anderson (1974) observed a clear seasonal pattern, with the density of parasites declining sharply during the summer months (Fig. 3.5). Indeed, Anderson (1974) observed a significant relationship between temperature and survivorship of *C. laticeps* in bream (Fig. 3.6). A striking difference in the results of Kennedy (1968) and Anderson (1974), however, was that in dace the parasite disappeared completely after June, not to reappear until the following December. In bream, on the other hand, gravid adults remained throughout the summer and the parasite was present throughout the year. The occurrence of gravid adults in the summer months was one of the key elements of Anderson's (1974) analysis regarding the dynamics of *C. laticeps*. Thus, it was during the summer months that the largest proportion of gravid worms was present, at the time when the overall densities of the worms were also the lowest. He speculated that this strategy by the parasite prevented a large number of intermediate hosts from becoming infected at the same time. In turn, he suggested this would keep the size of the adult population in the fish host at a low level while minimizing the potential for intraspecific competition and maximizing the efficiency of reproductive effort.

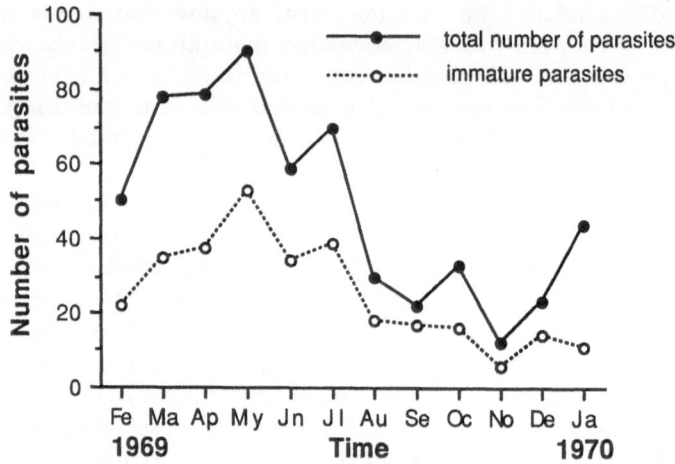

Figure 3.5 Seasonal variation in total parasite numbers and immature parasites (*Caryophyllaeus laticeps*) per sample of 30 fish (*Abramis brama*). (With permission, from Anderson, 1974.)

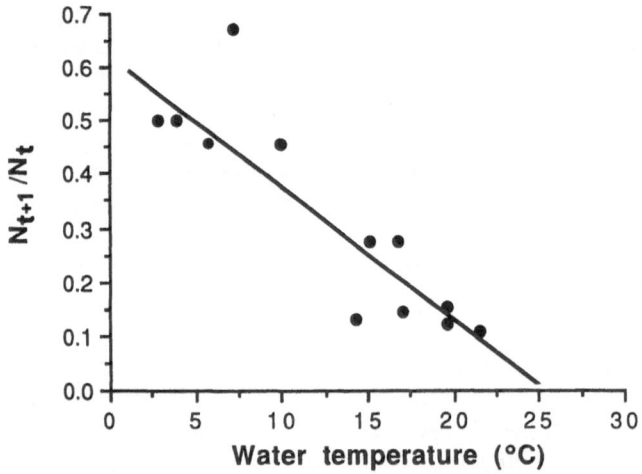

Figure 3.6 The relationship between the ratio N_{t+1}/N_t and water temperature, where N_t is the number of immigrant parasites at time t and N_{t+1} is the number of mature parasites at time t + 1. (With permission, from Anderson, 1974.)

Another component of Anderson's (1976) study involved the spatial distribution of infected intermediate hosts in combination with the feeding behaviour of bream and the frequency distribution of parasites within the definitive hosts. The parasites were overdispersed within the fish hosts and the distribution was best described by the negative binomial model (see Chapter 4 for additional discussion of this concept). He suggested that the overdispersed frequency distribution was generated by variation in the feeding behaviour of individual fishes within the population. Additionally, he noted that the age structure of the fish population was a significant factor in affecting the metapopulation dynamics of *C. laticeps*. Thus, both the age distributions and the feeding behaviours in different size classes of fishes varied and, in turn, affected the population biology of the parasite.

There are strong seasonal similarities between the results obtained by Anderson (1976) and those of Kennedy (1968, 1969, 1971). On the other hand, a basic difference was that *C. laticeps* was present in bream throughout the year, but not in dace or orfe. This would suggest that the latter species may not be the normal host for the parasite and could provide an explanation for the reported differences.

In summary, then, these studies serve to illustrate the impact of an abiotic factor, namely temperature, on the seasonal infrapopulation dynamics of the cestode, *C. laticeps*. Moreover, the efforts of Anderson (1976) also indicate how the spatial distribution of infected tubificids may influence the dispersion pattern of adult parasites in their definitive hosts.

3.2.2 *Bothriocephalus acheilognathi*

The Asian tapeworm, *Bothriocephalus acheilognathi*, was first described from cyprinid fishes in Japan (Yamaguti, 1934). It was introduced into the United States in the mid-1970s (Granath and Esch, 1983a) and into Great Britain at about the same time (Andrews *et al.*, 1981). The parasite has several unique characteristics. First, it has been reported from at least 40 different species of definitive hosts, so its host specificity is much broader than that of almost any other species of cestode. This, in combination with its broad range of copepod intermediate hosts, unquestionably accounts for its wide geographical distribution and for its ability to colonize rapidly once it is introduced into a new habitat. Another unusual feature of the parasite is its capacity for what appears to be indeterminate growth. For example, in the mosquitofish, *Gambusia affinis*, it will grow to a maximum length of approximately 3 cm, while in the common carp, *Cyprinus carpio*, it may reach 15–20 cm. The size of most helminth parasites at maturity is fixed within some narrow range, indicating determinate growth patterns for these species, but not for *B. acheilognathi*.

The life cycle of *B. acheilognathi* is simple (Fig. 3.7). Eggs are produced in the piscine definitive host and shed in the faeces. After a period of development, they hatch and a ciliated coracidium emerges. The motile coracidia swim free in the water column where they are consumed by cyclopoid copepods of several different species. On ingestion by a copepod, the parasites penetrate the gut wall and enter the haemocoele where they develop into procercoids. When the infected copepod is eaten by the definitive host, the procercoid develops first to the plerocercoid stage in the small intestine of the fish and, when provided with an appropriate stimulus, into an adult parasite.

Most of the investigations that comprise the present case study were conducted in Belews Lake, a cooling reservoir of 15.63 km^2 (1563 ha), located in the Piedmont region of North Carolina (USA). There are three reasons for focussing on these particular investigations. First, they were conducted over a 7-year period, giving them a reasonably long time frame. Second, because the lake is a cooling reservoir, there are ambient areas in which temperatures are 32–33°C during summer, while in thermally altered areas, water temperature rises to nearly 40°C. These differences provided an ideal opportunity to compare the effects of warm, and even warmer, temperatures on the population biology of the parasite. Finally, the lake was established in 1971 and had a normal fauna of fishes until 1974 when the

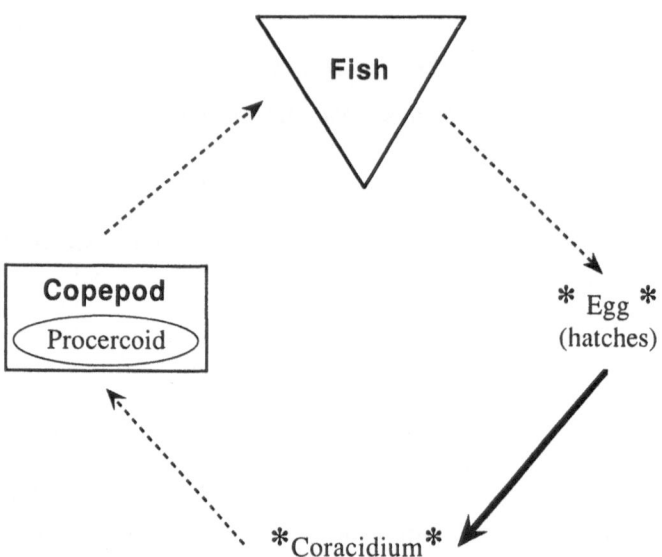

Figure 3.7 A schematic life cycle for the cestode, *Bothriocephalus acheilognathi*. The symbols are explained in the caption for Fig. 1.1 (page 7).

effects of selenium pollution were first observed. Then, over a period of 2 years, all but two of the original 26 piscine species became locally extinct; a cyprinid species was introduced into the lake in 1978, increasing the number of species present to three. By the time this series of studies was begun in 1980, those species present were the mosquitofish, *G. affinis*, the fathead minnow, *Pimephales promelas*, and the red shiner, *Notropis lutrensis*. Carp, *Cyprinus carpio*, and channel catfish, *Ictalurus punctatus*, were occasionally seen, but were apparently unsuccessful in spawning. Near the headwaters of the lake, there is an exceedingly strong chemocline that has effectively created two distinct, but contiguous, bodies of water, the main one of which is polluted by selenium and another that is not polluted. At the latter site, there is a fish community that is identical to the one that was present throughout the reservoir before the selenium pollution; it is also typical for the southeastern part of the USA. Among the other species present at the unpolluted end of the reservoir, is the carnivorous largemouth bass, *Micropterus salmoides*. The tapeworm is present in fishes in the unpolluted part of the reservoir. This unique physical boundary and the resulting biological structure in the lake also created an opportunity

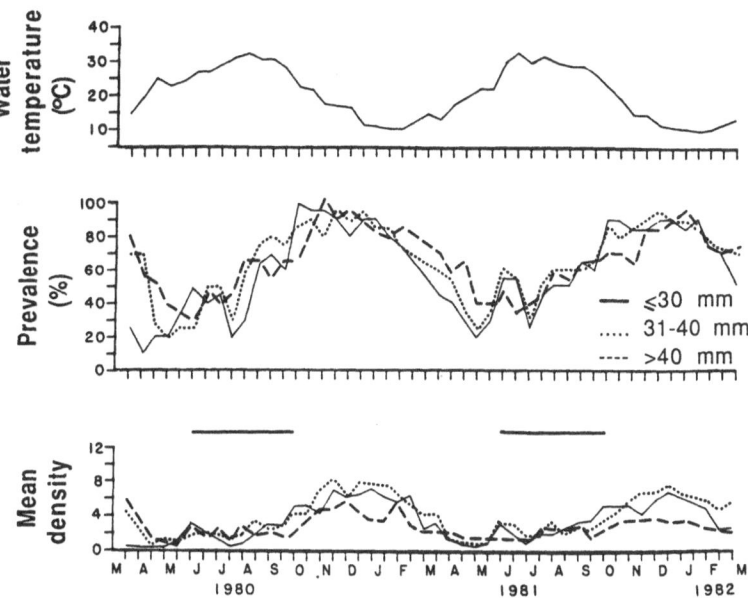

Figure 3.8 Biweekly changes in temperature, prevalence, and density of *Bothriocephalus acheilognathi* within three size classes of *Gambusia affinis* from an ambient-temperature site in Belews Lake, North Carolina, USA. The horizontal bar indicates when recruitment of the cestode occurred. (With permission, from Granath and Esch, 1983c.)

for comparing the presence and absence of piscine predation on the population biology of the parasite.

The studies, begun in 1980, were initially designed to compare the prevalence, density, and recruitment patterns of the parasite in mosquito-fish from thermally altered and ambient locations in the polluted part of Belews Lake (Granath and Esch, 1983a, b, c). The effects of temperature were striking (Figs. 3.8, 3.9). In both areas, parasite densities and prevalences were lowest during the summer months. Densities at both sites then increased sharply in the autumn, peaked in early winter, and declined again in winter. Recruitment of the parasite at the ambient location was also seasonal, beginning in late spring and continuing into October. At the thermal site, however, recruitment began 2 weeks earlier, lasted 2 weeks longer and was interrupted for several weeks in the summer when water temperatures exceeded 35°C. This observation illustrates how a parasite's metapopulation biology can be affected by seasonal changes in water temperature and that excessively high temperatures can fundamentally alter normal patterns of parasite recruitment.

Laboratory investigations confirmed the field observations regarding the relationship between the metapopulation biology of *B. acheilognathi* and

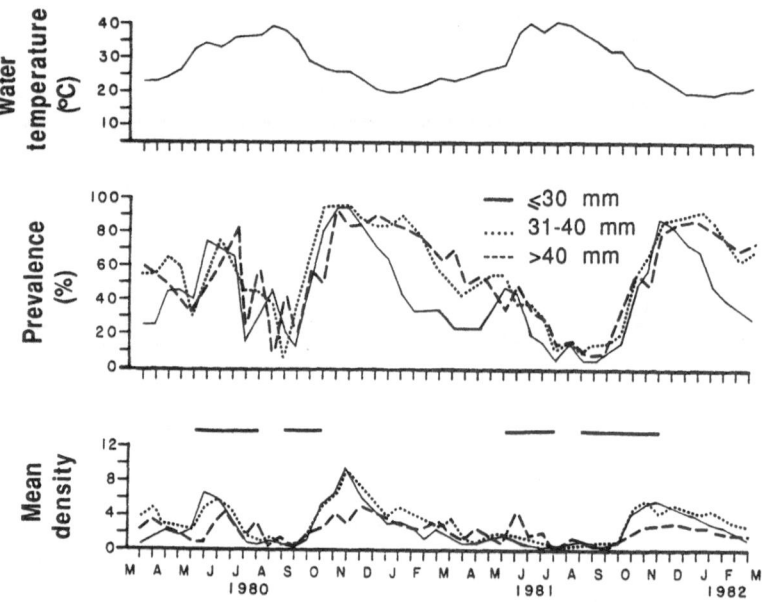

Figure 3.9 Biweekly changes in temperature, prevalence, and density of *Bothrio-cephalus acheilognathi* within three size classes of *Gambusia affinis* from a thermally altered site in Belews Lake, North Carolina, USA. The horizontal bar indicates when recruitment occurred. (With permission, from Granath and Esch, 1983c.)

water temperature (Granath and Esch, 1983b). They also analysed the infrapopulation structure with respect to the influence of water temperature on the parasite's developmental pattern. Beginning in October and continuing into May, more than 98% of the worms present were non-segmented. When water temperatures rose above 25°C, segmentation proceeded at a rapid rate (Fig. 3.10). The segmentation was accompanied by a sharp and significant decline in infrapopulation densities, an observation not unlike that of Anderson (1976) for *C. laticeps* in bream. Granath and Esch (1983b) speculated that a temperature-dependent rejection response could be responsible for the decline. However, they favoured the notion that intraspecific exploitative competition for limited spatial and nutrient resources was most likely responsible for the decline in infrapopulation densities within the mosquitofish during the summer months. In effect, they proposed that while a density-independent factor (temperature) was the driving force for the seasonal changes in infra- and metapopulation biology, a density-dependent factor (competition) was involved in regulating (*sensu* Scott and Dobson, 1989) infrapopulation densities.

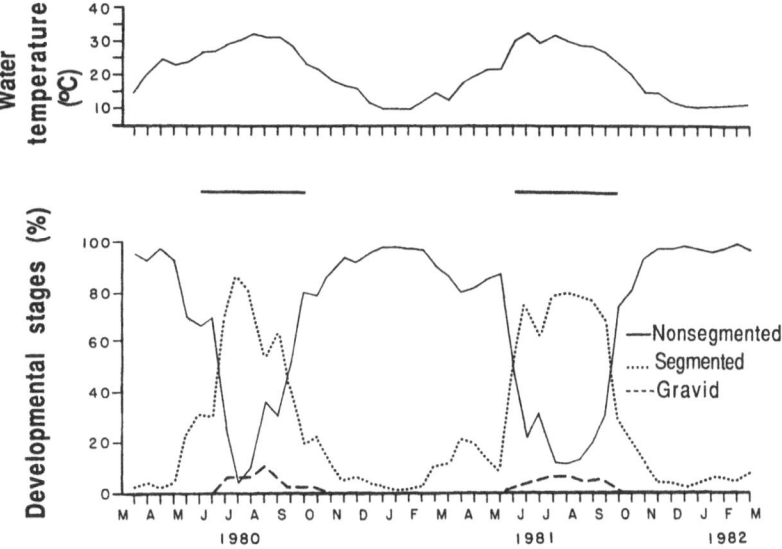

Figure 3.10 Seasonal changes in temperature and developmental stages of *Bothriocephalus acheilognathi* within *Gambusia affinis* from an ambient site in Belews Lake, North Carolina, USA. Each percent refers to the proportion of the total infrapopulation of the cestode at that stage of development. The horizontal bar indicates when recruitment of the cestode occurred. (With permission, from Granath and Esch, 1983a.)

Subsequently, a follow-up study of the *Bothriocephalus*–mosquitofish system in Belews Lake found that a striking change in the population biology of the parasite had taken place from the 1980–1982 period to 1984–1986 (Marcogliese and Esch, 1989). Thus, both the prevalence and densities of the parasite in mosquitofish were much lower in 1984–1986. Moreover, maximum prevalence and density in mosquitofish shifted from the autumn months in 1980–1982 to summer in 1984–1986. These observations would, at first glance, tend to raise questions regarding the conclusions of Granath and Esch (1983a, b, c) as to the influence of temperature on the population biology of the parasite. However, as mentioned earlier, a third species of fish was introduced into the lake in 1978. Between 1976 and 1980, the dominant species was the fathead minnow, *P. promelas*, a detritivore. Mosquitofish were also quite abundant in the early 1980s, ranging far from their normal littoral zone habitat into limnetic parts of the lake while foraging because of the absence of both piscine predators and potential competitors. The introduced species was the red shiner, *Notropis lutrensis*, which is a highly efficient planktivore and a competitor for mosquitofish. Three cyclopoids are known to transmit the parasite in Belews Lake, but the primary copepod host for *B. acheilognathi* in the lake at the earlier time was *Diacyclops thomasi* (= *Cyclops bicuspidatus*). From 1980 to 1982, it was abundant in both late spring and early summer, but underwent diapause from early summer to autumn. From 1984 to 1986, only the copepod *Tropocyclops prasinus* was abundant and *D. thomasi* densities were sharply lower than in the earlier period. Marcogliese and Esch (1989) proposed that with the introduction of the red shiner in 1978, the diversity of the copepod community was substantially altered between then and the 1984–1986 study period. Moreover, the population biology of the parasite was significantly changed as well. Although *T. prasinus* is a suitable intermediate host for the parasite, it is a much smaller species than *D. thomasi*, about half its size. As a result, it is less vulnerable to predation by visually oriented fish (Brooks and Dodson, 1965; Fulton, 1984) such as the dominant, planktivorous red shiner. Thus, according to Marcogliese and Esch (1989), '*T. prasinus* would not be as effective as *D. thomasi* in transmitting the parasite to the definitive host, simply because the former is smaller and suffers less predation. As a consequence, fewer parasites were transmitted to mosquitofish, with lower autumnal abundance (density) and prevalence in 1984–1986 than in 1980–1982'. Indeed, Riggs (1986) and Marcogliese and Esch (1989) reported that the diversity of the copepod community was strikingly altered by size-selective predation over the 10-year period from 1976 to 1986 and that, presumably, changes in the parasite's population biology within the reservoir reflect these changes.

Bothriocephalus acheilognathi is a unique cestode for several reasons, not the least of which is its lack of specificity at both the intermediate and definitive host levels. This accounts, in part, for its wide dispersal throughout Asia, Europe, and now North America. It is clear that the population biology of the cestode in its definitive host is affected by normal seasonal changes in water temperatures. However, in Belews Lake its suprapopulation biology also has been greatly impacted through the long-term alterations in the community structure of the copepod intermediate hosts brought on via selenium pollution in the reservoir and, in turn, by the selective predation of planktivorous fishes such as the red shiner.

3.2.3 Eubothrium salvelini

A 15-year study was conducted on the metapopulation biology of the pseudophyllidean cestode, *Eubothrium salvelini*, in sockeye salmon, *Onchorhynchus nerka*, from Babine Lake, Canada (Smith, 1973). Salmon spawn in tributaries that feed the lake. Fry of the salmon then migrate into the lake in May and June where they remain for one year as underlings. By the age of 1 year, they mature into smolts and are ready for their migration to the ocean. There are two populations within the lake, one residing in the northern basin and one in the southern. Migration to the sea is bimodal, with the northern population moving a few weeks sooner than the southern population. The stimulus for migration is apparently 'ice-off', or spring turnover.

Adult *E. salvelini* occur in the intestines of underlings and smolts. The life cycle is shown in Fig. 3.11. Eggs released from the fish are ingested by the copepod, *Cyclops scutifer*. On hatching, the parasite penetrates the gut wall and enters the haemocoele where it develops into a procercoid. These are ingested by planktivorous smolts and underlings where procercoids are freed upon digestion of host tissue; they then mature sexually. On migration to the ocean, the adult parasites are lost from the salmon.

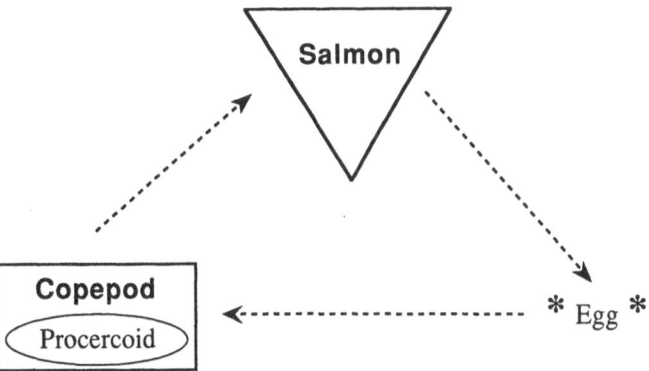

Figure 3.11 A schematic life cycle for the cestode, *Eubothrium salvelini*. The symbols are explained in the caption for Fig. 1.1 (page 7).

The parasite recruitment pattern generally was quite consistent over the 15 years of study. As smolts pass through the lake in May and June, they shed eggs that are eaten by *C. scutifer*. At about the same time, fry move into the lake from tributaries where they had been spawned and become underlings. As underlings, they begin to consume copepods and become infected in the process. Prevalence in the underlings reached between 30% and 40%. Interestingly, the prevalence did not go higher than this even though infected copepods remained in the lake. The explanation for the cessation in recruitment is related to a complete shift in prey preference from copepods to cladocerans in July; as a result, the fish stop recruiting *E. salvelini*. Moreover, *C. scutifer* is most abundant in May and June and then the population density declines rather precipitously at about the time the cladocerans become more abundant. This seasonal succession in the plankton community in Babine Lake is typical for most lake systems in temperate parts of the world.

The shift in diet from copepods to cladocerans is most important for the population biology of the parasite and probably the host as well since it is known from hatchery studies that high parasite densities will kill smolts. Thus, continued feeding of underlings on copepods after June could easily induce death of the fish in Babine Lake. Moreover, not only is there synchrony in dietary preference and transmission of the parasite, synchrony is also important to the survival of both the host and the parasite. The patterns of parasite recruitment and the level of parasites were consistent over the 15-year period with two exceptions, one lasting for two summers and another for three. During these years, the prevalence of the parasite was substantially lower than normal. The change was attributed to the asynchronous appearance of potential intermediate hosts (the copepods) and the presence of infected definitive hosts. No reasons for the asynchrony were assigned.

The metapopulation biology of *E. salvelini* in salmon is certainly influenced by seasonal temperature changes. However, there is also a behavioural element involved in affecting the population biology of the parasite when the underlings change their dietary preferences from one planktonic species to another. The dietary shift may be related to a natural, seasonal decline in the copepod densities, but more likely it is due to alterations in size-selection predation by the underlings. In other words, the underlings are better off foraging for larger cladocerans than on smaller copepods from the standpoint of energy expenditure and return for the effort. This is supported by the observation that the population density of *C. scutifer* is seasonally bimodal with a large increase in numbers in August as well as in the early summer. The underlings do not return to copepods as a dietary preference during August, remaining with the larger cladocerans.

The effects of temperature on the metapopulation biology of *E. salvelini* are clearly illustrated by the long-term study of Smith (1973). However, this abiotic influence has been modified by a shift in foraging behaviour by underlings during mid-summer away from copepod intermediate hosts to cladocerans which do not serve as hosts for the parasite. For this system, then, there is a combination of abiotic and ontogenetic biotic forces which affect the metapopulation biology of the cestode in the fish definitive host.

3.2.4 *Diphyllobothrium sebago*

Diphyllobothrium sebago is a common tapeworm in herring gulls, *Larus argentatus*, in the Rangeley Lakes area of northern Maine, USA (Meyer, 1972); the parasite has a three-host life cycle (Fig. 3.12). Eggs are shed from herring gulls in their faeces. The eggs hatch and free-swimming coracidia emerge. Planktonic copepods, *Diacyclops thomasi* (= *Cyclops bicuspidatus*) consume the coracidia; the parasite penetrates the gut wall, enters the haemocoele, and transforms into a procercoid. In turn, planktivorous smelt fry, *Osmerus mordax*, eat infected copepods. The procercoids then penetrate the gut wall and enter the tissues of the new host where they develop into plerocercoids. When an infected smelt is eaten by a gull, the cycle is complete.

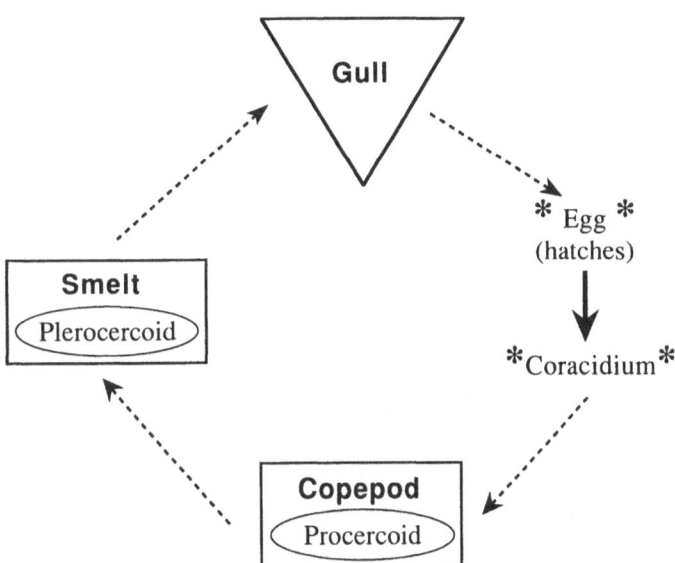

Figure 3.12 A schematic life cycle for the cestode, *Diphyllobothrium sebago*. The symbols are explained in the caption for Fig. 1.1 (page 7).

An unusual aspect in the life cycle of *D. sebago* is the elegant and complex temporal and spatial synchrony that has been evolved by the parasite and its hosts. Smelt infected with plerocercoids migrate from the lakes into feeder streams and spawn. They then return to the lake where they die. Subsequently, the young fry return to the lake where they migrate into deeper, hypolimnetic locations. They assume a planktivorous mode of feeding before they move back into the streams to spawn 2 years later.

Gulls become infected when they feed intensively on dead and dying smelt as they return to the lake after spawning. This corresponds with the time when gulls are raising their young so that young birds also become infected. Adults mostly favour the littoral zone in their early foraging. However, when the young birds fledge, both the fledglings and the parents move to the deeper, limnetic parts of the lake to feed. By this time, *D. sebago* has matured sexually and is producing eggs. After being shed from infected birds, the eggs hatch, releasing free-swimming coracidia. Coracidia are consumed by copepods where they develop into procercoids. Simultaneously, smelt fry are intensively feeding on copepods and thus become infected; in doing so, the parasite completes its life cycle.

In summary, the success of the parasite in this setting rests, in part, on the temporal sequence of smelt spawning and on the reproductive activity of the definitive host. There is also a spatial component in which infected birds move from the near-shore, littoral zone to the limnetic areas of the lakes. This ensures that eggs released from adult parasites are shed in open water; eggs sink into the hypolimnion before hatching, providing a much greater chance for the free-swimming larval stages to come into contact with appropriate intermediate hosts which are cold-water stenotherms. The suprapopulation biology of the parasite appears to be directly influenced by normal seasonal events that involve the behaviour of both definitive and intermediate hosts and, to a much lesser extent, with other density-independent abiotic factors.

3.3 DENSITY-DEPENDENT FACTORS: INTRODUCTION

Any density-dependent factor that influences the population biology of a free-living organism automatically implies regulation. The factors involved could include predation, competition, or parasitism. Among most parasite infrapopulations, predation can be excluded as a major regulator; the only predatory parasites are digenetic trematodes that have rediae as a part of their life cycle (Kuris, 1990; Sousa, 1990). Indeed, Kuris (1990) has developed an elaborate scheme of dominance hierarchies that applies to trematode infracommunities within the horn snail, *Cerithidea californica*.

For some of these trematodes, predation by parasitic rediae is an important mechanism for maintaining their position within the dominance hierarchy; these kinds of host–parasite interactions will be examined thoroughly in Chapter 6.

The immune responsiveness of some hosts is certainly a density-dependent phenomenon in certain systems. However, a serious problem in these cases is that field evidence, except in a few instances, is unavailable and conclusions can only be inferred from laboratory-based, experimental observations. Generally, immune capacities are only partially successful in regulating parasite infrapopulations; sterile or complete immunity, such as the kind induced by measles, has been reported for only a very few species. The complex regulatory system described by Wassom and his co-workers (Wassom, Guss and Grundmann, 1973; Wassom, DeWitt and Grundmann, 1974; Wassom *et al.*, 1986) for *Hymenolepis citelli* and *Peromyscus maniculatus* is unusual in that it is based on a combination of both host genetics and immunity.

Competitive interactions among parasites appear to result in both exclusion, or at least partial exclusion, and reductions in fecundity. The effects of competition have been viewed largely in terms of parasite infra- and component community dynamics and less so from an infra- or metapopulation standpoint (Holmes, 1973; Bush and Holmes, 1986b; Esch, Bush and Aho, 1990). In the present Chapter, only a few of the more classic cases of competition will be reviewed; most of the examples will be reserved for separate consideration within the context of parasite community dynamics.

The phenomenon of developmental arrest in nematodes (Schad, 1977) is, in part, influenced by density-dependent factors. Arrest, as defined by Michel (1974), is a temporary cessation in the development of parasites at a precise point in their early development, 'where such an interruption contains a facultative element, occurring only in certain hosts, certain circumstances, or at certain times of the year, and often affecting only a proportion of the worms'. Schad (1977) also emphasized that arrest is a temporary cessation of development. He pointed out that in some situations, it is a natural mechanism by the parasite to avoid harsh external environmental conditions and, in this sense, could be a density-independent process. However, he also noted that 'it is equally apparent that the entry of newly acquired parasitic nematodes into arrest, often at precisely the time when adult worm burdens are becoming particularly dense, is a highly adaptive mechanism for regulating [infra]populations of adult worms'. It is, therefore, an unusual mechanism for avoiding competition and clearly has density-dependent ramifications.

Finally, there is the interaction between parasite density and the potential for parasite-induced host mortality. Beginning with Crofton

(1971a, b), the relationship between overdispersion of parasite infra-populations and regulation has been examined by a large number of investigators, both in the field and in the laboratory. The contention is that parasites are able to cause host mortality under certain circumstances. If this could be established, then there would be a basis for inferring regulation. There is also the potential for mutual, regulatory interaction between host and parasite populations. This occurs when parasite density becomes high enough to cause mortality of a host either directly or indirectly. By causing the host to die, the parasite clearly is affecting the host population density. Simultaneously, the parasite is affecting its own density at both the meta- and suprapopulation levels. If the stage that is eliminated through host mortality is reproductive, then the rate at which new propagules are introduced into the environment could be affected and this would also affect parasite numbers. The extent of density-dependent regulatory interaction under natural conditions is still under debate and will be considered further in Chapter 4.

3.4 DENSITY-DEPENDENT FACTORS: CASE HISTORIES

3.4.1 Immunity

The impact of the immune response on parasite infrapopulation biology within humans is known but, from an experimental and strictly quantit-ative point of view (*sensu* Crofton, 1971a), it is not well understood or studied. In natural populations of other animals, much less is known. In the laboratory, the tapeworm, *Hymenolepis nana*, has been found to stimulate **sterile immunity** (complete and permanent immunity) in its mouse host (Heyneman, 1962; Weinmann, 1966), lasting for up to 120 days. Esch (1983) speculated that the duration of such an immune response under natural conditions could be enough to provide a density-dependent regulatory influence, but there are no data from field studies to support the assertion. A second tapeworm, *Taenia pisiformis*, has a tissue-dwelling cysticercus in its rabbit intermediate host, *Oryctolagus cuniculus* (Bull, 1964). In natural populations of rabbits in New Zealand, the frequency of the parasite increased with increasing host age, but mean density remained constant once a certain density threshold was reached. The number of eggs that were initially consumed by the rabbit apparently determined the long-term size of the infrapopulation since subsequent exposure to eggs of the parasite did not produce an expansion of the infrapopulation size. While the immune responsiveness in rabbits was not evaluated in the laboratory, the evidence certainly suggests density-dependent regulation via a strong immune response. Sterile immunity is also known for the haemoflagellate, *Leishmania tropica*, the causative agent for Oriental Sore. This parasite produces an open skin lesion in humans; after it has healed, the individual cannot be re-infected with the parasite. While this parasite is not regulated in terms of the classic density-dependent process, it is nonetheless regulated by the host immune system.

The prevalence of the malarial parasite, *Plasmodium falciparum*, within a human population where the parasite is being transmitted at a high rate, produces a curve that suggests transplacentally acquired immunity in newborn infants (Fig. 3.13). The immunity lasts for about 3 months before disappearing. Then, for approximately the next 5 years, children are totally susceptible to infection by the parasite; in Africa alone, about a million die from malaria each year. Humans are apparently capable of acquiring immunity to the parasite because, after 5 years, prevalence of the disease declines. It is impossible to assign any real significance to the effects of the immune response in humans on the overall population biology of the parasite or the host since field studies are not generally designed to answer such questions.

Schistosomiasis is another disease about which a relationship between immunity and parasite density is known to occur and for which there are some field data available. If the prevalence or density of schistosomes is plotted against age in a village where the parasite is endemic, peak

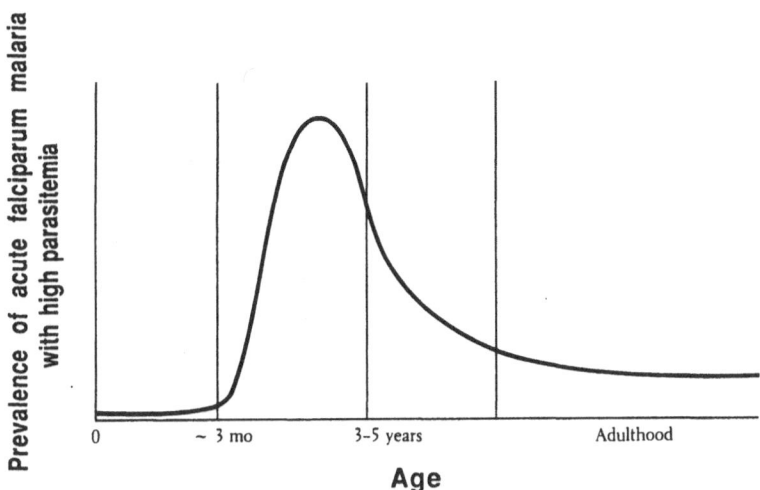

Figure 3.13 An age–prevalence curve for the malarial parasite, *Plasmodium falciparum*, in a hypothetical African village. Newborns have transplacentally acquired immunity to infection. After 3 to 6 months of age, individuals become highly susceptible to infection. After several years exposure, immunity is re-acquired and parasitemias are low.

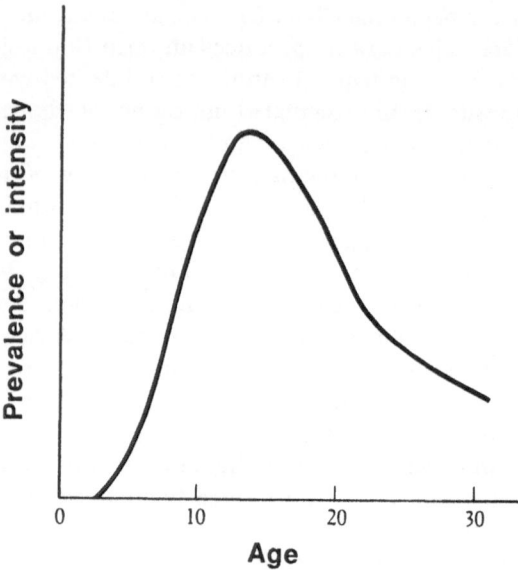

Figure 3.14 An age-prevalence curve for *Schistosoma* spp. in a hypothetical African village. The decline in prevalence with age may be attributable to a slow, spontaneous mortality of adult worms combined with acquired resistance, or a reduced level of exposure, or both.

numbers in both parameters will be seen in the early teenage years (Fig. 3.14). After the peak, both prevalence and parasite density decline. The exact explanation for the decline is not known, but it may be due to any of several factors. It could be related to natural senescence and mortality of older worms. Perhaps it is affected by acquired resistance, or by a change in behaviour of the potential host with age so that there is a reduction in exposure to the parasite with passing time. Based on animal models in laboratory settings, it is known that concomitant immunity will keep parasite infrapopulations at a lower level by effectively preventing the super-imposition of new infections on pre-existing ones.

The immune capacity of an individual host obviously plays a significant role in the infrapopulation biology of many parasites. However, as has been noted, documentation of this effect in natural systems is not extensive and, therefore, requires substantial additional study. Moreover, it would be of interest to extend these efforts to the level of the intermediate host and, in particular, to molluscan intermediate hosts since there is mounting laboratory evidence that snails have a reasonably powerful immune capacity.

3.4.2 Host genetics, immunity, and *Hymenolepis citelli*

One of the most unusual forms of host–parasite interaction occurs between the deer mouse, *Peromyscus maniculatus*, and the tapeworm, *Hymenolepis citelli*. Adult parasites are found within the intestines of deer mice where they produce eggs that are shed to the outside in faeces (Fig. 3.15). Eggs are then accidentally ingested by camel crickets, *Ceuthophilus utahensis*. They hatch, a larval stage emerges, penetrates the gut wall, and migrates to the haemocoele of the cricket. In the haemocoele, the parasite develops into a cysticercoid. When eaten by a deer mouse, the cysticercoid develops into an adult cestode and the cycle is complete.

Laboratory-reared deer mice were 100% susceptible to infection with *H. citelli* (Wassom, Guss and Grundmann, 1973). However, most of the animals eliminated the worms before they were able to produce mature proglottids. When mice were subsequently challenged with a new infection, they demonstrated a marked resistance to the parasite. In a few mice, however, there was no resistance to either the stimulating or the challenge infection. The resistance established in mice was density-dependent, being directly related to the dose of cysticercoids presented to the mice in the challenge infection.

Subsequently, it was shown that the acquired resistance by *P. maniculatus* to *H. citelli* was under the influence of a single, autosomal dominant gene and that susceptible hosts were genetically homozygous for the recessive gene (Wassom, DeWitt and Grundmann, 1974). Moreover, the immunity could be transferred to uninfected hosts via what they termed 'immune' lymphoid cells, but not with serum from infected hosts.

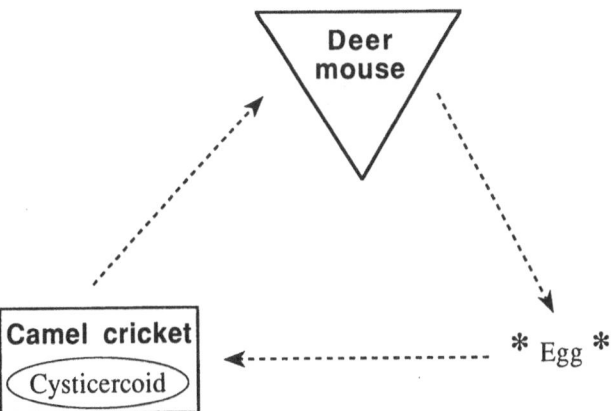

Figure 3.15 A schematic life cycle for the cestode, *Hymenolepis citelli*. The symbols are explained in the caption for Fig. 1.1 (page 7).

The susceptibility of outbred deer mice to infection was compared between the Utah (USA) strain of *H. citelli* and one isolated from California (USA) deer mice. The Utah mice responded much more rapidly to the Utah strain of the parasite than to the one from California. Field research revealed that the parasite also was patchily distributed, occurring only in small foci that favoured the co-existence of both the definitive and the intermediate hosts. It was speculated that the parasite was sustained in such foci by susceptible hosts that represent about 25% of the total population.

In field populations of deer mice, Wassom, Guss and Grundmann (1973) and Wassom, DeWitt and Grundmann (1974) found that the parasite was highly overdispersed and suggested that the contagion was influenced by resistance-related genetic factors. Wassom *et al.* (1986) tested and confirmed this hypothesis using controlled laboratory experiments. By using such an approach, most of the ecological variables that otherwise could have been invoked in contributing to the parasite's over-dispersed frequency distribution were eliminated. It was also stressed, and correctly so, that host population genetics should be considered more frequently in evaluating other parasite frequency distributions.

The metapopulation biology of *H. citelli* also was examined in the white-footed deer mouse, *Peromyscus leucopus*, in Wisconsin (USA) (Munger, Karasov and Chang, 1989). They also found *H. citelli* in very low prevalences (2–3%), but the parasite in *P. leucopus* behaved very differently than in *P. maniculatus*. For example, although up to 100% of *P. leucopus* could become infected with *H. citelli* under laboratory conditions, the parasites also developed to maturity before being rejected 28 days after infection. Some of the laboratory-infected mice were returned to the field. These mice retained their infections much longer than in the lab, with some animals still shedding eggs up to 100 days post-infection. When *P. leucopus* were challenged with new infections, the majority resisted infection. Another interesting observation was that when laboratory-infected mice were introduced into the field, they apparently stimulated transmission of the parasite to mice that had been previously uninfected, thereby increasing prevalence of the parasite in the natural population. While genetically based expulsion of worms may be important in the Utah system, the rejection phenomenon was not the overriding regulatory factor at the Wisconsin study site. Overdispersion and low prevalence at the Wisconsin site were attributed to heterogeneous distributions and low densities of the intermediate host in conjunction with the abbreviated life span of the parasite.

While the studies of Wassom, Guss and Grundmann (1973), Wassom, DeWitt and Grundmann (1974), Wassom *et al.* (1986) and Munger, Karasov and Chang (1989) differ in conclusions regarding parasite population dynamics, they serve to illustrate two important points regarding regulation. First, in the future, the population genetics of hosts must be assessed in more depth than it has been. Second, the population biology in two closely related systems, while seemingly very similar in pattern, is not necessarily affected by similar forces. This provides a clear warning that the conclusions reached in a single study, or even a series of studies, cannot be extrapolated to other systems without great care in the analysis of all events and factors associated with the population biology of both the parasites and their hosts.

3.4.3 Competition

'Competition occurs whenever two or more organismic units use the same resources and when those resources are in short supply' (Pianka, 1983). The competing species are usually at the same trophic level. **Exploitation competition** is the interaction between two species or individuals that is mediated indirectly through the use of a common resource. **Interference competition** occurs when two individuals or species are attempting to acquire a common resource and there is direct confrontation or interaction that reduces access to the resource by one or more of the individuals involved.

The outcome of competition can be considered in three ways. The first is **exclusion** which forms the basis for the so-called Gausian Principle (Gause, 1934). In essence, it states that two species cannot co-exist simultaneously in the same space and time. According to Holmes (1973), competitive exclusion among helminth parasites is a common phenomenon. **Interactive site segregation** refers to the specialization or segregation of niches by two species in which the realized niche of one, or both, is reduced by the presence of a second species (Holmes, 1973). **Selective site segregation** occurs within an evolutionary context. It is non-interactive and implies the absence of current competition. Thus, it is suggested that, within an evolutionary time frame, genetic changes occurred among one or both formerly competing species and the resource requirements of the competitors diverged.

As discussed earlier, most of the investigations on competition that followed those of Holmes (1961, 1962a, b) have largely focussed on the structure and organizing forces involving parasite infra-, component, and compound communities. A review of these studies will occur subsequently within the framework of parasite community ecology. This Chapter will review Holmes' reports (1961, 1962a) because these were seminal efforts in this area and because these can be most closely identified with the concept of parasite population regulation.

The cestode, *Hymenolepis diminuta*, has a two-host cycle. Adults are found in the small intestine of the rat, *Rattus rattus*. Eggs are shed in the faeces and then consumed by the grain beetle, *Tribolium confusum*. On hatching, an oncosphere emerges, penetrates the gut wall, and enters the haemocoele where it develops into a cysticercoid. After consumption by the rat definitive host, the parasite becomes a sexually mature adult. A second parasite employed by Holmes was the acanthocephalan, *Moniliformis dubius*. This parasite also uses the rat as a definitive host and, in fact, occupies the same section of the small intestine as *H. diminuta*. Eggs are released from sexually mature females (all acanthocephalans are dioecious) and are shed with the host faeces. On ingestion by cockroaches, eggs hatch and a spined larva called an acanthor penetrates the gut wall; it enters the haemocoele and develops into an infective stage known as a cystacanth. When the infected cockroaches are eaten by a rat, the parasite matures sexually, completing its life cycle.

Holmes (1961) conducted a series of experimental infections in laboratory rats. In one protocol, the rats were intubated with known numbers of cysticercoids and, after a period of maturation, the rats were killed and the precise linear location of the attachment site of the adult's scolex was noted. The attachment site by the tapeworms was always in the first 35% of the intestine. Infections of *M. dubius* were obtained by intubating rats with known numbers of acanthocephalan cystacanths. The *M. dubius* likewise became attached in the first 35% of the intestine. When the two parasites were introduced simultaneously, however, *M. dubius* continued to occupy the same site, while *H. diminuta* was forced back in the intestine into a section that it did not normally occupy.

Another set of experiments was designed to determine the effects of single and concurrent species infections on several physical parameters of the worms (Holmes, 1961). These included weight, length and weight–length ratios. The data (Table 3.2) show that the number of individual parasites which became established was not affected by the presence of the other species, but that weight, length and weight–length ratios of *H. diminuta* were substantially affected in the rats with double infections. *Moniliformis dubius* also was influenced by the presence of *H. diminuta*, but not to the extent that the cestode was affected.

He further assessed the impact of crowding and concurrent infection by *M. dubius* on the linear intestinal distribution of *H. diminuta*. The reverse of the experiment was also attempted for the two parasites. As the numbers of *H. diminuta* increased, the effects of crowding could be clearly seen, with the points of attachment being forced more and more posteriorly in the intestine. In single-species infections with *M. dubius*, basically the same pattern was observed, with the site of attachment moving further back in the intestine as parasite densities increased. In concurrent infections though, the impact was not as great as in crowding from heavy infections with a single species. Based on the role of carbohydrate in

Table 3.2 Effects of concurrent infection on the number, wet weight, length and weight:length ratios of *Hymenolepis diminuta* and *Moniliformis dubius*. (From Holmes, 1961).

	Single Infection			Concurrent Infection			F	Significance
	Rats	X̄ ± S.E.	Range	Rats	X̄ ± S.E.	Range		
H. diminuta								
Numbers	14	4.0 ± 0.3	2–5	14	4.7 ± 0.2	4–5	2.7	n
Weight (mg)	14	664.6 ± 30.7	496–918	14	251.4 ± 28.0	122–434	134.0	HS
Length (mm)	14	529.4 ± 21.3	388–662	14	323.5 ± 14.6	261–458	23.9	HS
Weight:length	14	1.272 ± 0.056	1.03–1.77	14	0.766 ± 0.055	0.46–1.04	38.3	HS
M. dubius (aggregate)								
Number	14	7.3 ± 0.6	3–10	14	6.4 ± 0.6	2–9	0.8	n
Weight	—	—	—	—	—	—	19.9*	HS*
Length	—	—	—	—	—	—	3.7*	B*
Weight:length	—	—	—	—	—	—	20.2*	HS*
M. dubius males								
Weight	14	79.5 ± 2.3	65–88	14	59.1 ± 2.4	40–74	*	*
Length	13	93.8 ± 1.2	85–103	13	88.4 ± 2.4	77–101	*	*
Weight:length	13	0.862 ± 0.098	0.64–1.04	13	0.697 ± 0.038	0.51–0.90	*	*
M. dubius females								
Weight	14	363.7 ± 15.1	285–430	13	235.8 ± 14.7	141–332	*	*
Length	14	207.4 ± 3.8	188–231	13	188.4 ± 4.7	151–208	*	*
Weight:length	14	1.751 ± 0.052	1.34–1.97	13	1.248 ± 0.072	0.88–1.77	*	*

*Data on *M. dubius* males and females were combined in the analysis of variance (with segregation of a highly significant variance due to worm sex) to give a single test of significance that takes into account the difference between sexes because of concurrent infection, and also the concordance of direction of change between the sexes.

F = variance ratio; n = not significant at the 10% level; B = significant at the 10% level but not the 5% level; HS = significant at the 1% level.

crowding and competition, Read (1959) and Holmes (1961) concluded that their observations were consistent with the hypothesis that competition occurred because of limited nutrient resources.

As has been noted, the role of competition in influencing the infrapopulation dynamics of parasites in natural systems has not been well documented. However, there is evidence that infracommunity structure is affected by interspecific competition in certain host species, discussed in Chapter 6, indicating that parasite infrapopulation densities could also be affected.

3.4.4 Developmental arrest

Developmental arrest is a wide-spread form of diapause that frequently occurs among nematode species infecting mammals (Schad, 1977; Adams, 1986). Schad (1977) identified three sets of factors that are believed to induce arrest. The first includes external environmental factors, causing a condition in the parasite which resembles diapause after it reaches the host. Second, there are factors associated with the host that will influence its capacity to either stimulate the development of the parasite or cause it to stop development when external conditions become adverse. And third, there are genetic and density-dependent factors related to the parasite that are involved with the induction of arrest in some species.

Environmentally induced developmental arrest appears to be fairly common among certain species of gastrointestinal nematodes in mammals from temperate regions of the world. Arrest among these species typically occurs during seasons of the year when external environmental conditions would be most detrimental to the survival of free-living larval stages. The primary stimuli for arrest thus vary from species to species. Examples of stimuli include low temperature, declining temperature, changing photoperiod, or combinations thereof (Fernando, Stockdale and Ashton, 1971; Hutchinson, Lee and Fernando, 1972; McKenna, 1973; Michel, Lancaster and Hong, 1975a, b; Eysker, 1981).

Host-induced developmental arrest also may be related to either natural or acquired resistance (Dineen, Donald and Wagland, 1965). Arrest under these conditions is additionally influenced by host age, sex, or species. Adams (1986) suggested that fourth-stage larvae of the sheep nematode, *Haemonchus contortus*, included a number of different subpopulations in a given host. He speculated, and then confirmed, that some of these subpopulations could be stimulated to undergo arrest in response to immunological stimuli while, in others, arrest was genetically inherent to the parasite. In the canine hookworm, *Ancylostoma caninum*, larvae may remain in a state of arrest in the musculature of bitches until parturition, at which time the third-stage larvae migrate to the mammary glands and are transferred to suckling pups via milk. (Stoye, 1973).

It is also clear that arrest in some species, even among strains of the same species, is probably genetically controlled as is the case for *H. contortus* (Adams, 1986). Nawalinski and Schad (1974) suggested that a strain of *Ancylostoma duodenale* (a human hookworm), isolated by culturing the eggs from a single female worm in West Bengal, was genetically fixed with respect to arrest. Frank *et al.* (1988) switched populations of the nematode, *Ostertagia ostertagi*, between northern Ohio (USA) and southern Louisiana (USA) and examined responses of the parasites to new environmental stimuli. They reported that 'the transplanted northern isolate (Ohio) had not adapted to respond to environmental stimuli in the south (Louisiana), whereas the southern isolate continued to respond to spring stimuli in both the north and the south, with no adaptation to autumn stimuli in the north'. This is clearly an indication of genetically based arrest.

Developmental arrest is a density-dependent phenomenon among species of *Ostertagia*, *Cooperia*, *Nematodirus*, *Graphidium*, *Obeliscoides*, and *Haemonchus* (Schad, 1977). The nematode, *Obeliscoides cuniculi*, was intensively studied in rabbits by Russell, Baker and Raizes (1966) and Michel, Lancaster and Hong (1975b). Russell, Baker and Raizes (1966) infected rabbits with doses of larvae ranging from 2500 to 25 000. Subsequently, all rabbits were killed and the larvae were counted. They observed a direct relationship between dose size and arrest in the larval infrapopulation up to a certain point; the density of arrested larvae then reached a plateau and levelled off. This plateau was referred to as a 'biomass or immunogenic threshold'.

Michel, Lancaster and Hong (1975a, b) extended these observations and concluded that both environmental conditioning and density-dependent factors were involved in many of the observed dose effects. These conclusions were based on a combination of laboratory and field observations which clearly add a dimension of credibility.

3.5 SUPRAPOPULATION DYNAMICS: INTRODUCTION

The complexity of abiotic and biotic factors involved in regulating or controlling infrapopulation densities should be apparent by this point. The complexity of these factors at the suprapopulation level is obviously compounded by the multi-host, life-cycle patterns of many parasitic organisms. The occurrence of free-living, life-cycle stages complicates the situation even further. Moreover, many parasitic organisms have more than one species of intermediate or definitive host and this places even greater demands on the investigator who chooses to undertake the study of a suprapopulation. The numbers of investigations at the suprapopulation level are, therefore, limited. For the most part, however, whenever such studies have been attempted, the results that were generated were well worth the effort.

3.6 SUPRAPOPULATION DYNAMICS: CASE HISTORIES

3.6.1 *Schistosoma japonicum*

The first comprehensive study on the suprapopulation biology of a helminth parasite was conducted by Hairston (1965) on the trematode, *Schistosoma japonicum*, in the Philippines. This digenetic trematode occurs in the superior mesenteric venous system that drains the small intestine of its definitive host. The life cycle is shown in Fig. 3.16. It infects a range of mammals, including man, where it causes a chronic and debilitating disease. Eggs produced by the parasite are pushed from small venules into the tissues of the intestinal wall (it is the pathology that is produced by the eggs in the tissues that is the primary cause of the disease syndrome). The eggs are then forced by the contraction of smooth muscle through the tissues into the lumen of the gut where they are eventually shed in the

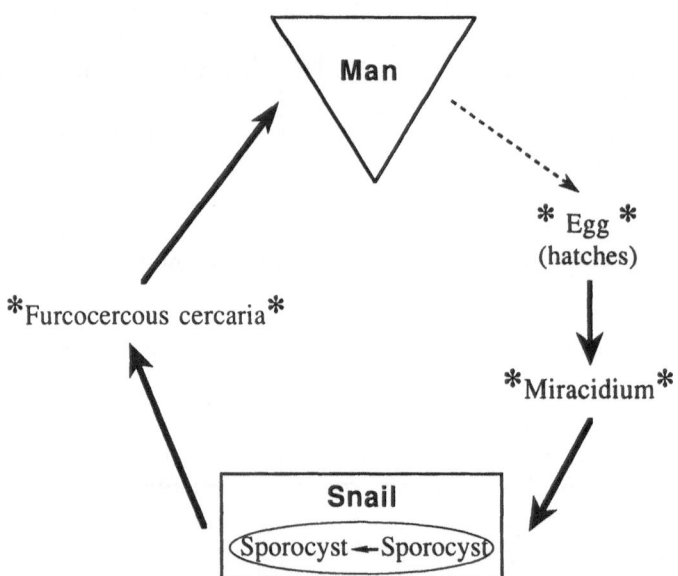

Figure 3.16 A schematic life cycle for the trematode, *Schistosoma japonicum*. The symbols are explained in the caption for Fig. 1.1 (page 7).

faeces. On reaching water, the eggs hatch immediately and free-swimming miracidia emerge. The miracidia locate and penetrate planorbid snails where they move to the hepatopancreas and develop into sporocysts. After two generations, daughter sporocysts produce and release large numbers of free-swimming, furcocercous cercariae. When cercariae make contact with the skin of a definitive host, they drop their tails and penetrate, becoming schistosomules in the circulatory system. The schistosomules ultimately find their way via the circulatory system to their final site of infection, usually in the superior mesenteric veins.

One of the critical observations made by Hairston (1965) relates to the importance of reservoir hosts in maintaining the parasite infrapopulations at high levels. He states that if the parasite could be eliminated completely from a given village, within a year it would return to the same levels as before its elimination. This assertion was based on its high prevalence in reservoir hosts such as dogs, pigs, and field rats. Indeed, it was suggested that metapopulations of parasites in humans or dogs could not be maintained except by the parasite flow through field rats. This high flow rate was accounted for by the intimate contact between rats and snails; he even argued that in the Phillipines, field rats are the primary hosts, not humans. This was despite the fact that in humans, egg-producing females had both their longest life span and highest total egg output. In field rats, the prevalence was equal to that in humans, but parasite densities were only about a third of that in humans and the egg output was less than 0.1% of that in humans.

The consequences of a study such as Hairston's (1965) are clear from an epidemiological point of view. Thus, they can be designed to generate baseline data which can then be used in developing strategies for appropriate control measures. Without certain basic information regarding critical aspects of the suprapopulation biology of the parasite, however, the design of control efforts for *S. japonicum* would have been totally inadequate.

3.6.2 *Metechinorhynchus salmonis*

The suprapopulation dynamics of the acanthocephalan, *Metechinorhynchus salmonis*, was examined in several species of fishes within Cold Lake, Alberta, Canada and in the parasite's intermediate host, the amphipod, *Pontoporeia affinis* (Holmes, Hobbs and Leong, 1977). Adults of the parasite occur in the fish intestine and produce eggs that are shed to the outside via faeces. The eggs hatch when ingested by the amphipod; an acanthor penetrates the amphipod gut wall and develops into a cystacanth within the haemocoele. The cycle is complete when the infected amphipod is eaten by the fish.

On the surface, this would seem like a reasonably simple system for a comprehensive study at the suprapopulation level because the parasite has a simple two-host life cycle. This is not the case, however, because at least 10 species of fishes in the lake can become infected by the parasite. Some species are infected directly by ingesting infected amphipods as is the case for whitefish, *Coregonus clupeaformis*. Others, such as lake trout, *Salmo trutta*, or coho salmon, *Oncorhynchus kisutch*, are infected indirectly; they acquire the parasite by feeding on smaller, infected fishes that, in effect, become transport hosts in the transmission process. The complexity of the definitive host system became somewhat less·complicated when two species of sucker (*Catostomus* spp.), northern pike (*Esox lucius*), stickleback (*Pungitius pungitius*), burbot (*Lota lota*) and walleye (*Stizostedion vitreum*) were eliminated from the study because, although they can recruit the parasite, it seldom or never matures sexually in these hosts. This left the four salmonid species for further consideration.

The prevalence, density, and percent of gravid worms was determined in several hundred salmonid hosts from Cold Lake (Holmes, Hobbs and Leong, 1977). The proportion of gravid worms is important for the assessment of relative parasite flow rates through the different species of definitive host and, therefore, for the suprapopulation dynamics of the parasite in the lake. According to their results, the parasite densities in whitefish are constrained by density-dependent forces. This conclusion was based on observations that:

1. the mean number of gravid worms remained constant with increasing density of worms in older age classes of fishes;
2. the mean number of gravid worms remained the same seasonally, despite radical and irregular shifts in overall parasite densities from month to month, and;
3. there was a significant negative correlation between the percentage of gravid females and the density of infrapopulations within individual fish.

In cisco, *Coregonus artedii*, the pattern was completely different to whitefish. Cisco are planktivores and the acquisition of infected benthic amphipods is purely by chance. There would, therefore, be little opportunity to evolve effective feedback mechanisms for regulating *M. salmonis* infrapopulations in cisco. Lake trout acquire the acanthocephalans by preying on fishes that are already infected with the parasite. While the percentage of gravid females in lake trout was relatively low, Holmes, Hobbs and Leong (1977) indicated that there was a negative correlation between gravid *M. salmonis* and densities of the cestode, *Eubothrium salvelini*. This suggested the possible influence of intra- or interspecific competition on parasite maturation and, therefore, parasite fecundity.

Coho salmon were introduced into Cold Lake in the summers from 1970 to 1972; results of the introduction were difficult to assess as far as the parasite's suprapopulation biology was concerned. They suggested, however, that the acanthocephalan may have been killing large numbers of young coho salmon and thereby could have had a significant impact on parasite flow within the system as well as on the population biology of the salmon.

Using data on parasite prevalence, densities, numbers of gravid worms, and the relative abundance of different fish species in the lake, the relative output of eggs from each host species was computed. The resulting estimate indicated that most of the flow through the system was via whitefish. It was hypothesized that regulation of flow through whitefish was 'sufficient to regulate the entire system'. After modelling the system mathematically, they concluded that regulation at the suprapopulation level may operate through individual infrapopulations in only one of several host species. Moreover, it was emphasized that the host species in which regulation occurs need not have the highest numbers or even be the one through which the largest parasite flow is occurring. They admonished that 'those working with the population dynamics of parasites having alternative definitive hosts, investigate relative flow rates through those hosts, and keep their populations in perspective when studying potential regulatory mechanisms'.

3.6.3 *Haematoloechus complexus* (= *H. coloradensis*)

In a very elegant study, Dronen (1978) examined the suprapopulation dynamics of the digenetic trematode, *H. complexus*, focussing on the flow rates and transmission efficiencies at different steps in the parasite's life cycle rather than on the nature of potential regulatory processes. The parasite's definitive host is the frog, *Rana berlandieri*, where it occurs in the lungs (the life cycle is shown in Fig. 3.17). Eggs produced by adults are coughed up and swallowed; they pass unaffected through the intestine and are released in the faeces. Eggs do not hatch, and must be eaten by the pulmonate snail, *Physa gyrina*, where they hatch, releasing miracidia that migrate to the hepatopancreas and give rise to two sporocyst generations. Daughter sporocysts produce cercariae that emerge from snails and penetrate nymphs of three different species of odonates where they encyst as metacercariae. When odonates are consumed by *R. berlandieri*, the metacercariae migrate to the lungs and develop into adults.

For two years, Dronen (1978) assessed the density and prevalence of *H. complexus* in different size classes in each host of the life cycle. The parasite's seasonality was examined in all hosts in the life cycle and the efficiencies with which each step in the life cycle could be completed were

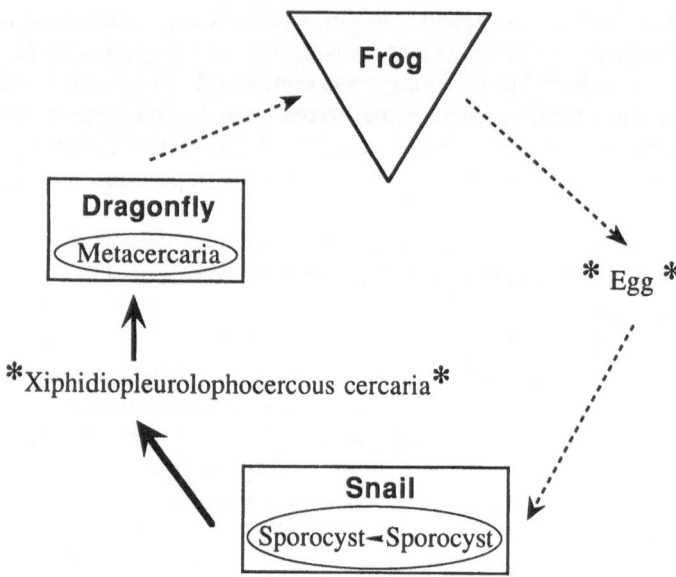

Figure 3.17 A schematic life cycle for the trematode, *Haematoloechus complexus*. The symbols are explained in the caption for Fig. 1.1 (page 7).

determined. The dynamics of the parasite's suprapopulation were monitored in 11 permanent ponds located in Sierra County, New Mexico (USA).

Haematoloechus complexus was highly seasonal, increasing rapidly in both prevalence and density in early summer. Overall prevalence of the parasite was lowest in snails (5–7%), next in odonates (15–17%) and highest in frogs (69–76%). The parasites were contagiously distributed in both odonates and frogs, and the distributions could be described by the negative binomial model. No firm conclusions were reached regarding the manner in which the distributions were generated although he noted that the frequency distributions did conform to the Poisson distribution in frogs of certain sizes. This suggested that perhaps contagion was being produced by the compounding of Poisson variates. Parasite prevalences increased with increasing size of frogs and snails, peaked in middle-size groups and then declined as snails and frogs continued to grow. There was a positive correlation between nymph size and parasite prevalence throughout the aquatic phase of the dragonfly life cycle.

Density estimates (numbers per cubic metre) were made for each host in the cycle. Using data from estimates of egg and cercariae production in the laboratory, data regarding parasite prevalence and density in the field and data for host densities, flow-number diagrams were developed to characterize efficiencies for each step in the life cycle. In 1972, these efficiencies were 0.02% for eggs becoming successful miracidia, 4.4% for cercariae production from successful eggs, 0.08% for metacercariae from successful cercariae, and 0.9% for new adults from metacercariae. The next year, these efficiencies were 0.01%, 3.3%, 0.1% and 0.2%, respectively.

Based on these numbers, not surprisingly it was concluded that transmission efficiency was very low, but that reproductive efficiency was very high. As pointed out by both Holmes, Hobbs and Leong (1977) and Dronen (1978), there is a very delicate balance between reproductive and transmission efficiencies in parasite life cycles. On the one hand, if transmission efficiency drops too low, the chances of local extinction are increased. If the parasite's transmission efficiency becomes too great, there is the risk a host will recruit too many parasites which, in turn, could produce host morbidity and even mortality. As Dronen (1978) noted, 'it should be kept in mind that this equilibrium between reproductive potential and efficiency has evolved under the influence of selective pressures on the gene pools of both the parasite and its host; and although these selective pressures were probably numerous, host density and the selective pressures that determined the expression of these densities probably played a major role'.

3.6.4 *Bothriocephalus rarus*

The life cycle (Fig. 3.18) of the pseudophyllidean cestode, *B. rarus*, was determined by Thomas (1937) and confirmed by Jarroll (1979). The definitive host for the parasite is the red-spotted newt, *Notophthalmus viridescens*. Eggs are shed in the faeces of newts and hatch, releasing ciliated coracidia. The coracidium is ingested by a copepod in which the parasite develops to the procercoid stage in the haemocoele. When newts feed on copepods, the life cycle is completed. The parasite may be acquired by both larval and adult newts. The latter are also known to cannibalize their larvae. If an infected larval newt is eaten by an adult, its parasites will be successfully transferred to the predator. Larval newts migrate from the pond for periods of time lasting from 2 to 7 years, carrying adult worms with them when they leave the pond and when they return.

The suprapopulation dynamics of *B. rarus* was studied (Jarroll, 1980) in a small, permanent pond in Ritchie County, West Virginia (USA). All parts of the parasite's reproduction and transmission were examined experimentally in the field or in the laboratory. The seasonal dynamics of the parasite in copepods and newts was also assessed. Based on the results of his investigation, the probability of an egg successfully surviving to the procercoid stage was 2.2%. The probability of a procercoid surviving to become a worm in an adult newt was 3.8% and to a worm in a larval newt, the probability was 2.9%.

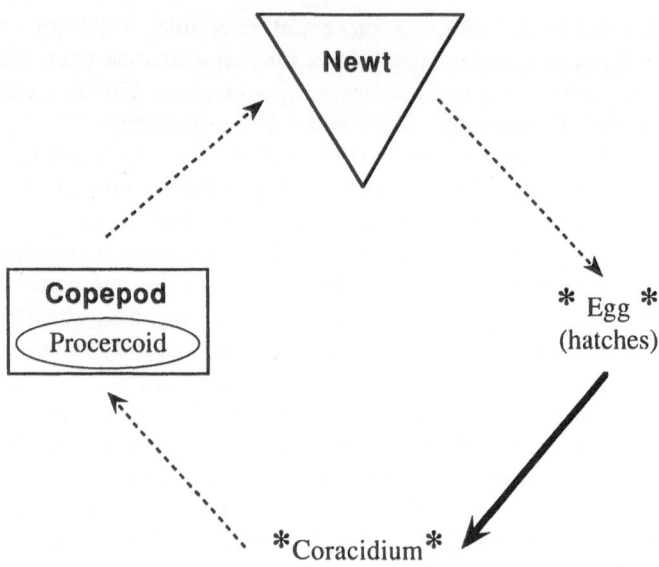

Figure 3.18 A schematic life cycle for the cestode, *Bothriocephalus rarus*. The symbols are explained in the caption for Fig. 1.1 (page 7).

He indicated that the period of greatest loss was from the egg to the procercoid stage. This pattern was similar to that seen by Dronen (1978) for *H. complexus*, but not by Hairston (1965) for *S. japonicum* where the greatest losses in the life cycle were in the cercariae-to-definitive-host phase of the cycle. Jarroll (1980) observed a much higher efficiency in going from the procercoid to the adult parasite. This was attributed to two factors, one associated with the strong spatial overlap between the copepods and newts and the second associated with a strong predator–prey relationship between larval newts and copepods. It was suggested, although no evidence was provided, that infected copepods may in some way be more vulnerable to predation.

Finally, larval newts had the largest parasite densities and exhibited the highest recruitment rates. However, when they became efts and left the pond there was a tendency to lose worms, so much so that mean infrapopulation densities came to resemble those of adult newts in the pond. He speculated that 'if one assumes that a dynamic equilibrium exists in the newt and worm populations, then the returning efts and their worms may serve to replace the lost newts and their worm infrapopulation'. In this way, the returning infected efts would serve as a reservoir for the parasite and act to prevent its local extinction. This is also an excellent mechanism for the dispersal of the parasite to other ponds.

3.6.5 *Bothriocephalus acheilognathi*

The life cycle of *B. acheilognathi* was presented in section 3.2.2 of this Chapter. Briefly, it includes a free-swimming coracidium, a procercoid in several species of copepods, and an adult in at least 40 species of fish. Aspects of the infra- and metapopulation biology of the parasite have been discussed earlier in this Chapter in terms of responsiveness to temperature.

Certain aspects of the parasite's suprapopulation biology were invest-igated (Riggs and Esch, 1987; Riggs, Lemly and Esch, 1987) in Belews Lake, a large cooling reservoir in the Piedmont area of North Carolina (USA). As was described previously, the lake had undergone severe selenium pollution, to the extent that the main body of water was completely devoid of piscine predators, such as largemouth bass. Indeed, at the beginning of the study, only three fish species were present in the lake. These included mosquitofish (*Gambusia affinis*), fathead minnows (*Pimephales promelas*), and red shiners (*Notropis lutrensis*), all of which could be infected by *B. acheilognathi*. Over a 30-month period beginning in 1980 and continuing into 1983, the density, prevalence, and dispersion patterns of the parasite were followed at three distinct sites in the reservoir. One was in the polluted part of the lake (without piscivorous predators). A second was in the headwaters of the lake where no pollution had occurred, but where there was a diverse fish community, including piscivorous largemouth bass. The third was in an interface zone where predation was probably occurring, but its effects were clearly less than at the unpolluted site. There was some pollution at the intermediate site, but selenium concentrations were comparatively low.

The approach taken by Riggs and Esch (1987) was different from those of Holmes, Hobbs and Leong (1977), Dronen (1978) and Jarroll (1980). The *Bothriocephalus* study was designed to examine the interaction of season, site, host size and host diet on changes in parasite prevalence, density, dispersion, and fecundity in three species of definitive host. There were seasonal changes in prevalence of the parasite at all three sites, among all three definitive host species, and in three, arbitrarily established, host-size classes. Interestingly, the variations observed were strongly related to seasonal changes in species diversity within the copepod community, to size-related changes in prey preferences among the three host species, and to the influence, or absence, of predation pressure.

They also observed striking differences in the ability of the parasites to mature sexually in the three host species. Holmes (1976, 1979) defined three classes of hosts in terms of their capacity to allow successful maturation. The so-called 'required hosts' are those in which the majority of gravid worms are found. In 'suitable hosts', the parasite will mature sexually, but will not possess gravid individuals in large numbers.

'Unsuitable hosts' may recruit the parasites, but the parasites will not become gravid. In Belews Lake, both fathead minnows and red shiners are clearly required hosts while, for the most part, mosquitofish fall into the suitable category. In mosquitofish, worms appeared to be stunted and occupied only a very narrow part of the gut. Moreover, Granath and Esch (1983b) observed, and Riggs (1986) confirmed, that growth of even a few worms to sizes larger than 40–50 proglottids caused mortality in mosquitofish.

One of the more striking observations was the difference in fecundity and egg viability among the worms from polluted and unpolluted sites, as well as between red shiners and fathead minnows. Thus, fecundity and egg viability were higher in fathead minnows; the same two parameters were also higher at the unpolluted sites. The former observation is a clear indication of differences in physiological compatibility of the parasite for these two host species. The latter finding suggests that the power plant effluent was definitely having an impact on the parasite's fecundity throughout the polluted part of the reservoir. Selenium toxicity halts reproduction in fishes and was responsible for the negative impact on the fish community in Belews Lake (Cumbie and Van Horn, 1978). The selenium concentration was examined in the tissues of all three fish species in the polluted and unpolluted parts of the reservoir as well as from tapeworms removed from the same fishes. The results were clear. The concentration of selenium in tapeworm tissues containing gravid proglottids was 10 times higher than in muscle from the host fish and was six to eight times higher than in the scolices of the same worms. The study also showed that selenium was being concentrated in the host gonadal tissue, but was still far less than half what it was for the gravid proglottids from the tapeworms. When the concentration of selenium in host and parasite tissues from the unpolluted sites was examined, it was found to be inconsequential and up to four orders of magnitude less than comparable tissues from fishes and parasites from polluted areas. It was apparent that the parasites were acting as selenium 'sinks' and that both fecundity and egg viability were being affected as a result.

Riggs and Esch (1987) concluded their study by saying, 'factors that affect, directly and indirectly, the dynamics of the intermediate hosts and the frequency and intensity of predation on them by definitive hosts seem to be of primary importance' in influencing the prevalence, density, and aggregation of *B. acheilognathi* in Belews Lake. It was emphasized, however, that the one variable in the transmission process not associated with predator–prey relationships was parasite fecundity in required hosts. In the end, then, it is the unique combination of biotic and abiotic components within the host's ecosystem that affects the suprapopulation dynamics of *B. acheilognathi*.

3.6.6 *Cystidicoloides tenuissima*

Aho and Kennedy (1987) examined the circulation pattern and transmission dynamics of the nematode, *C. tenuissima*, in all its intermediate and definitive hosts in the River Swincombe, a third-order stream located in Dartmoor National Park in Devon, England. Definitive hosts for the parasite include brown trout, *Salmo trutta*, and juvenile Atlantic salmon, *Salmo salar*. Intermediate hosts included 18 species of insects, but the parasite was able to develop into infective third-stage larvae only in the mayfly, *Leptophleba marginata*.

Even though trout and salmon were suitable definitive hosts, the former species was the primary host because the mayfly was a major component in its diet. It was estimated that 99% of the parasite's egg production originated in the trout, but that only 10% of the eggs were ingested by insects. Most of the eggs were apparently ingested by two species of insects that could not serve as suitable intermediate hosts; nearly 80% of these intermediate host populations were infected by *C. tennuissima*, while only 10% of *L. marginata* were infected. The overall transmission rate of eggs to larval *L. marginata* varied from 0.25 to 0.87%; transmission rates from insect to fish were much higher, ranging from 10.8 to 39.8%.

While most of the parasite eggs were apparently consumed by unsuitable hosts, the parasite could not become established perhaps because of physiological incompatibility or because of an immunological response. It was concluded that the differences between trout and salmon as a potential definitive host were not related to a physiological/ immunological response, but to an ecological factor. In the River Swincombe, trout were restricted primarily to pools and salmon to riffles. During periods of the year when parasite transmission occurred, *L. marginata* was also confined to the pools thereby bringing the mayfly into the foraging arena of the trout. The parasite could not, therefore, be transmitted to salmon since they did not typically feed on the mayfly host. Also of interest was the comparison made by Aho and Kennedy (1987) of the suprapopulation dynamics of *C. tenuissima* with those of the acanthocephalans, *Pomphorhynchus laevis* and *Metechinorhynchus salmonis* (Hine and Kennedy, 1974; Holmes, Hobbs and Leong, 1977). All three parasites have a two-host life cycle, with the definitive hosts being fishes. Aho and Kennedy (1987) pointed out that there was a strict specificity at both the intermediate and definitive host levels for all three species of helminths. However, in the case of the nematode, large numbers of eggs were consumed by insects in which the parasite could not develop properly while the acanthocephalan eggs were consumed only by hosts in which proper development could occur. Conversely, acanthocephalan intermediate hosts were consumed by a wide range of potential fish hosts, but could mature in only one.

As noted earlier, the suprapopulation dynamics of parasitic organisms have not received extensive attention. However, in those studies which have been conducted, there is ample indication of their worth. Thus, Hairston (1965) used the approach in identifying a critical epidemiological factor for the potential control of schistosomiasis in the Philippines. Other investigators (Holmes, Hobbs and Leong, 1977; Dronen, 1978; Jarroll, 1979; Granath and Esch, 1983a, b; Aho and Kennedy, 1987; Riggs, Lemly and Esch, 1987) employed several different host–parasite systems to characterize reproductive and transmission efficiencies at various life-history steps and to emphasize the delicate balance between the two processes in parasite life cycles. Studies such as these are quite valuable and should be extended to other systems as well.

4 Influence of parasites on host populations

4.1 INTRODUCTION TO THE CONCEPT OF REGULATION

In the previous Chapter, an effort was made to assess the nature of those factors, both biotic and abiotic, that are known to affect the biology of parasite populations. The purpose of this Chapter is to reverse that thrust and to determine if, and under what circumstances, parasites can affect host population dynamics.

In order to identify the nature of regulatory influence on host population dynamics, it is first necessary to define what is meant by regulation. Scott and Dobson (1989) described it as a phenomenon involving processes that will 'reduce the per capita survival or reproduction' within a population as the population's density increases. The key element in the definition is the reduction in host survivorship, or fecundity, as a function of parasite density. In other words, there must be the clear implication of a density-dependent effect on host mortality or reproductive fitness for there to be regulation. As they pointed out, however, the measurement of density-dependent regulation is not easily accomplished since it is not always feasible, or even possible, to manipulate host populations in the field or in the laboratory. For this reason, some of the examples cited in this Chapter as being regulatory in character will be inferential only.

4.2 CROFTON'S APPROACH

4.2.1 Introduction

Before considering Crofton's (1971a) analysis of the host–parasite system with which he worked, or the mathematical model he developed (Crofton,

1971b) for better understanding the relationship between hosts and parasites, it is instructive to know why he undertook such an effort. As explained previously, he was disenchanted with the various definitions of parasitism, mainly because they were too qualitative in character. He was evidently impressed with the approaches taken by Kostitzin (1934, 1939) and Lotka (1934) who both were familiar with frequency distributions, but lacked the sophisticated computational technologies necessary for fitting theoretical models to observed distributions. After considering several applications to the problem, Crofton (1971a) became convinced that a quantitative methodology could best be applied through the use of the negative binomial model.

The negative binomial model is given by the expression:

$$(q - p)^{-k}$$

where $q = 1 + p$ and k is positive (Crofton, 1971a). The parameter p is the probability of an event occurring one way (i.e. presence of parasites), q is the probability of the same event occurring in an alternative way (i.e. absence of parasites) and the exponent k is related to the spatial distribution of the organisms. Frequency distribution curves that can be described by the negative binomial model for various values of k and μ (the population mean) are shown in Figure 4.1.

It was suggested that 'the Negative Binomial distribution is a "fundamental model" of parasitism in so far as it describes the distribution of parasites among hosts' (Crofton, 1971a). The elegance of the negative binomial rests with the observation that distributions which can be described by the model have hypothetical bases for inferring the manner in which they can arise. With an hypothesis in place, it then can be tested and either supported or rejected. Most other models do not offer such an advantage. Thus, for example, a distribution that can be described by the negative binomial might be generated through the compounding of Poisson variates. In other words, if a host acquires parasites through a series of random waves of infection so that the chances of infection by each succeeding wave is independent of the previous ones, then the final distribution can be described by the negative binomial model.

It is emphasized that the transmission processes used by some parasites will not always result in overdispersed frequency distributions and, therefore, the negative binomial model is not always applicable to host–parasite systems. Most protozoans, as well as the sporocysts and rediae of digenetic trematodes in snails, for example, will not generate overdispersed frequency distributions. This is because once infected, intramolluscan larval stages, for example, reproduce asexually thereby generating frequency distributions that are highly skewed. The same observation holds for most protozoans in, or on, their hosts.

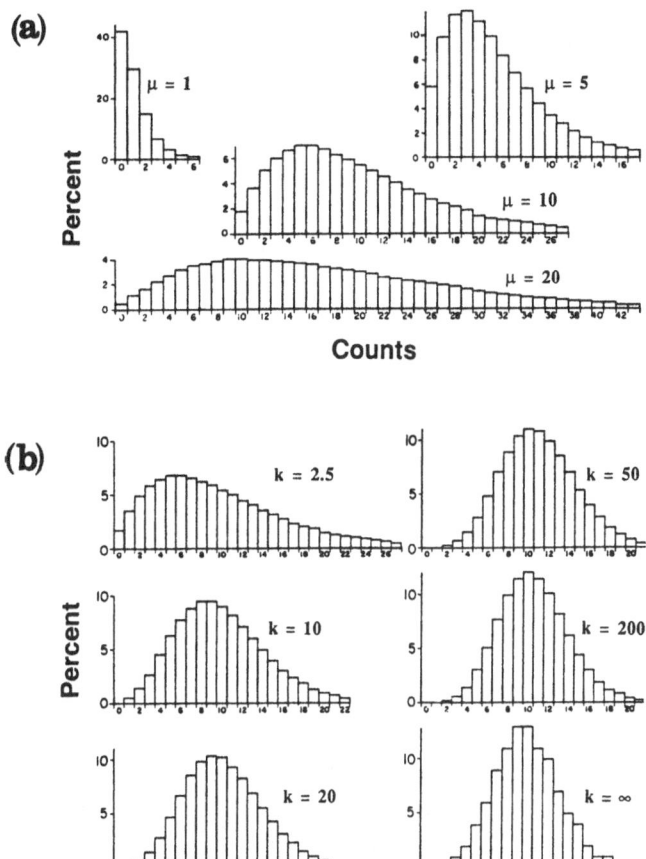

Figure 4.1 Negative binomial frequency distributions: (a) Frequency distributions for $k = 2.5$ and various values of μ, from 1 to 20; (b) Frequency distributions for $\mu = 10$ and various values of k, from 2.5 to ∞. Frequency of each count is expressed as percentage of total count. (With permission, from Elliot, 1983.)

If the infective stages of a parasite are contagiously distributed in space, the likelihood is that the parasite will be overdispersed within the population of the next host in the life cycle. This is, in part, due to the fact that most organisms are not randomly or regularly distributed in space, but are themselves overdispersed. It is also due to the manner in which infective stages themselves are dispersed from an infected host; they are frequently shed in 'bunches', e.g., eggs in faeces, creating microgeographic infection foci.

As already noted, some parasites have the ability to alter a host's behaviour. If infection increases the transmission probability for a parasite, then overdispersion will be a likely outcome. If, within a population, the chances for infection are unequal because of some form of variation in individual hosts, then overdispersion may be the result. The individual variability in hosts may be related to genetic differences in susceptibility–resistance traits, to physiological differences, to age differences within a host population, etc. Similarly, individuals within a population change over time and such changes could also create conditions that could lead to overdispersed frequency distributions.

The reason for emphasizing the nature of overdispersion is that, because of it, the majority of parasites within a host population will be found in only a few hosts. This suggests that if the parasite has any lethal qualities, then mortality may be restricted to but a few hosts, those that are heavily infected. The risk of parasite-induced mortality, and morbidity, is thus spread unevenly within a host population. Presumably, in the course of co-evolution by the various host–parasite systems, efficiencies in life-cycle patterns have developed to the point that transmission strategies involving parasite-induced host mortality are such that even local extinction is a low probability except under very exceptional conditions. There are other evolutionary implications regarding the dispersion patterns of parasites, but these will be discussed subsequently.

4.2.2 The *Polymorphus–Gammarus* system

The data for Crofton's (1971a) quantitative approach were provided by Hynes and Nicholas (1963) who examined the metapopulation biology of the acanthocephalan, *Polymorphus minutus*, in its amphipod intermediate host, *Gammarus pulex*. The definitive hosts for the parasite are ducks. In their study area, ducks were kept in cages next to a stream into which faeces (and parasite eggs) from the ducks were washed. The amphipods were collected at several stations, each one located at a distance further away from the source of eggs. The amphipods were returned to the lab where they were dissected and the larval cystacanths counted.

Not surprisingly, parasite densities were highest at the station immediately adjacent to the duck cages. At each station down stream from the duck pens, the parasite densities declined in the amphipods. These data were analysed by Crofton (1971a); both the Poisson and negative binomial models were used to describe parasite frequency distributions in the amphipods, with the latter clearly giving the best fit. However, he also observed that the fits for zero, and one, parasite per host classes were unsatisfactory, especially at sites closest to the duck pens. He reasoned that an explanation for the poor fit was that heavily infected hosts had been killed by the parasite and that by removing these hosts from the population, the resulting distribution had been skewed.

This hypothesis was then tested using a truncated negative binomial model to describe the distributions. According to Crofton (1971a), 'truncation is usually applied when the negative and sometimes the zero terms of a frequency distribution are rejected or accumulated at the zero frequency (Brass, 1958)'. It was concluded that there was little or not mortality at stations located the greatest distance from the egg source. As the stations came closer and closer to the duck pens (and the source of parasite eggs), parasite losses due to host mortality appeared to become greater and greater. Losses at stations closest to the duck pens were estimated at between 5 and 60%.

The real significance of Crofton's (1971a) analysis was that it combined the concepts of parasite frequency distributions and parasite-induced host mortality in a practical manner for the first time. However, it was vigorously emphasized that the precise form of the frequency distribution is unimportant as long as the distribution is clumped. 'It is the overdispersion and the relationship of parasite density to lethal factors that produce a disparity in parasite and host deaths. This disparity, offset by the greater reproductive rate of the parasite, can produce the dynamic equilibrium of host and parasite populations that is essential to the continued association of the host and parasite species. In effect, the parasite acts as a regulator of the host population, the intensity of the regulatory function being related to both host and parasite population dynamics' (Crofton, 1971a).

4.2.3 Crofton's model

In conjunction with his study of the *Polymorphus–Gammarus* system, Crofton (1971b) developed a mathematical model to describe some of the dynamic qualities of host–parasite interactions. The deterministic model that was developed employed several basic assumptions regarding certain characteristics of the host and parasite populations. For example, the host population was considered to have varying rates of logistic growth. In the model, the population never achieved its carrying capacity and was, therefore, always in exponential growth.

The parasite was described as always having a higher rate of reproduction than the host, a component of his quantitative definition of parasitism (Crofton, 1971a). In the model, the reproductive rate of the parasite and its potential to infect a host were combined into a single term called the Achievement Factor. Since the probability of infecting a host is always less than unity, the Achievement Factor for the parasite was always less than its reproductive rate. The frequency distribution of the parasites within the host population was overdispersed and could be described by the negative binomial model, another essential element in his quantitative approach to parasitism. The concept of the Lethal Level (L) was also incorporated into the model. It was 'defined in terms of the number of parasites required to kill the host before it can reproduce' (Crofton, 1971b). After using several different approaches, he was concerned about the nature of transmission rates in relation to host density, and decided to make the model's transmission rates directly correlate with host density. The efficiency of the model was expressed in terms of an equilibrium between densities of hosts and parasites. Equilibrium was assumed to have been reached when constant levels of parasites and hosts were attained or when there were persistent oscillations in parasite and host densities.

After developing the basic model, simulations were conducted in order to assess the effects of changing various parameters. For example, k is a measure of overdispersion in the negative binomial distribution. As k was increased in the model, the amount of overdispersion decreased. The influence of k was found to be greatest on the equilibrium levels when it was less than two; when it was more than two, only small changes in densities were observed. Population densities of both hosts and parasites were a function of the Achievement Factor. As the Achievement Factor increased, population densities of both parasites and hosts fell because of host mortality. The effects of three different L values were then simulated. With low pathogenicity, the equilibrium levels of the parasite population were high. As L increased, the mean density of parasites also increased. Crofton (1971b) pointed out, 'this is due not only to an increase in the number of parasites but, perhaps surprisingly, to a decrease in the number of hosts'. Apparently, as transmission success increased, more and more hosts acquired parasite numbers that were greater than L and were thus eliminated. Host sterilization by parasites was found to be similar to host death in terms of the equilibria that could be established. Another observation worth noting is that when dispersion was reduced, a low L produced instability, causing the entire system to collapse. Finally, when the effects of immunity were introduced, the results were inconclusive for the most part. When overdispersion was high, immunity produced a stable, oscillating system. If immunity was strong enough to increase host survivorship and produce higher host population densities, parasite densities will also be higher, not lower as might be expected.

Based on this modelling effort, Crofton (1971b) concluded that his quantitative description of parasitism (Crofton, 1971a) was a 'functional' one because parasite pathogenicity was found to be a primary factor in establishing host–parasite population equilibria. The significance of the model is not so much that it attempted to replicate with precision the 'real world', but that it provided the basic hypothesis from which subsequent modellers could develop more sophisticated methods for interpreting host–parasite interactions. In effect, it became both the baseline and the initial stimulus for future effort in this area of study.

4.2.4 Crofton revisited (May, 1977)

As pointed out by May (1977), the Crofton (1971b) model for the dynamics of host–parasite relationships was overly simplistic, but it 'retains pedagogical value as *the* (his emphasis) basic model'. According to May (1977), there were two essential flaws in the original model. First, Crofton (1971b) used a fixed value for his Lethal Level (L). In other words, all hosts with a parasite density greater than L parasites were eliminated. This feature of Crofton's model was regarded as too deterministic. Lethality should have been considered in more probabilistic terms. Thus, some hosts with more than L parasites might die and others would not, a more plausible assumption. When the deterministic effects of parasite-induced host mortality were exchanged for probabilistic expressions, May (1977) found that the key parameters in stabilizing the system were k, that describes parasite overdispersion, and $\propto$, that refers to the growth potential of the host population. Another objection was that Crofton assumed parasite transmission was a direct function of host density. Accordingly, May included 'saturation effects in the parasite transmission factor' in his modification of the model. He indicated that Crofton devised a parasite transmission factor (F) that was 'linearly in proportion to the number of hosts'. The problem with this assumption is that transmission will not increase linearly with host density indefinitely; instead 'it must saturate to unity at high host population levels' (May, 1977). Because these saturation effects were not included by Crofton (1971b), they were developed by May (1977) and appeared to generate more stability in the model.

A basic similarity also was noted between the stabilization of the host–parasite relationship by overdispersion and related interactions involving prey–predator and host–parasitoid systems (May, 1977). He further emphasized that relationships involving predators and prey 'can be stabilized by differential aggregation of predators or by explicit refuges for the prey'. Finally, it was observed that host–parasitoid interactions (Hassell, 1976, 1978) and those described by Crofton's (1971b) model for host–parasite dynamics, are similar because 'the underlying interaction processes are similar'.

4.3 OVERDISPERSION AND REGULATION: INTRODUCTION

There is a direct correlation between overdispersion and regulation even though Keymer (1982) has said that the statistical distribution of parasites within a host population is not, in and of itself, a regulatory factor. None-theless, regulation cannot operate in the absence of overdispersion. The effects of overdispersion may be influenced by a number of other factors, e.g., diet, stress, host age, immunity, etc. In some of these situations, overdispersion may be exacerbated and, in others, it may be constrained.

The effects of overdispersion may be measured in two different ways. First, it may induce host mortality, either directly or indirectly. Second, some parasites have the capacity to cause a reduction in host fecundity, even castration, and this is also regulatory in character (also recall that the impact of castration at the host population level is the same as host death). On the other hand, in most of the cases involving trematodes in molluscan hosts, the parasite is not overdispersed. The effects of overdispersion and other factors on host mortality and fecundity will now be examined through consideration of a series of case studies.

4.4 OVERDISPERSION AND REGULATION: CASE HISTORIES

4.4.1 *Hymenolepis diminuta* and its intermediate host

Keymer (1981, 1982) examined the effects of the cestode, *Hymenolepis diminuta*, on its intermediate host, the common grain beetle, *Tribolium confusum*. Experimental protocols were designed to assess various aspects of the relationships between parasite establishment, growth and infectivity, as well as density, and the host's population biology.

Results show that there were no density-dependent constraints on the establishment of infrapopulation within the grain beetles, at least within the range of exposure employed. However, there was an inverse relation-ship between infrapopulation densities and cysticercoid size, suggesting intraspecific competition for food or space, or both. No relationship was observed between infrapopulation densities and infectivity although there was a marked decline in cysticercoid infectivity with the passage of time.

The transmission of the cysticercoid to the definitive host is a much more complex process than of an egg to the intermediate host. Trans-mission dynamics in both cases are, in part, related to feeding behaviour, but in the case of the definitive host there is also an apparent link between infrapopulation size and overdispersion of cysticercoids in the beetles.

Changes in beetle population size were followed over time, with and without the parasite. The populations were begun with 400 beetles of uniform size, half of which were exposed to parasite eggs and half of which were not. Beetles were counted and re-exposed to additional parasite eggs at regular intervals. The results are illustrated in Fig. 4.2. Under these conditions, densities of infected beetle populations were reduced up to 50%

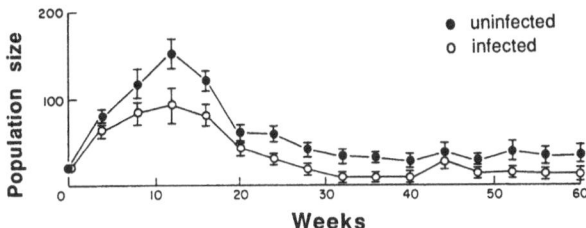

Figure 4.2 The influence of infection with *Hymenolepis diminuta* on the population growth of *Tribolium confusum*. The points represent observed values of total population size; the vertical bars indicate the 95% confidence limits of the means. (With permission, from Keymer, 1981.)

as compared with uninfected populations. Keymer (1981) pointed out that while the parasite is regulatory under laboratory conditions, it is probably ineffective as a regulator under field conditions where predation and competition are more likely to be the primary constraining factors for the grain beetle. On the other hand, Evans (1983) examined the cestode, *Hymenolepis tenerrima*, that uses ostracods as an intermediate host and ducks as definitive hosts. In a natural pond in West Sussex, England, the cysticercoid was found to cause substantial mortality in heavily infected hosts and fecundity in the ostracod population was reduced by nearly 10%.

In summary, Keymer (1981) indicated that adult parasites are probably incapable of inducing mortality of rat definitive hosts, except under conditions of nutritional stress (see also Goodchild and Moore, 1963; Dunkley and Mettrick, 1977). Regulation of *H. diminuta* dynamics at the suprapopulation level is probably via density-dependent effects on fecundity and survival of adult worms in the definitive host.

4.4.2 *Uvulifer ambloplitis* and bluegill sunfishes

The life cycle of *U. ambloplitis*, the causative agent for 'blackspot' disease in centrarchid fishes, is shown in Fig. 4.3. Briefly, the adult parasites are enteric in kingfishers, *Megaceryle alcyon*. Adults shed eggs that are passed in the faeces. Miracidia hatch from eggs and penetrate the pulmonate snail, *Helisoma trivolvis*, where intramolluscan development occurs. Cercariae released from snails locate centrachid fishes and penetrate, becoming encysted as metacercariae within the flesh. When a kingfisher eats an infected fish, the cycle is completed. The metapopulation biology of *U. ambloplitis* was studied within the snail and fish intermediate hosts in a combined laboratory and field effort that lasted for approximately three years (Lemly and Esch, 1983, 1984a, b, c, 1985).

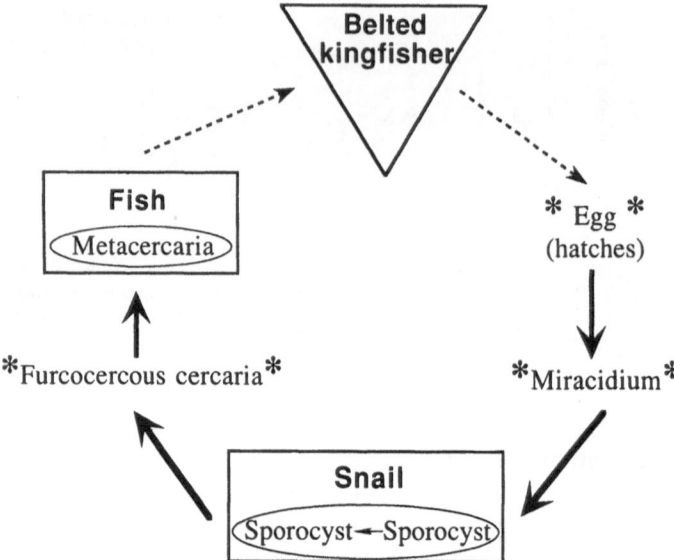

Figure 4.3 Life cycle of the trematode, *Uvulifer ambloplitis*. The symbols are explained in the caption for Fig. 1.1 (page 7).

Snail densities and parasite prevalences had a striking annual peak in July of each year, followed by a decline (Fig. 4.4). It was speculated that a combination of the life cycle of the snail and the visitation periodicity of the kingfishers to the pond (most frequently in spring and early summer) were responsible for the seasonal pattern of parasite prevalence in snails. The snail thus recruited the parasite when kingfishers were foraging in the pond in the spring and early summer. The decline in prevalence then corresponded with the annual period of senescence and mortality within the snail population and the recruitment of a new snail cohort in midsummer.

Seasonal changes in prevalence and density of metacercariae were observed in juvenile (< 70 mm) bluegill, *Lepomis macrochirus*, and largemouth bass, *Micropterus salmoides*, in the pond over a 3-year period. Prevalences in three, arbitrarily established, size classes of bluegill showed a marked seasonal pattern; highest percentages occurred from spring to the mid-winter. There were then decreases in prevalence. Peak metacercariae densities were in September, followed by declines into the following spring and then again by increases until the next September. Fishes were also maintained in liveboxes within the littoral zone and the rates of parasite recruitment were followed throughout the year. Acquisition of parasites began in April, peaked in July and stopped completely by

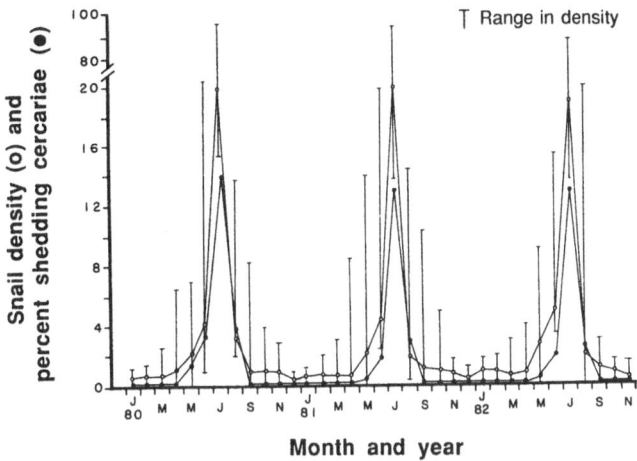

Figure 4.4 The density (number of individuals per square metre) of the snail, *Helisoma trivolvis*, in the littoral zone of Reed's Pond, North Carolina, USA, and the percentage of the snails shedding cercariae of *Uvulifer ambloplitis*. (With permission, from Lemly and Esch, 1984a.)

October. All of the parasite recruitment coincided with those periods in which snail shedding also occurred. The parasite frequency distributions were highly overdispersed in each of the years and were adequately described by the negative binomial model.

The variance-to-mean ratio ($S^2/\bar{x}$) for *U. ambloplitis* in bluegills, increased continuously in all three size classes from early spring to September and then stopped (Fig. 4.5). During that time period, the parameter appeared to be tracking parasite recruitment by the fish hosts. Lemly and Esch (1984c) reasoned that if host or parasite mortality did not occur and if no fish moved in or out of the pond, then the $S^2/\bar{x}$ ratios should remain constant from September until the following spring, but they did not. Instead, a huge decrease occurred so that within 3 months following the September peak, the ratio was 10- to 100-fold smaller. Under similar circumstances, other investigators had suggested parasite-induced host mortality as an explanation for changes in these ratios (Crofton, 1971a; Pennycuick, 1971; Lester, 1977; Gordon and Rau, 1982). However, direct observation of host mortality was not made in any of these studies.

Lemly and Esch (1984c) hypothesized that blackspot was causing mortality of heavily infected fishes in the pond. A series of experiments was then designed to examine the relationship between parasite density and host mortality. Bluegill with a wide range of parasite densities were kept in unheated outdoor aquaria and within liveboxes in the pond during

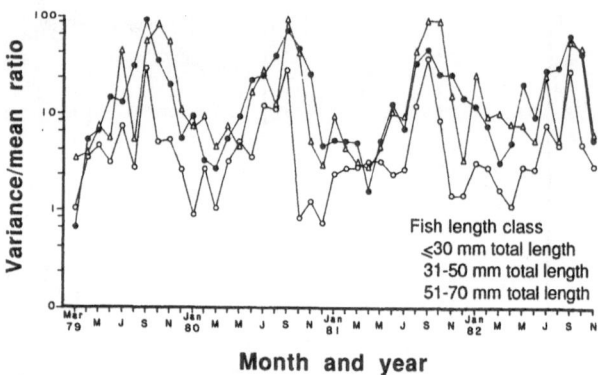

Month and year

Figure 4.5 Variance-to-mean ratios for metacercariae of *Uvulifer ambloplitis* in bluegill sunfishes, *Lepomis macrochirus*, from Reed's Pond, North Carolina, USA, from March 1979 to November 1982; n = 10 fish per data point. (With permission, from Lemly and Esch, 1984b.)

the winter months. A set of aquaria was placed in a controlled-temperature room and fishes were exposed to temperature and light regimens that simulated those of a typical winter, while another set was placed in a controlled-temperature room where temperature was held constant throughout the winter. When fishes died in any of the aquaria or in the pond, they were dissected and the metacercariae counted. In several experiments, fishes that died were also frozen for subsequent lipid analysis.

In most of the field and laboratory experiments, the coefficient of body condition was calculated for each fish in which metacercariae were counted. The coefficient is expressed as:

$$K = (100 \times \text{weight (g)}) / (\text{standard length (cm)})^3$$

Body condition is a standard method by which fisheries biologists measure individual robustness within a given population (K in this case is not to be confused with k in the negative binomial model).

All of the heavily infected, juvenile fishes (with more than 50 metacercariae) in both the field and the laboratory experiments died when water temperatures fell below 10°C (Fig. 4.6). If fishes were longer than 70 mm, high parasite densities had no effect. Similarly, high parasite densities had no effect on fishes held under constant temperature (22°C) and fed *ad libitum* in the laboratory. Based on these experiments, in combination with the field observations on changes in $S^2/\bar{x}$ ratios, it was concluded that parasite-induced host mortality was occurring among juvenile bluegill in the pond. Moreover, it was proposed that it was also directly involved with density-dependent regulation of the bluegill population. Based on population estimates for juvenile bluegill and on the

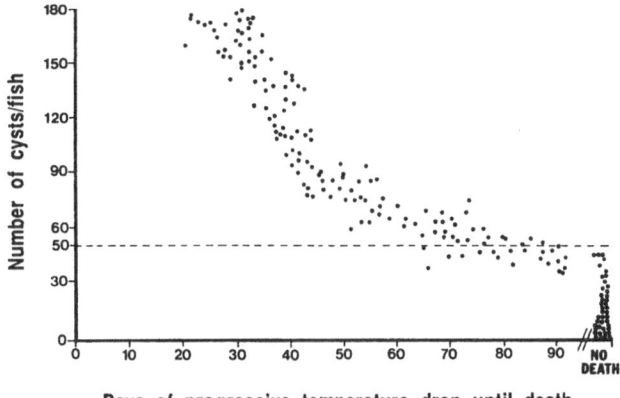

Days of progressive temperature drop until death

Figure 4.6 Association between parasite density and survival, following onset of declining water temperature for bluegill sunfishes, *Lepomis macrochirus*, from Reed's Pond, North Carolina, USA, during the autumn and winter of 1980, 1981 and 1982. Data are for fish held in outdoor aquaria at ambient temperature (n = 18) or in liveboxes in the littoral zone (n = 174). All fish were naturally infected with *Uvulifer ambloplitis* (≤ 70 mm total length). Correlation coefficient $(r_S) = -0.94$ (P < 0.001). The dotted line indicates the maximum density observed for bluegill that overwintered successfully in Reed's Pond. (With permission, from Lemly and Esch, 1984c.)

incidence of heavily infected fishes in the pond during the late autumn, it was suggested that 10–20% of the bluegill were eliminated by parasitism annually.

During the course of the study, metacercariae were counted in a large sample of juvenile bluegill taken from the pond. A striking and significant correlation was observed between body condition and parasite density. Thus, as parasite densities increased, body conditions declined (Fig. 4.7). Three bluegill are shown in Figure 1.2 (page 9). The body conditions of the uninfected fishes at the top and bottom are both greater than 7.0 while the infected fish in the middle is close to 5.0; all three fishes are exactly the same standard length. It was reasoned that, in some way, the decline in body condition contributed to a fish's inability to survive the winter months. Measurements of total body lipid revealed that it was positively correlated with body condition and negatively correlated with parasite density. Finally, in one last set of experiments, a group of uninfected fishes was starved in order to reduce their body conditions to levels similar to those caused by heavy infections. When these fishes were subjected to temperatures that simulated winter conditions (<10°C), they all died (Fig. 4.8).

It is clear that the energy demands on hosts during the infection process by certain parasites are substantial. When strigeid cercariae enter the flesh

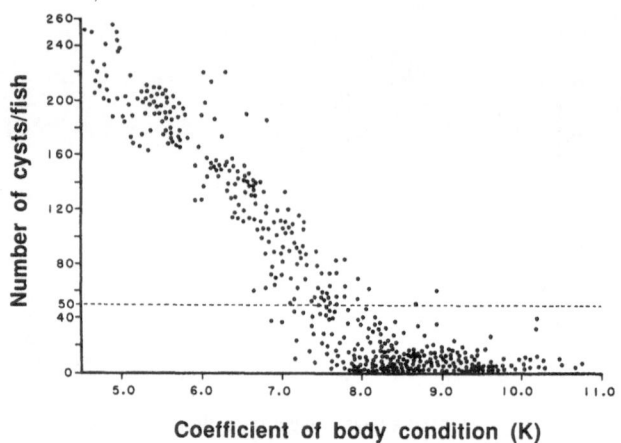

Figure 4.7 Association between parasite density and body condition of bluegill sunfishes, *Lepomis macrochirus*, from Reed's Pond, North Carolina, USA, March 1979 to November 1982. All fish (31–50 mm total length) were naturally infected with *Uvulifer ambloplitis*, n = 440, correlation coefficient (r_S) = −0.933   (*P* < 0.001). The dotted line indicates the maximum density observed for bluegill that successfully overwintered in Reed's Pond. (With permission, from Lemly and Esch, 1984c.)

of a fish, for example, there is a striking host reaction (Davis, 1936; Lewert and Lee, 1954a, b; Erasmus, 1960). It includes inflammation, localized bleeding, cell destruction, and oedema. With *U. ambloplitis*, the initial reaction is similar but, as it progresses, it actually becomes more and more intense. Thus, in addition to the localized tissue response, the fish host produces a large fibrous capsule around the parasite; this is followed by the migration of melanocytes in the wall of the cyst creating the characteristic appearance of the 'blackspot' (Hunter and Hamilton, 1941). Lemly and Esch (1984c) concluded that the metabolic demand involved with the encystment of dozens of parasites significantly decreased a fish's lipid reserves and because bluegills do not feed during the winter months, these lipid reserves are critical to their survival through the winter.

There are two important points that should be emphasized by this series of studies on *U. ambloplitis*. First, parasite-induced host mortality was inferred from changes in $S^2/\bar{x}$ ratios of parasite metapopulation within fish from field studies. Hypotheses regarding mortality of heavily infected hosts were then proposed, tested, and confirmed or rejected through both field and laboratory experiments. Second, further analysis of the data provided a framework for interpreting the physiological basis for the parasite-induced host mortality. It was thus not surprising to find that the mortality was related to a massive pathological response of the host to the

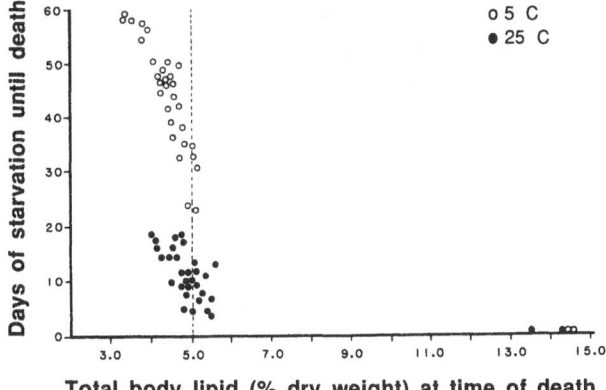

Figure 4.8 Total body lipid present at the time of death for uninfected bluegill sunfishes, *Lepomis macrochirus*, from Reed's Pond following starvation, n = 32 for each water temperature; all fishes were ≤ 70 mm total length. Correlation coefficient (r_S) = –0.211 for 5°C (P > 0.05); r_S = –0.364 for 25°C (P > 0.05). (With permission, from Lemly and Esch, 1984c.)

parasite. Among most tissue-dwelling helminth parasites, it is the trauma of the tissue reaction and the parasite-induced pathology that produces the serious morbidity problems for the host. A somewhat similar conclusion was reached by Gordon and Rau (1982) in their study of the strigeid trematode, *Apatemon gracilis*, in brook sticklebacks, *Culaea inconstans*. Although this parasite does not encyst, the metacercariae are quite large and capable of inducing considerable damage if they occur in large numbers. Moreover, Gordon and Rau (1982) also invoked the notion of parasite-induced host mortality based on seasonal changes in the $S^2/\bar{X}$ ratios, but they did not conduct experimental studies to confirm or reject the hypothesis.

4.4.3 *Heligmosomoides polygyrus* and regulation

Heligmosoides polygyrus is a common intestinal nematode in mice. It has a direct life cycle in which eggs shed in the faeces hatch and free-living larvae become infective within 7 days. The larvae are accidentally ingested by mice where they migrate into the intestinal wall, stay in the area of the serosa for about a week, and then migrate back into the lumen of the intestine before maturing sexually. Scott (1987a) designed a series of experiments in which the aim was to compare the population dynamics of infected (with *H. polygyrus*) and uninfected mouse colonies. The colonies were established using outbred animals of the same strain.

In the uninfected colony, reproduction and mortality stabilized at a point where densities were about 320 mice per square metre. At these high densities, there was density-dependent mortality among young mice, some cannibalism, and a decline in reproductive success over time. As Scott (1987a) indicated, such mechanisms are considered as 'intrinsic regulatory mechanisms in mouse populations where emigration is impossible (MacKintosh, 1981)'.

On introduction of the parasite into immunologically naive mouse populations, the impact was dramatic (Fig. 4.9). In the three arenas housing the exposed colonies, mouse densities were reduced by 50% within 7 weeks after introducing the parasite and, by 9 weeks, 90% of the mice were dead. Parasite densities in necropsied mice from these three arenas averaged from 213 to 892 per animal, with some mice harbouring up to 1500 nematodes. Mice in one of the three arenas were then transferred into a habitat in which high transmission rates were prevented. Mortality of mice continued at a high rate. However, when these mice were treated with an anthelmintic, the population growth rate increased significantly.

Based on these results, it is clear that *H. polygyrus* is a potential pathogen of considerable magnitude for mice. The study also demonstrated that the parasite is capable of regulation in the sense that it reduced population equilibria to levels lower than those that existed without the parasite. Scott (1987a) noted, however, that the experiments were conducted using naive populations and that the high mortality rates she observed could have been a result of the lack of prior exposure to the parasite. She concluded by emphasizing that 'the relative importance of parasitism in relation to other regulatory factors including predation, competition and dispersal will need to be assessed'. This admonition is certainly a clear warning for ecologists working on free-living organisms and that is: parasites can kill, they can castrate, and they can alter host behaviour.

4.4.4 *Gyrodactylus turnbulli*, guppies, and aggregation

The metapopulation biology of *G. turnbulli*, a monogenetic trematode of guppies, *Poecilia reticulata*, was examined in a series of studies by Scott (1982, 1985a, b, 1987b), Scott and Anderson (1984), Scott and Nokes (1984), Scott and Robinson (1984), and Madhavi and Anderson (1985). In addition to evaluating the parasite's metapopulation biology, the report by Scott (1987a) is of particular interest because it provides a useful discussion of the value k in the negative binomial model. Perhaps more importantly, it makes a careful comparison between the use of k and the variance/mean ($S^2/\bar{x}$) ratio in evaluating parasite aggregation.

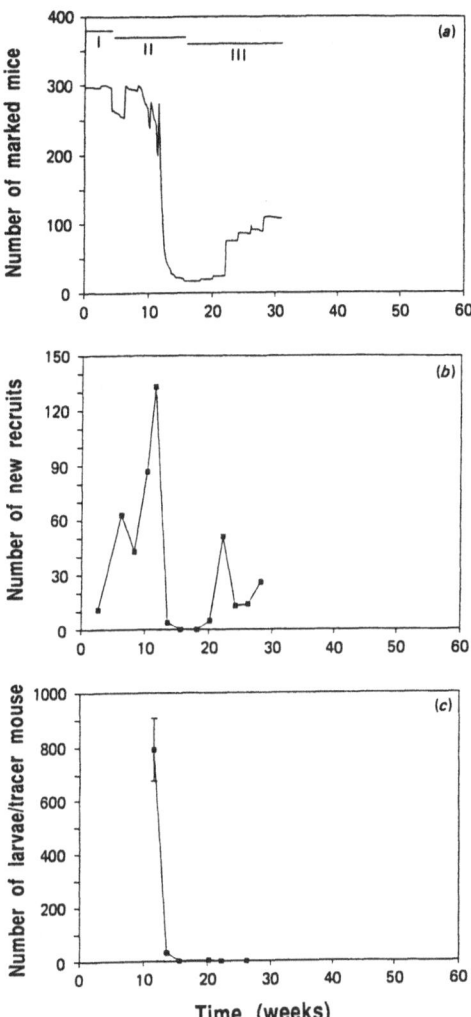

Figure 4.9 Effect of the introduction of *Heligmosomoides polygyrus* into an immuno-logically naive mouse population: (a) Change with time in the number of marked mice (> 2 weeks old) during three phases of the experiment. Phase I represents the mouse population prior to parasite transmission. Phase II represents the mouse population during the initial period of *H. polygyrus* transmission. Phase III represents the mouse population during continued low-level transmission of *H. polygyrus*; (b) Numbers of newly recruited mice (2 weeks old) at each census; (c) Mean (± SE) numbers of *H. polygyrus* larvae in tracer mice exposed to contaminated arena for 24 h following each census. (With permission, from Scott, 1987a.)

The system on which she based her assessment was described in detail by Scott and Anderson (1984). In brief, *G. turnbulli* is an ectoparasitic, monogenetic trematode that lives on the fins and other body surfaces of guppies. It has a short generation time, a direct life cycle, and transmission between fishes is rapid. Another attractive feature of the system is that fishes can be anaesthetized with ease and the parasites can be counted using an ordinary dissecting microscope. In other words, the more traditional 'kill and count' method of enumerating parasite infrapopulation density was not necessary.

The aims of Scott's (1987b) investigation were to examine the temporal aggregation of the parasite and to determine if changes in the level of clumping would provide any insights into the biological processes that produced the aggregation. The results of several carefully designed experiments revealed that the parasite undergoes repetitive epizootic cycles on guppies. In that part of the epizootic where parasite numbers are increasing, parasites became more aggregated presumably because of direct reproduction by the parasites on individual fish. Then, as parasite prevalences and densities peaked, aggregation began to fall and continued downward as the prevalence and densities declined. According to Scott (1987b), 'this is thought to be a function of density-dependent death of infected hosts, and density-dependent reduction in parasite survival and reproduction on hosts that recover from infection'. The first part of her suggestion thus is not unlike that of Lemly and Esch (1984c) for *U. ambloplitis* in bluegill.

Scott and Anderson (1984) had previously discussed and evaluated the same data set and concluded 'that the degree of aggregation decreased while the prevalence and abundance of infection increased, based on the positive correlation between k and abundance'. Scott (1987b) subsequently rejected that interpretation of their data. Thus, while k did rise as density increased, it would not necessarily also indicate that over-dispersion was simultaneously decreasing since both the density and the percentage of infected fish were changing. She stated, 'it is suggested that this relationship is simply a function of the interaction between mean burden (=density) and k in the negative binomial distribution, and does not indicate anything *per se* about the changes in the degree of aggregation'. The term k was considered a 'mathematical convenience' and it was urged that it should not be interpreted outside of that consideration. The $S^2/\bar{x}$ ratio is the appropriate term of biological sensitivity, especially if prevalence or density, or both, are in flux, or if the tail of the distribution is the focus of evaluation. While k can be useful in describing a given frequency distribution, it must be considered with caution when drawing conclusions having any sort of biological implications. As she declared, the value of k in the negative binomial is a theoretical description while 'the variance to mean ratio is an absolute measure of the degree of aggregation'.

4.5 EPIDEMIOLOGICAL IMPLICATIONS

4.5.1 Aggregation and predisposition

Despite the availability of suitable anthelmintic drugs, the human intestinal helminths continue to represent an undesirable and unnecessary global scourge. Most are concentrated within the so-called Third World countries where their prevalences continue to increase. *Ascaris lumbricoides*, for example, was estimated to occur in about one billion individuals in 1986 (Crompton, 1989). Stoll (1947) had placed the number at 645 million, meaning that the prevalence of this parasite has increased by about 50% in just under 40 years. The prevalences of several of the other debilitating parasite diseases of man have increased in a similar manner (Table 1.2 on page 8).

The life cycles of these parasites and many of their epidemiological characteristics have been known for years. However, it was not until the early 1980s that the emergent ecological doctrine, developed by Crofton (1971a, b) was applied to human helminth parasites. At about the same time, some epidemiologists and ecologists recognized that there was a problem in Crofton's assertion that parasites have the capacity to kill. It is known that many parasites, if not most, do not cause host mortality. Accordingly, Anderson and Gordon (1982) qualified Crofton's mortality position by stating 'that the probability of a host dying in a given time interval is some function of its parasite burden'. Their revised approach appears much closer to reality, certainly as it applies to the large group of enteric helminths affecting man.

The capacity of the enteric helminths to cause morbidity or mortality is clearly influenced by an enormous array of environmental factors that operate in conjunction with nutritional status, immunocompetency, and the physiological condition of the host. Moreover, the latter three factors frequently are so intertwined as to be virtually inseparable. Consider hookworm disease as an example. A single adult hookworm attached to the intestinal wall of a human host will consume approximately 0.2 ml of blood per day. Assume that a host has an infrapopulation density of 100 worms. In a single day, the host will lose 20 ml of blood. If this individual is on a high protein diet, the probability of morbidity is low and mortality is nil. On the other hand, if the host is living under exceedingly poor socioeconomic conditions and the diet is meagre and low in protein, the loss of 20 ml of blood per day may very well produce morbidity and perhaps even contribute to the individual's death.

The insidious nature of hookworm disease exacerbates the problem. The loss of blood from the individual living under poor socioeconomic conditions is compounded by a typically low protein diet, meaning that lost haemoglobin cannot be readily replaced. The immunocompetency of the individual is further compromised through loss of gamma globulin and this translates into antibody loss. Thus, the poor diet will influence the host's immunological capacity while the worms affect both the O_2 carrying capacity of the host and its immune system. The host's size, age, sex, genetics, and cultural background also will influence the situation regarding morbidity or mortality. These sorts of host–parasite relation-ships are exceedingly complex and understanding the epidemiological characteristics of the diseases caused by these parasites requires multi-variate approaches.

There are two ways of estimating infrapopulation densities of enteric helminths in human populations. One is to count parasites in stools immediately after treating the individual with an appropriate anthelmintic. This procedure has its problems, one of which is the cost of the treatment itself. Another difficulty is that parasite expulsion can be extended over a period of several days so that the reliability of faeces collections may be questionable, especially when dealing with children (L. Kightlinger, personal communication). The second procedure for estimating parasite densities in humans relies on faecal egg counts; however, a primary difficulty with the technique is related to the tremendous variability in the number of eggs shed per day over a given time interval. Moreover, there are also density-dependent constraints created by intraspecific parasite competition, causing even greater variability in daily egg output. In many of these diseases, egg output will increase as parasite densities increase, but only until a threshold is reached and then it may decline or fluctuate radically. A number of investigators have attempted to resolve these problems and the reader is referred to these papers for a more thorough review of the issues and solutions (Croll *et al.*, 1982; Anderson and Schad, 1985; Holland *et al.*, 1988; Elkins and Haswell-Elkins, 1989).

One of the initial observations on the epidemiology of human infections using the approach developed by Crofton (1971a, b), was made in the early 1980s by Croll and Ghadarian (1981). They worked in a number of Iranian villages generating data on the frequency distributions of several enteric helminths, including the nematodes, *Ancylostoma duodenale*, *Necator americanus*, *Ascaris lumbricoides*, and *Trichuris trichiura*. The first two (both hookworms) employ soil-dwelling, filariform larvae that penetrate the skin of their hosts. The latter two species of worms are transmitted directly, through the accidental ingestion of eggs.

All four parasites were found to be overdispersed in their frequency distributions within the village populations. All of the villagers were then de-wormed using appropriate anthelmintics. After about a year, the same people were re-examined for parasites. The investigators found that those individuals who had been wormy the first time were not wormy the second time. In other words, 'wormy persons' from the initial screening were not necessarily predisposed to re-infection. As Croll and Ghadarian (1981) stated, 'in this population of mixed, ages, sexes, and social histories, the "wormy persons" prior to treatment were not reliable predictors of subsequent intensities of infection'.

These findings suggested two important conclusions. First, host genetic predisposition did not appear to play a role in the acquisition of these helminth parasites, or in creating overdispersion. Second, the worms did not appear to have stimulated the host's immune responsiveness to the extent that it could prevent re-infection. They concluded that 'risk factors are much more subtle than the classic categories we are considering or that "wormy persons" in these communities result from the super-imposition of otherwise random events'.

The study by Croll and Ghadarian (1981) was a watershed in modern epidemiology, in part because of their use of Crofton's (1971a, b) approach. Their results and conclusions regarding the absence of genetic predisposition were, however, viewed somewhat sceptically by some investigators. Since the time of their report, there have been a large number of studies in different parts of the world that have attempted to determine if predisposition is, or is not, an epidemiological cornerstone (Bundy *et al.*, 1987; Haswell-Elkins, Elkins and Anderson, 1987; Anderson, 1989).

Generally, the notion of predisposition is considered within one of three contexts. The first relates it to non-genetic factors such as personal hygiene, defecation practices, socioeconomic status, religion, water resources, etc. Any of these factors could influence transmission dynamics in such a manner as to create the impression of predisposition, as well as to produce parasite distributions that are overdispersed. The second context relates parasite frequency distributions to either host or parasite genetics, or both. For example, there could be genetically determined resistance that would create differential levels of immunity within a community and thus contribute to parasite overdispersion. Genetically based resistance has been documented for parasites in a number of animal models, such as mice, cattle and guinea pigs, but its occurrence in humans is less well understood. Finally, the presence or absence of predisposition could also be affected by differential levels of nutrition within a population and this could be a confounding factor in assessing the resistance/susceptibility and, therefore, genetic component in parasite transmission dynamics.

The significance of overdispersion and predisposition as they affect the enteric helminths of humans has several practical implications when it is considered within the framework of mathematical models (Anderson and May, 1979; May and Anderson, 1979; Schad and Anderson, 1985). Thus, the latter authors, especially, argued that treatment of only the heavily infected (and predisposed) humans would be effective in reducing overall infection within a community, or possibly in eliminating the parasite entirely. They based their assertions on predictions regarding transmission dynamics generated from mathematical models and on the number of parasites necessary to sustain a given parasite species within a host population. In other words, it may be necessary to apply therapy to only a few people within a community in order to effectively control the parasite. As they suggested, cost-effective treatment becomes a clear possibility except for the expense involved in identifying the predisposed individuals within a population. Recently, however, Keymer and Pagel (1990) reported that under 30% of paired hosts had the same ranked infrapopulation density order at the first and final sampling times. This caused Guyatt and Bundy (1990) to suggest that 'fixed, long-term predisposition is modified by transient changes during the relatively short course of the observation (usually less than two years). Equally it may indicate that the phenomenon of predisposition is a reflection only of short-term factors that remain constant for some individuals during the study period'. In any event, the idea of predisposition through genetic mechanisms is certainly a debatable issue at the present time.

4.5.2 Non-genetic predisposition

Defecation practices are among the most important factors in creating the differential frequency distribution patterns of certain enteric helminths. This behaviour is directly related to the transmission patterns associated with these parasites. Haswell-Elkins *et al.* (1988), for example, observed in an Indian coastal village that women were much more heavily infected with hookworms because they used shaded tree groves in which to defecate while men used the sandy beaches. Hookworm larvae require special soil and shade conditions in order to develop successfully; such conditions were optimal in the shaded tree groves, but not on the sandy beaches. Hookworm disease was much more common 60–70 years ago, than it is now, in the southeastern United States and there was a distinct, age-related pattern of infection. Teenagers were likely to have more hookworms than young children or adults because they would commonly return to a previously established outdoor site to defecate, while young children tended to defecate more randomly and adults would use indoor facilities when they were available.

The wearing of shoes prevents filariform larvae of hookworms and *Strongyloides stercoralis* from penetrating the skin and thus is a considerable socioeconomic deterrent to transmission in certain parts of the world. Holland *et al.* (1988) reported that Panamanian children living under poverty conditions, such as in houses with dirt floors, were much more likely to have enteric helminths than children living in concrete block houses. Crowding was also reported as a factor conducive to increasing prevalence of enteric worms among children on the island of St. Lucia.

4.5.3 Genetic predisposition

Long before the study by Croll and Ghadarian (1981), a basis was proposed for genetic predisposition to parasite infection. Thus, Keller, Leathers and Knox (1937) demonstrated that the frequency of hookworm disease could be separated along racial lines, with non-Caucasians having significantly lower prevalences of the disease than Caucasians. Cram (1943) reported the same racial trend for the nematode, *Enterobius vermicularis*, which is directly transmitted via the ingestion of eggs.

Schad and Anderson (1985) examined hookworm infections in several villages in West Bengal, India, prior to, and after, drug therapy was administered. They were able to show that the frequency distributions were overdispersed both before and after drug therapy and, moreover, that certain individuals appeared to be predisposed to infection. They stated, 'we suspect that both heterogeneity in exposure to infection within human communities and genetically determined host resistance mechanisms play important roles as determinants of parasite aggregation and host predisposition to infection.' More recently, some efforts have been made to correlate resistance to infection with certain types of known genetic markers, but the results have not been very conclusive (Wakelin, 1987; Bundy, 1988). However, animal models have been successfully employed to establish relationships between genetically based predisposition and infection (for review, see Wakelin, 1984, 1987). For example, Sher, Smithers and Mackenzie (1975) provided a clear indication of a single locus control in the protective immunity against the trematode, *Schistosoma mansoni*, in laboratory mice. Enriques, Zidian and Cypess (1988) reported that resistance to infection of *Heligmosomoides polygyrus* in rats could be under the control of several genes, some of which may belong to the major histocompatibility complex.

4.5.4 Concluding remarks on epidemiology

Based on the preceding review, it should be clear that 50–60 years of study on the epidemiology of parasitic diseases, especially those involving enteric helminths, have yielded an enormous amount of information. Yet, these diseases are on the rise in most Third World countries (see Table 1.2 on page 8). There are several reasons for this discouraging observation. Part of it is the result of the divisive social and political pressures in many of these countries that have resulted in greater and greater socioeconomic repression of already poor populations. The example of the widespread migration of the Kurds in northern Iraq following the Gulf War illustrates this point clearly. Another problem is the disproportionate emphasis for research on diseases affecting much smaller segments of the world's population. Many of these diseases are horribly unpleasant and deadly, and because of this, they attract enormous public attention and financial treasure. This has produced the apparently intractable paradox of spending large sums of money on some of the 'lesser', though no less deadly, diseases of mankind.

A second issue is that the amount of money allocated by certain governments, for malaria research as an example, almost always increases during the time of war when their soldiers are exposed to the parasite. After the war is over, the money spent on these research programmes declines, leaving many unfinished. In effect, national research priorities change rapidly and this too is most unfortunate.

A question of some significance relates to the importance of density-dependent constraints, or regulation, as applied to the enteric helminths that infect man. The position taken by Quinnell, Medley and Keymer (1990) is that many of these species are regulated at the infrapopulation level by density-dependent factors, primarily nutrient and spatial resources, and immunity. A problem in their argument is that their conclusions for the human enterics are largely extrapolated from data generated in the laboratory on non-human, host–parasite systems. They, however, recognized these limitations and recommended further effort to alleviate this shortcoming.

Guyatt and Bundy (1990) urged that 'future research should focus on longitudinal studies that monitors an individual's infection and variables related to exposure or susceptibility through time'. They are concurring with a point made earlier regarding the almost complete lack of long-term studies on host–parasite interactions. This point was also clearly made by Quinnell, Medley and Keymer (1990) who commented, however, on the ethics of long-term studies as they apply to human populations. As they indicated, 'There are, however, noticeable gaps in the available data [regarding long-term epidemiology]. These arise mainly from the ethical requirement to treat people where possible, thus making long-term observational studies unacceptable and the necessity to spend considerable effort persuading subjects to collaborate, thus reducing sample sizes'.

Even though the epidemiologies of disease produced by enteric helminths are well known, they are still incompletely understood. For example, the present concept of genetic predisposition to infection is relatively new and, accordingly, the basic or more subtle nuances of the concept have not yet been clearly evaluated. If genetic predisposition to infection can be demonstrated, it may be necessary to treat only the 'wormy persons' in order to effectively reduce the prevalence and density of most enteric helminths (Schad and Anderson, 1985). The problem with such an approach, however, is that it depends on the validity of the genetic predisposition concept. The question persists, is genetic predisposition the core of a realistic paradigm, or is it simply a smaller segment of a much larger landscape?

4.6 MODELS

'It will be acknowledged that a predictive mathematical model of the epidemiology of any disease is desirable, both from the standpoint of intellectual satisfaction and from the standpoint of the usefulness in planning measures to control the disease' (Hairston, 1965). With this statement, the *raison d'être* for the mathematical model in epidemiology is certainly made clear. There are, however, two other important issues that need to be raised with respect to the broader development and use of mathematical models. First, any model is only as good as the assumptions required to make it and, moreover, the assumptions are only as good as the quality of the data provided to make them. Second, any model, as a research tool, is a sophisticated mathematical **hypothesis**. Models are attempts to predict reality in mathematical terms. In biology, one of the greatest problems with modelling is associated with the vagaries of biological systems; another is the extraordinary complexity of the same systems. Despite these almost intractable difficulties, the mathematical models developed for host–parasite population dynamics have become valuable tools in understanding how these systems function, and for developing important and useful applications in the treatment and control of parasitic diseases.

Hairston's (1965) study was a clever effort to apply the life-table concept to understanding the efficiencies of transmission dynamics in the life cycle of the digenetic trematode, *Schistosoma japonicum*, in the Philippines. Two important findings emerged from his studies. First, he found that the then existing information regarding the epidemiology of *S. japonicum* was inadequate for the approach he wanted to use in developing his mathematical model. This observation serves as a critical object lesson in that one cannot assume that existing data, even though they may be accurate, will provide the sort of information necessary to develop a realistic model. Second, he made the determination that reservoir hosts were much more important to the transmission dynamics of *S. japonicum* than had been previously suspected. This led him to the conclusion that humans were not the primary hosts for the schistosome in the Philippines. Thus, as previously noted, if the parasite was eliminated from a local population, it would quickly return because of its prevalence in many reservoir hosts.

The deterministic model developed by Crofton (1971b) and subsequently critiqued by May (1977) has already been discussed in section 4.2. The latter author correctly viewed Crofton's effort as 'seminal' and as having 'pedagogical value'. Crofton's model also provided a natural connection to his (Crofton, 1971a) quantitative approach in defining parasitism; moreover, it clearly established a relationship between pathogenicity and population stability. Another positive contribution rests with the reinforcement of Hairston's (1965) observation regarding a lack of understanding of the transmission dynamics for many parasitic organisms. Thus, even though a parasite's life cycle may be well known and studied, subtleties in the flow of parasites through a series of hosts can make the efficacy of mathematical modelling questionable.

The use of modelling for host–parasite interactions was extended when Anderson (1974, 1976) focussed attention on the metapopulation biology of the cestode, *Caryophyllaeus laticeps* (Fig. 3.3 outlines the life cycle). He developed an immigration–death model to assess various components of the parasite's life cycle, but found the model inadequate because of the complexity of the parasite's life cycle. Specifically, the problems were associated with too many time lags, the age structure of the host population, and the periodicity associated with the maturation of the parasite.

Subsequently, Anderson (1978) developed a more sophisticated model for the analysis of host–parasite interactions. In it, he suggested that stability in host populations will be affected by parasite overdispersion, by density-dependent restrictions on the growth of parasite infrapopulations, and by non-linear, parasite-induced host mortality. On the other hand, he also determined that host and parasite infrapopulation growth can be destabilized when parasites are able to reduce the host's reproductive potential as with the castration of molluscs by larval digenetic trematodes. Thus, some parasites that reproduce within their hosts can have a destabilizing effect on host population growth without causing mortality. Finally, time delays can affect population growth as well as the stability between host and parasite interactions.

Anderson and May (1979) and May and Anderson (1979) presented a most comprehensive treatment of mathematical models as applied to what they called microparasites (viruses, bacteria and protozoans) and macro-parasites (helminths and parasitoids). They related the patterns of disease caused by microparasites to:

1. the host providing a suitable environment;
2. the extent of pathogenicity caused by the parasite or its effectiveness in reducing the host's reproductive capacity;
3. the parasite's capacity for the induction of acquired immunity; and
4. the requirement that a parasite be transmitted from one host to the next.

Macroparasites were considered within two contexts, those with direct life cycles and those with indirect life cycles. In the former group, patho-genicity, host resistance, and production of transmission stages all affected parasite densities within the host population. The phenomenon of parasite overdispersion also was re-emphasized. They predicted that overdisper-sion would influence both pathogenicity and parasite transmission dynamics. For macroparasites with indirect life cycles, they emphasized the difficulty in modelling systems with multi-host life cycles. Based on these models, however, they provided several generalizations regarding the evolution of host–parasite relationships. Assuming no genetic change in the parasite, for example, it was suggested that selection in an evolu-tionary time-frame would ultimately lead to a reduction in the numbers of susceptible hosts. In this way, if a host population is initially regulated by a parasite, it was predicted over the long term that it may escape regulatory influences altogether.

They also predicted that parasites with indirect life cycles could be identified with a number of distinctive characteristics in their host populations. For example, pathogenicity in the definitive host would be low, prevalence high, and the expected life span of the host would be long. Conversely, for the first intermediate host, pathogenicity should be high, prevalence low, and the host's life span should be short. To support this contention, they cited as evidence the *Haematoloechus coloradensis* study of Dronen (1978) that was described in section 3.6.3. In that system, prevalence was highest in frogs, but no apparent pathology was produced. In dragonflies, where prevalence was moderate, there was likewise only moderate mortality caused by the parasite. In snails, prevalence was low, but mortality was high. Generalizations such as this, however, cannot be extended to all systems. Consider the *Halipegus occidualis*–ranid frog combination (Crews and Esch, 1986; Goater 1989) described in section 2.3.1. *Halipegus occidualis* occurs in approximately 60% of the frog definitive hosts and no apparent pathology is produced. On the other hand, up to 60% of the molluscan first intermediate hosts can have patent infections by the end of their life span and, in all cases, the snails are totally castrated. As has been stressed, while castration is not death, it is the equivalent from the standpoint of host reproductive fitness.

A long-standing discussion concerns the extent of parasite-induced host mortality in natural populations. Empirical evidence to support this notion is not extensive, although some does exist. The best case for supporting the idea of mortality caused by parasites is provided by several of the digenetic trematode–molluscan combinations, two of which were just cited (Dronen, 1978; Crews and Esch, 1986; Goater, 1989; Fernandez and Esch, 1991a, b). Other intermediate hosts are also known to be adversely affected by certain protozoan and helminth parasites (Crofton, 1971a; Keymer, 1980, 1981; Evans, 1983). In many of these systems, one of the overriding factors supporting the case for parasite-induced host mortality is the overdispersed frequency distribution. Anderson and Gordon (1982), in a series of Monte Carlo simulations, examined parasite dispersion patterns within the context of probability models for increasing and decreasing host and parasite populations. They showed 'that, for certain types of host–parasite associations, convex curves of mean parasite abundance in relation to age (age–intensity curves), concomitant with a decline in the degree of dispersion in the older classes of hosts, may be evidence of host mortality by parasite infection'. In a very real sense, the laboratory studies by Scott (1987b) on the monogenean, *Gyrodactylus turnbulli*, tend to confirm the interpretation of Anderson and Gordon (1982) regarding the effects of aggregation on host mortality within a temporal framework.

The consequences of parasite-induced changes in host behaviour have been reviewed by Holmes and Bethel (1972), Moore (1984a, b) and Barnard and Behnke (1990). Dobson (1985) attempted to quantify the population dynamics of hosts exhibiting behavioural modifications resulting from parasitism by developing a set of mathematical models. In each case, he estimated the basic reproductive rate (R_o) of the parasite at the outset of its introduction into the host population. The R_o was then used to predict the threshold values for intermediate and definitive hosts that were required to maintain the parasites at levels where local extinction would not occur. In each system examined, when the parasite caused changes in host behaviour, the R_o would increase and the threshold numbers necessary to maintain the parasite were reduced. Using phase plane analyses, he showed that intermediate host and parasite infrapopulations had a tendency to oscillate over time. When intermediate host densities were low, changes in host behaviour increased the frequency of oscillations. In contrast, when intermediate host densities were high, parasite densities were affected more by interactions with definitive hosts with intermediate host behaviour becoming less significant in influencing parasite infrapopulation sizes. He concluded by emphasizing that since host–parasite systems are patchily distributed in space, any change in host behaviour would clearly benefit the parasite. It would act in this manner by serving to reduce the density threshold necessary to sustain the host over time and increase the basic reproductive rate of the parasite. These adaptive permutations were viewed as highly advantageous in exploiting host populations that occur in small pockets or in situations when hosts are erratic in their site visits to a particular habitat.

In a recent review regarding the extent of regulation by parasite species in nature, Scott and Dobson (1989) identified several host–parasite systems in which regulation was clearly demonstrated, or at least could be strongly inferred. They even referred to the possibility that parasites may function as 'keystone species' at the community level. They quickly qualified this statement by saying 'we should curb our enthusiasm and understand that just because parasites were largely ignored in ecological contexts in the past, an understanding of host–parasite interactions will not explain everything about host population regulation and community structure'. This exhortation regarding less heat and more light should provide a strong indication of where the areas of parasite population biology and mathematical modelling are at the present time. As the literature indicates over and over, there is a severe shortage of appropriate field and laboratory data and, moreover, not enough are of a long-term nature. When more of these studies are completed and more information regarding the epidemiological or epizootiological nature of parasitic disease is acquired, then, and only then, will there be a reasonable basis for effectively considering parasitic diseases from the standpoint of mathematical models. As Cohen (1977) states in a caveat lector in his paper regarding mathematical models of schistosomiasis, 'some models contain so many internal inconsistencies that they should be perused with extreme caution, if at all'.

5 Life history strategies

5.1 INTRODUCTION

It has been indicated in section 1.4.1 that the residual component in the life cycles of many species of parasitic organisms represents an important strategy for local transmission as well as dispersal. Among those parasites with a residual component in their life cycle, many have larval stages that can remain in a habitat long after the definitive host has deposited the parasite's propagules and disappeared from the scene. On the other hand, not all parasites have complicated life cycles. Many have direct life cycles where an egg, for example, may be directly ingested and the host thus infected. Despite the route of infection, whether direct or indirect, many parasites have co-evolved with their hosts to the extent that their survival strategies involve mechanisms that increase their reproductive fitness. Some strategies include those that ensure spatial or temporal overlap with a potential host. Indeed, as stated by Janovy and Kutish (1988), some 'studies suggest that temporal and spatial aggregation of infective stages may have *equivalent* effects on the structure of parasite populations' (emphasis added); in other words, time and space are of equal importance in affecting the aggregation of parasites. Other parasites may repress a host's immune responsiveness, allowing the parasite to successfully establish and thrive. Whatever the strategy, the outcome permits fecundity levels to remain high enough to sustain the parasite's suprapopulation density above some threshold level.

Modern life-history theory for free-living organisms has its origins with Cole (1954). Since that time, attempts have been made to both broaden and consolidate his approach into a viable concept to explain life-history patterns of various species (Lewontin, 1965; MacArthur, 1972; Wilbur, Tinkle and Collins, 1974; Pianka, 1976; Stearns, 1976, 1977; Parry 1981; Sibly and Calow, 1986; Southwood, 1988). More recently, Hughes (1989) has reviewed the ideas regarding life-history strategies as they apply to animals that are capable of cloning, including many parasite species. Using models developed variously by Calow (1983a, b), Caswell (1985) and Sackville Hamilton, Schmid and Harper (1987), Hughes (1989) proposed that if gametic reproduction is negatively affected by extended embryological development and high death rates among juveniles, then there will be an advantage to agametic reproduction if high population increases are possible. The opposite is true if the conditions are reversed. Thus, if growth is slowed, if development of an embryo is rapid, and if juvenile mortality is limited, then gametic reproduction will be favoured. These ideas, and related ones such as *r*- and K- selection (MacArthur, 1972) and S and G selection (Sibly and Calow, 1985, 1986) will be reviewed more thoroughly at the end of this Chapter.

5.2 REPRODUCTIVE STRATEGIES

5.2.1 Agametic strategies

Agametic reproduction by parasites is most common among the intra-molluscan larval stages of digenetic trematodes. From a single, penetrating larval stage, the parasite may give rise to as many as four successive larval stages inside the snail, i.e., mother sporocyst, daughter sporocyst, mother redia, and daughter redia. Depending on the species, one or both redia stages may be skipped. Internally, each larval stage is filled with primitive reproductive cells that are capable of giving rise to offspring either by **polyembryony** or by **cyclical, apomictic parthenogenesis** (partheno-genesis lacking meiosis). According to Hughes (1989), however, the exact mechanism of reproduction is still under debate.

Some authors claim that the reproductive process in digenetic trema-todes represents **cyclical parthenogenesis** (mictic reproduction alternat-ing with parthenogenetic reproduction) because most flukes that are **monoecious** (hermaphroditic) also have the ability to cross-fertilize if more than one sexually mature adult is present. The initial division by the zygote of a trematode results in the production of both a somatic and a reproductive cell (Hughes, 1989). Division continues with the somatic cell line producing only somatic cells. The reproductive cell line at each division gives rise to one more somatic and one reproductive cell. When the larva matures, the single reproductive cell then gives rise to a pool of reproductive cells. The question revolves around whether the reproductive cells represent a germ line or if the tissue is embryonic in character. If the former and they give rise to germ cells by meiosis, then they are parthenogenetic. If the latter and they give rise to new cells by mitosis, then they are embryonic. Khalil and Cable (1969) have described meiosis in the trematode, *Philopthalmus megalurus*, but Hughes (1989) has encouraged verification of this observation.

The life cycle of digenetic trematodes continues with the production of cercariae. Again, reproduction is either by parthenogenesis or poly-embryony. Since some 40 000 species of digenetic trematodes have been described to date, it is not surprising that the numbers of cercariae produced, their sizes and shapes, and the strategies employed in reaching the next host in the life cycle are exceedingly variable. These and other considerations regarding colonization will be discussed later in this Chapter.

Another group of parasitic organisms is known to engage in both sexual and **homogonic** (asexual) reproduction; it includes a number of species of Cestoda (tapeworms), primarily those belonging to three families in the order Cyclophyllidea. These include the Taeniidae, Dilepididae, and Mesocestoididae, although Beaver (1989) and Conn (1990) have raised some interesting questions regarding this form of reproduction in the latter family.

Most adult tapeworms undergo **strobilization**. This term simply refers to the process of growing individual **proglottids** (segments) from a region in the neck immediately behind the **scolex** that is the organ of attachment for the worm in the gut of its definitive host. The neck possesses a large population of primitive stem cells that give rise to all of the somatic and reproductive tissues within each proglottid. Except for the reproductive cells in the mature proglottids, stem cells are the only ones in an entire strobila that have the capacity to divide. Hughes (1989) refers to strobilization as **modular iteration**, or the production of repeating modules that are suitable for sexual reproduction. Indeed, the individual proglottid is an extraordinary reproductive machine. In almost all cestode species, each proglottid has a complete set of male and females repro-ductive organs and is capable of producing thousands of eggs that are shed to the outside in the faeces.

Some cestodes are **apolytic**. That is, eggs are retained in the proglottid when it drops off the strobila. Faecal examinations of stools for eggs of species of cestodes that employ apolysis would not be very productive. This kind of egg-shedding strategy will affect the dissemination pattern of eggs. Other cestodes are **anapolytic**. In these species, eggs are lost from the proglottid through a uterine pore while the proglottid is still attached to the strobila and then they are shed free of the proglottid in the faeces. When egg production ceases in the proglottid, it is released from the strobila and usually totally disintegrates by the time it is passed to the outside in the faeces.

There are several exceptions to this pattern, two of which are interesting enough that they deserve attention at this point. The first is *Haplobothrium globuliformae*, a pseudophyllidean cestode of the primitive bowfin fish, *Amia calva*. The life cycle is shown in Figure 5.1. The bowfin becomes infected when it eats a piscine second intermediate host possessing plerocercoids of the parasite. In the gut of *A. calva*, the plerocercoid undergoes strobilization, but genital primordia do not form in the proglottids. Once the strobila reaches a certain size, the proglottids begin dropping off, one by one, and new ones are formed to take their place. The proglottids, now free from the primary strobila, migrate forward in the gut; they develop a typical pseudophyllidean scolex and undergo secondary strobilization. The proglottids on the secondary strobila develop sexually, produce eggs and release them in typical pseudophyllidean style, via anapolysis. Morphologically, the scolex on the primary scolex is much more similar to that of species in the order Tetraphyllidea, parasites of elasmobranchs (sharks, skates and rays); the scolex possesses four spinose tentacles. The tetraphyllidean scolex on the primary strobila, in combination with the presence of a pseudophyllidean-like scolex on the secondary strobila, the occurrence of the tapeworm in the primitive

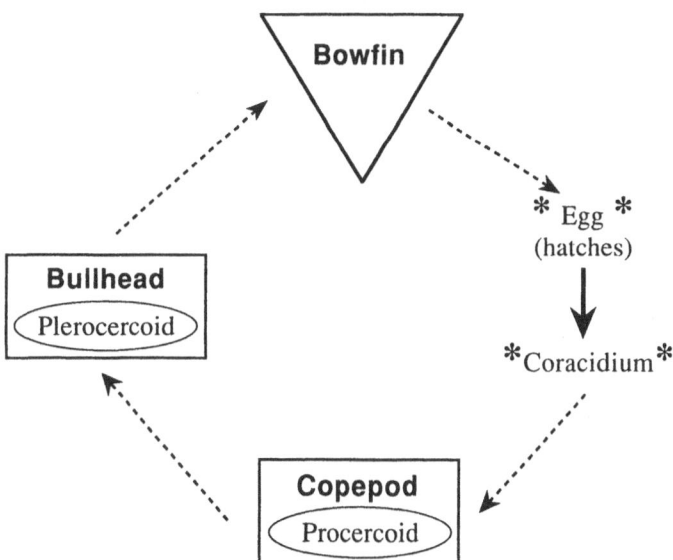

Figure 5.1 A schematic life cycle for the cestode, *Haplobothrium globuliformae*. The symbols are explained in the caption for Fig. 1.1 (page 7).

bowfin fish, and its pseudophyllidean life-cycle patterns, have collectively suggested to some that it represents an evolutionary 'link' between the orders Tetraphyllidea and Pseudophyllidea.

A second, bizarre developmental pattern is associated with the cestode, *Amurotaenia decidua*, a nippotaeniid that occurs in common bullies, *Gobiomorphus cotidianus*, in Australia. As long as the proglottids remain attached to the strobila of *A. decidua*, the reproductive system remains in a rudimentary state. However, after proglottids drop off the strobila, they then mature sexually and produce eggs; presumably, therefore, each proglottid acts as an unattached and independent reproductive unit. Moreover, there appears to be a density-dependent restraint on the numbers of freely existing proglottids that can become established within the gut of an infected fish (Weekes, 1990).

As mentioned above, some members of at least three cestode families are capable of cloning at a comparable stage in their life cycle. Within the intermediate hosts of some species in these families, the larval stage has the capacity to reproduce through the process of 'budding'. In some taeniid cestodes, primitive stem cells are involved in the development of a larval stage identical in form to the parent; they may remain attached to the parent or they may detach. In microtine rodents, *Taenia crassiceps* buds **exogenously** (to the outside) and each daughter bud completely disengages from the parent before undergoing maturation to form a new **cysticercus** (bladder worm). In others, e.g., *Echinococcus granulosus*, budding within the **hydatid cyst** is **endogenous** (to the inside, or **unilocular**) and the number of protoscolices produced is staggering. The size of a hydatid cyst is also enormous. A single hydatid cyst of *E. granulosus* may contain up to 12 litres of fluid and as many as 1×10^6 protoscolices (the fluid within bladder worms of all taeniid cestodes is a transudate from host tissue fluids). In microtine rodents, *Echinococcus multilocularis* produces protoscolices that remain inside the hydatid cyst, but the bladder buds exogenously, producing what are called **multilocular** cysts. In others, e.g., *Taenia multiceps*, typically there is no budding and the newly formed scolices remain attached to the inside of the bladder, or **coenurus**, which also may grow to enormous sizes in the lagomorph intermediate host.

Budding within the family Mesocestoididae widely has been assumed to be a common phenomenon (Cheng, 1986; Schmidt and Roberts, 1989; Noble *et al.*, 1989) but its significance has recently been questioned by Beaver (1989) and Conn (1990). *Mesocestoides corti* was originally described by Hoeppli (1925) from specimens taken from a house mouse, *Mus musculus*, collected by W.W. Cort. Subsequently, Specht and Voge (1965) isolated **tetrathrydia** from a fence lizard, *Sceloporus occidentalis*; tetrathrydia are larval forms of *Mesocestoides* spp., but the entire life cycle of the parasite has not yet been completely determined. When inoculated directly into the body cavity of laboratory mice, the tetrathrydia reproduced asexually by binary fission. When tetrathrydia were intubated into the stomachs of laboratory mice, they penetrated the gut wall and reproduced asexually in the liver and abdominal cavity. When fed to cats, the larvae developed into adult worms that were identified as *M. corti*. Eckert, Vonbrandt and Voge (1969) inoculated tetrathrydia into the abdominal cavity of dogs; some reproduced asexually, but others migrated into the intestine where they strobilated and became sexually mature adults. It appears, however, that asexual reproduction by tetrathrydia may be an anomalous phenomenon, restricted to an aberrant laboratory strain of *M. corti* (Conn, 1990). In a recent commentary, Conn (1990) pointed out that verification of the budding process has been attempted by several investigators (James and Ulmer, 1967; Loos-Frank, 1980; Conn and Etges, 1983), but to no avail. He also noted that Voge (1969) 'stated clearly that her diagram represented a cycle that is unusual even for *M. corti* and was established only experimentally in the laboratory' (Conn, 1990). The occurrence of asexual reproduction by larvae in the family Mesocestoididae must, therefore, be viewed with caution.

The developmental processes involving the budding stages of several species of taeniid and mesocestoidid cestodes offer, however, some interesting prospects for future investigation. For example, the definitive hosts for *Taenia multiceps* are canines, typically coyotes and dogs (Figure 5.2 shows the life cycle). The asexually reproducing larval stage, or coenurus, is normally found subcutaneously and intramuscularly in jack-rabbits, *Lepus californicus*. The same parasite (see Esch and Self, 1965) is known to cause gid, or staggers, in sheep because coenuri will develop in their brains if they accidentally ingest eggs. For a number of years, one of us (GWE) tried unsuccessfully to obtain experimental infections of *T. multiceps* in laboratory rabbits in which accidental infections had been previously reported (Clapham, 1942). Surprisingly, however, when laboratory mice were intubated with eggs of the parasite, the larval stages developed successfully in intramuscular and subcutaneous sites. When fed to a dog, these coenuri were found to be infective, producing normal strobila and eggs that, in turn, were infective for mice. Moreover, some laboratory mice developed signs of gid and, when necropsied, were found to have small, intracerebral coenuri (Esch and Self, 1965; Larsh, Race and Esch, 1965). *Taenia crassiceps* also normally uses canines as definitive hosts (Figure 5.3). Recently, however, Kitaoka *et al.* (1990) fed cysticerci of *T. crassiceps* to cortisone-inoculated golden hamsters, *Mesocricetus auratus*, and obtained sexually mature adults that produced viable, infective eggs. This particular strain had been originally isolated from microtine rodents in Japan in 1985 and maintained in typical fashion by serial intraperitoneal

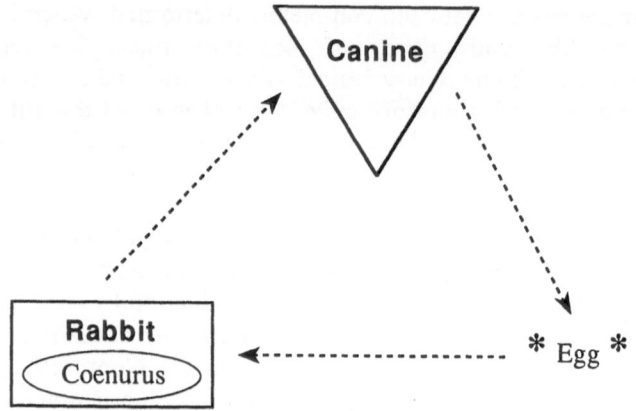

Figure 5.2 A schematic life cycle for the cestode, *Taenia multiceps*. The symbols are explained in the caption for Fig. 1.1 (page 7).

passage of cysticerci through Mongolian gerbils, *Meriones unguiculatus*. The results obtained from studies like these suggest that the keys for triggering asexual reproduction and for strobilization apparently are not significantly different. The results also indicate that ecological isolation of the parasite may play an important role in what otherwise appears to be a high degree of host specificity for these cestodes under natural conditions.

Williams (1975) proposed that environmental unpredictability and geographic dispersal were key elements in the evolution of alternating sexual and asexual generations. He suggested that by creating diverse pro-

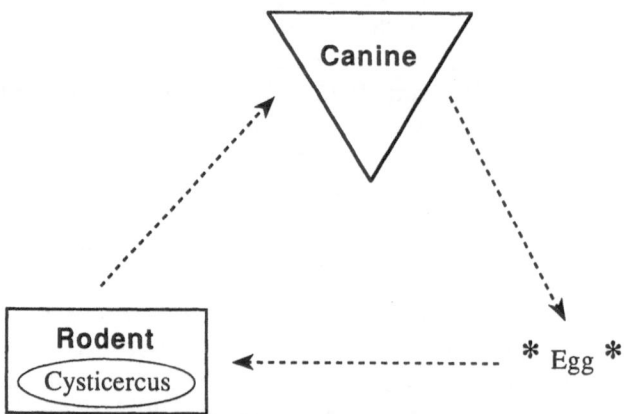

Figure 5.3 A schematic life cycle for the cestode, *Taenia crassiceps*. The symbols are explained in the caption for Fig. 1.1 (page 7).

pagules through genetic recombination at the time of sexual reproduction, survivorship probabilities would be maximized when these infective agents are dispersed into relatively more unpredictable environments. He attempted to apply these predictions to parasites and said that 'where there is more than one obligate host, the final host (where sexual reproduction takes place) will be the most mobile and disperse the parasite most widely' (Williams, 1975).

Moore (1981) took exception to Williams (1975) hypothesis and used members of the cestode family, Taeniidae, to show that it probably did not apply to parasites. The taeniids are an appropriate group of tapeworms to examine for this purpose because they include some species that have larvae that reproduce asexually and some that do not. Moreover, all of them use homeotherms as both definitive and intermediate hosts and all depend on predator–prey interactions to complete transmission to a definitive host. After thoroughly reviewing several basic components in the biology of these cestodes, Moore (1981) concluded that 'the occurrence of asexual reproduction in the life cycle is independent of possible indicators of predictability such as host size, immunological response, anatomical location, site/host-specificity and definitive host'. It was speculated that factors such as competitive interactions, genetic drift, group selection, and mortality/fecundity schedule, may have been the primary selection forces in the evolution of alternating sexual and asexual reproduction, at least among taeniid cestodes. But, she also pointed out that the data available are presently incomplete for identifying the relative importance of these various influences.

5.2.2 Gametic strategies

Asexual, but nonetheless gametic, reproduction has been reported for several nematode species, most prominently those from the Order Rhabditida. Perhaps the best known of these species is *Strongyloides stercoralis*, the so-called 'threadworm', that infects man (Figure 5.4). Adults of the parasite live in the mucosa of the small intestine where infrapopulations consist only of females that produce eggs by apomictic parthenogenesis. The mature eggs are released from the adult and hatch within the mucosa. The larvae move into the lumen of the intestine. Some larvae pass to the outside via the faeces, while others moult rapidly to become **filariform** larvae (L_3 stage) and penetrate the mucosa of the large intestine before they can be shed in the faeces. These larvae migrate to the lungs via the circulatory system. After moulting once in the lungs, they are coughed up, swallowed and move into the mucosa of the small intestine where they will become sexually mature females. This mode of infection is referred to as **internal autoinfection**.

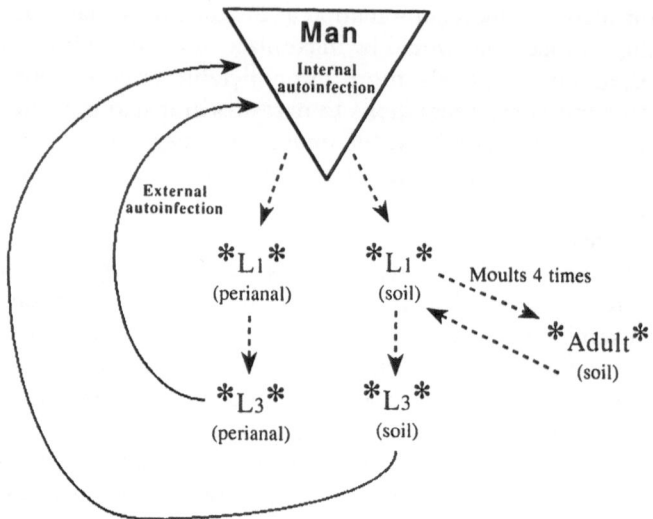

Figure 5.4 A schematic life cycle for the nematode, *Strongyloides stercoralis*. The symbols are explained in the caption for Fig. 1.1 (page 7).

If shed to the outside, some of the **rhabditiform** larvae (L_1 and L_2) may become infective (L_3) filariform larvae. These L_3s are then capable of penetrating the skin of humans and, if they do so, they will follow the same migratory route to get back to the intestine as larvae that infected the host via internal autoinfection. Those *S. stercoralis* that infect humans by either of these methods are said to be homogonic, or unisexual.

Some of the larvae shed from infected individuals may not become infective L_3s, but instead remain in the soil as non-infective rhabditiform stages where they moult four times, maturing into free-living adults of both sexes. The stimulus for development in this direction is apparently conducive environmental conditions, i.e., moist soil, warmth, shade, etc. Free-living females reproduce by meiotic parthenogenesis. Insemination of free-living females by males does occur, however. Ova are actually penetrated by sperm, but nuclei of the gametes do not fuse. Instead, nuclei from the egg's second division fuse to restore the diploid number. The cycle may continue in the free-living phase indefinitely. It will usually be halted by environmental adversity in which case rhabditiform larvae, after hatching from eggs, will moult twice and become infective L_3s before penetrating the skin of a host. This type of life cycle is **heterogonic** in character, and represents cyclical parthenogenesis. For a complete review of parthenogenesis and asexual reproduction, the reader is referred to Hughes (1989) and to Whitfield and Evans (1983) which is part of a larger volume (Whitfield, 1983) that considers the reproductive biology in a wide spectrum of parasitic sporozoa and helminths.

Another rhabditid nematode with some peculiar quirks in its reproductive cycle is the common lung nematode of frogs, *Rhabdias ranae*
(Figure 5.5). The adults of this parasite are **protandrous hermaphrodites**. That is, the male gonads develop first and produce sperm that are
stored in a seminal receptacle. The male system is resorbed and replaced
by a female system. Eggs are fertilized by the sperm stored in the
receptacle. The eggs are then coughed up from the lungs, swallowed and
hatch (to become rhabditiform, L_1 larvae) while still in the frog's gut
before being shed in the faeces. Once outside, the rhabditiform larvae
moult four times and become free-living males and females, making this
cycle heterogonic. The free-living females are ovoviviparous. After the
eggs hatch, the female retains the developing larvae that consume her as
they grow and moult (a parasitoid life-style exhibited by a soon-to-be
parasite). The filariform, L_3s free themselves from the adult female,
penetrate the skin of a frog, and migrate to the lungs via the lymphatic
system.

Gametic reproduction among most other parasitic helminths is hetero-
sexual. Acanthocephalans and nematodes (except those mentioned above,
along with some related species) are dioecious. Most of the cestodes and
trematodes, on the other hand, are monoecious. Even though they are
hermaphroditic, there is certainly strong evidence to indicate that cross-
fertilization and, therefore, genetic recombination occurs within species in

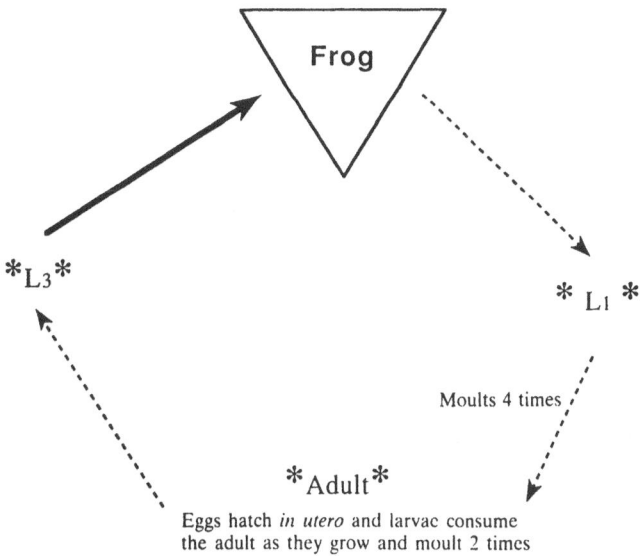

Figure 5.5 A schematic life cycle for the nematode, *Rhabdias ranae*. The symbols
are explained in the caption for Fig. 1.1 (page 7).

both groups of animals. However, there are also indications that there are tendencies by some species to self-fertilize as well (Williams and McVicar, 1968). Nollen (1983) reviewed several of these issues and indicated that a humoral link between hermaphrodites of the same species or males and females of the same species induced 'growth and reproductive development' or attracted a partner for copulation (Fried and Roberts, 1972; Kolzow and Nollen, 1978).

5.2.3 Fecundity

Fecundity among parasitic organisms is high, usually much higher than among their free-living counterparts (Table 5.1), but it is also variable. It runs as high as 700 000 eggs per day in *Taenia saginata* (Moore, 1981) and 200 000 per day in *Ascaris lumbricoides* (Chandler and Read, 1961). Many of these same parasites are not necessarily long-lived (Table 5.1). Some, however, like the trematodes, *Schistosoma japonicum* and *Schisto-*

Table 5.1　Fecundity values for some representative parasitic helminths.

Parasite	*Eggs/day*	*Life span*	*Source*
ACANTHOCEPHALA			
Macracanthorhynchus hirudinaceous	260 000	10 mo	(Kates, 1944)
CESTODA			
Taenia saginata	700 000	10 yrs	(Moore, 1981)
Taenia solium	300 000	–	(Moore, 1981)
Taenia hydatigena	60 000	–	(Moore, 1981)
Hymenolepis diminuta	250 000	2–3 yrs	(Keymer, 1980)
Hymenolepis microstoma	45 600	2 yrs	(Jones and Tan, 1971)
Diphyllobothrium latum	36 000	–	(Chandler and Read, 1961)
TREMATODA			
Fasciola hepatica	25 000	–	(Chandler and Read, 1961)
Schistosoma japonicum	1200	2 yrs	(Pesigan *et al.*, 1958)
Schistosoma mansoni	300	15 yrs	–
NEMATODA			
Haemonchus contortus	6000	–	(Kelley, 1955)
Ascaris lumbricoides	200 000	9 mo	(Chandler and Read, 1961)
Ancylostoma duodenale	20 000	5 yrs	(Chandler and Read, 1961)
Necator americanus	10 000	5 yrs	(Noble *et al.*, 1989)
Toxocara canis	200 000	4 mo	(Noble *et al.*, 1989)
Trichinella spiralis[1]	1500	–	–
Litosomoides carinii[2]	15 000	–	(Kennedy, 1975)

[1]Average number of larvae produced per female.
[2]Microfilariae released from a female.

soma mansoni, are long-lived, but produce only a few eggs per day. Egg production by helminth parasites can vary because of immunological, nutritional, and other various effects. Production within a single host also can vary from day to day, making it difficult to obtain accurate estimates of parasite density using this method.

An oft-cited example of high fecundity is the production of cercariae by *Cryptocotyle lingua* that infect the marine snail, *Littorina littorea*. Meyerhoff and Rothschild (1940) observed a single snail infected with this parasite for 5 years and estimated cercariae production at 1.5 million per year with a range in daily production of 3000 in the first year to 800 in the last. Shostak and Esch (1990a) measured cercariae production by *Halipegus occidualis* in its snail host, *Helisoma anceps*, and reported that, at 28°C it averaged approximately 1000 per day; at 22°C it was about 700 per day; and at 16°C it was nearly 550 per day. Based on unpublished computations, it is estimated that *H. anceps* in Charlie's Pond, in the Piedmont area of North Carolina (USA), will shed for about 180 days out of the year, with a mean daily temperature during the shedding period of 22°C. If these assumptions are accurate, then each infected snail will produce approximately 126 000 cercariae on an annual basis. Compound these cercariae production figures by egg production from adults in the definitive hosts (several thousand per day per individual) and the massive extent of fecundity in this species can be seen. What makes this even more interesting is that up to 60% of the *H. anceps* population in the pond will be shedding cercariae during late spring.

According to Calow (1983a), 'the traditional explanation for these enormously high fecundities is that parasites have to compensate for massive mortality in the transmission phase of their life cycles, but there are neo-Darwinian problems with this interpretation'. He referred to this as the **'balanced mortality hypothesis'** (Smith, 1954; Price, 1972). Neo-Darwinians would contend that natural selection will opt for genetically based characters that favour fecundity and survival, while simultaneously reducing generation time. At the population level, these traits will spread. Calow (1983a) argued that parasites are capable of higher fecundity because they have a greater resource base than their free-living counter-parts, although he acknowledged that resource availability is subject to question (Boddington and Mettrick, 1981). For the same reason, Calow (1983a) further conjectured that parasites using homeotherms as hosts should be reproductively 'better off' than those using poikilotherms. His argument in terms of resource base relied on the assumption that enteric parasites would be exposed to ample and consistent quantities of nutrients and, indeed, would have access to the resource before the host did. It was further argued that the resource base in the gut of a homeotherm is more ample and more consistent than in a poikilotherm and that, therefore, it should provide a larger resource base for reproduction.

A variety of factors are believed to affect the fecundity of parasites. Among these are genetic effects, differential immunity, host diet, intra- and interspecific competition, etc. Genetic effects can be approached from either the host or the parasite's perspective. Thus, variations in host genetics may influence parasite fecundity while genetic variations among parasites may also affect fecundity at the infra- and metapopulation levels. Consider the variability of cercariae production in different species of snail hosts as an example. *Schistosoma bovis* exhibits the broadest range of snail host specificity among species of schistosomes belonging to the terminal-spined-egg complex (Southgate and Knowles, 1975). When snails of three different species were exposed to infection by *S. bovis* and subsequent cercariae production monitored, Mouahid and Théron (1987) observed substantial differences between *Bulinus* species and *Planorbarius metidjensis* (Figure 5.6). Considerably less difference was observed between the two *Bulinus* species. The variation in production was attributed to differential larval demographics and biotic capacities among the three host species. It was speculated that differences in the 'quality' of hosts could affect sporocyst replication and, in turn, perhaps the rate of cercariogenesis.

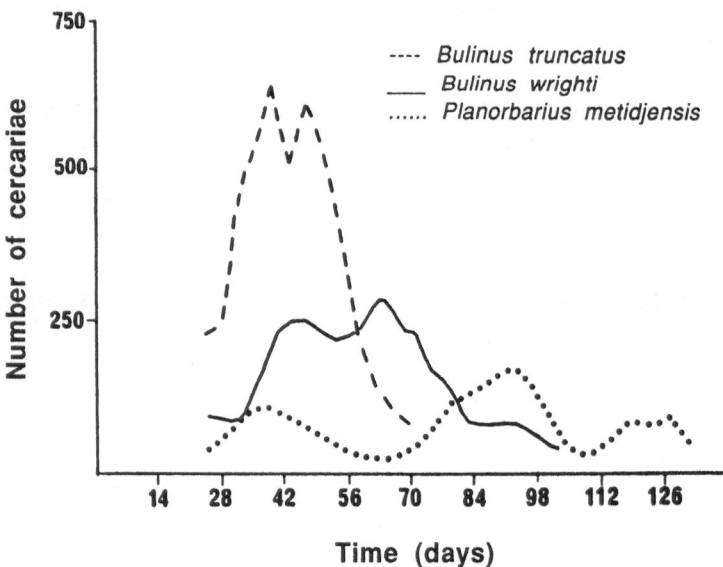

Figure 5.6 Comparative cercariae production dynamics of *Schistosoma bovis* developing in *Bulinus truncatus*, *Bulinus wrighti*, and *Planorbarius metidjensis*. Curves were obtained after smoothing the daily data by movable average methods in 10 steps. (With permission, from Mouahid and Théron, 1987.)

The pattern of host variability in affecting cercariae production is paralleled by similar genetic variability in the pattern of cercariae production from snails of the same strain. Cohen and Eveland (1988) removed single sporocysts of *Schistosoma mansoni* from individual *Biomphalaria glabrata* and transferred them microsurgically to other individual hosts of the same snail strain. Cercariae production was then monitored along with several other biological characteristics following introduction into the new hosts. For the traits measured, there was 'consistency within clones and significant differences between them were detected'. There was, for example, an order of magnitude in difference between production for the highest and lowest cercariae producers. They asserted these observations clearly demonstrated that differences were due to the parasite's genetic variation rather than phenotypic plasticity.

The ability of hosts to respond immunologically to most somatically-located adult helminths is well known. The impact of the immune response on the fecundity of these parasites is less clear, although there are indications that fecundity will be reduced in secondary hosts. Whether this is caused by the host's immune responsiveness or some other adverse physiological condition is unclear. There is also some evidence which suggests, that at least in *S. mansoni*, a host immune response is necessary for expulsion of eggs. Thus, if the immune response is weak, egg counts in faeces are low; if the immune response is strong, egg output is high (Damian, 1987). The ability to respond to other tissue-dwelling enteric helminths, e.g., *Trichinella spiralis*, is also known to occur (Larsh and Weatherly, 1974). Whether these immunological effects diminish fecundity is not known in most cases. Until recently, it was not known whether hosts were capable of responding immunologically to most lumen dwelling adult cestodes. Evidence has been available for a number of years that certain enteric cestodes could provoke a strong immune response (Weinmann, 1966; Befus, 1975; Hopkins, 1980). More recently, Ito and Smyth (1987) have indicated that the ability of adult enteric cestodes to stimulate immunity is probably more common than previously believed. If so, then the potential on the part of the host to immunologically affect parasite fecundity must be examined more carefully. Ito *et al.* (1988) and Ito, Onitake and Andreassen (1988) provided evidence that the immune response stimulated by *Hymenolepis nana* in euthymic mice was not only capable of reducing fecundity and biomass of the parasite, but prevented internal auto-infection from occurring as well. The latter action also would have a profound effect on egg production by an infrapopulation of *H. nana* from a given mouse. Wakelin (1987) placed great emphasis on the potential of the immune response to regulate parasite infrapopulation densities. Indeed, he related overdispersion to genetic variability in host populations. He contended that hosts within a population are constantly re-exposed to infective agents and those that accumulate large numbers of parasites do so because they are less competent immunologically than hosts with fewer parasites. The problem here is that there are not enough data from field studies to support this line of thinking.

Studies by Read (1959), Roberts (1966, 1980) and Dunkley and Mettrick (1969) have collectively demonstrated that the quality of host diet can have a negative impact on the fecundity of certain cestodes under laboratory conditions. Most of the dietary studies have focussed on the effects created by decreases in carbohydrate intake such as starch and several different monosaccharides and disaccharides. Crompton (1987) reviewed the responses of some 20 species of protozoan and helminth parasites to experimental manipulations in host diet and feeding programs. The effects were varied and ranged from changes in the rate of asexual reproduction to modification of the intensity of infection; these dietary changes also frequently modified parasite fecundity. If there is a problem here, however, it is that these kinds of laboratory experiments have not been paralleled by many field observations, so it cannot yet be said that host diet can have an effect on fecundity under natural conditions. On the other hand, as Crompton (1987) noted, 'in wild mammals, parasites of several species will be acquired and lost continuously, and will be expected to interact with each other and with the host immune response, in addition to the host diet, that may be more erratic in supply than in the laboratory'. While Crompton was emphasizing the situation in mammals, the same probably could be said for most hosts, both intermediate and definitive.

Intraspecific interactions among parasites may negatively impact on fecundity. This phenomenon has long been recognized and was early on termed, the 'crowding effect', by Read (1959). It was originally described as the result of competition for limited resources, primarily carbohydrates. The model for most of these studies was the rat tapeworm, *Hymenolepis diminuta*, although it also has been associated with several other species of helminth parasites. Boray (1969), for example, exposed sheep to varying densities of the trematode, *Fasciola hepatica*; in heavy infections, the percentage of patent parasites declined sharply and so did individual fecundity. The same kind of density-dependent effects were noted by Fleming (1988) in lambs infected with *Haemonchus contortus*, a particularly pathogenic nematode. While carbohydrate deprivation traditionally has been identified as an important component in the 'crowding effect', the stunting and fecundity reductions that result from high infrapopulation densities have also been attributed to growth inhibitors (Roberts, 1980), spatial constraints (Keymer, 1982), and immunological factors (Kennedy, 1983).

Interspecific competitive interactions may also affect the fecundity of certain parasites. The best model to illustrate the effects of inter-specific, density-dependent constraints has already been reviewed within another context in section 3.4.3. Referral is made to the competition studies by Holmes (1961, 1962a) that focussed on the combination of *Hymenolepis diminuta* and *Moniliformis dubius* in laboratory rats. In these experiments, the cestode was forced by the acanthocephalan to occupy an apparently undesirable position in the gut so that reduced biomass, and probably fecundity, for the cestode were the outcome of the interaction.

5.3 COLONIZATION STRATEGIES

The concept of colonization by parasites may be approached in several ways, all of which can be considered within the framework of temporal or spatial scales, or both. Thus, colonization of an individual host will involve strategies that may be linked to the behaviour or dietary preferences of one or most hosts in a life cycle. For example, some parasites depend on predictable predator–prey relationships to complete their life cycle. In the context of a more global scale, colonization of a given habitat or eco-system by a parasite will depend not only on the behaviour and dietary preference of a transporting host, but on the nature of the physical and biotic characteristics of the habitat. The fact that a migratory shore bird defecates in a pond and seeds it with the eggs of a particular digenetic trematode does not mean that the parasite will become established in the pond. An appropriate snail intermediate host may not be present. Col-onization can also be considered within a zoogeographic context that helps to make the concept even more difficult to address. At this point, focus will be on the smallest scale, reserving review of the other scale for Chapter 8, on biogeography.

In colonizing a particular host, a parasite is faced with several basic problems. The first is finding and getting to the host. As was pointed out earlier, recruitment of a parasite can be either passive or active. If the former, then locomotory capabilities are unnecessary. If the latter, then the parasite must be able to move and locate the host. The second problem in colonization for parasites is gaining internal access once the host has been reached. Finally, once the parasite has entered the host, it must be able to migrate to its proper site of infections, and then develop to maturity.

Parasites have evolved several different mechanisms for locomotion. For some trematodes and cestodes that rely upon aquatic media to complete their life cycles, ciliated miracidia and coracidia (Figure 5.7), respectively, are involved. Ciliated coracidia resemble planktonic protozoans and, presumably, the copepod that consumes a coracidium is deceived by its movement and is thus 'accidentally' infected. Miracidia of most digenetic

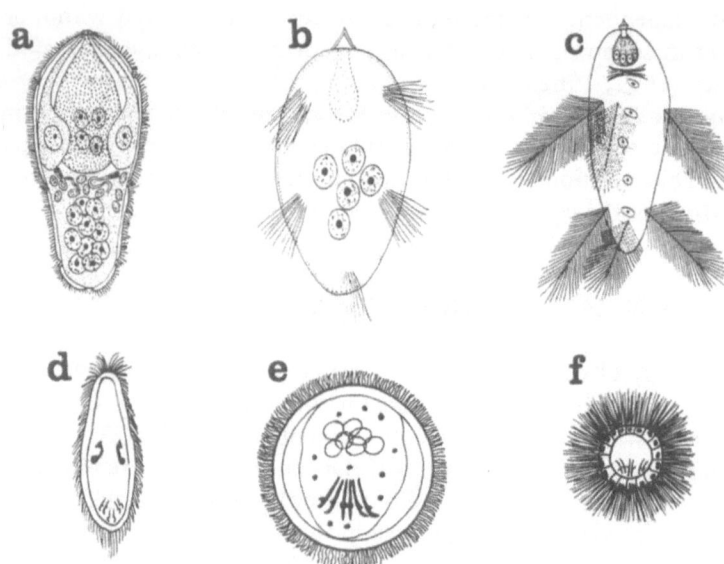

Figure 5.7 Morphology of trematode miracidia and cestode coracidia: (a) opecoelid miracidium, with ciliated epithelium; (b) miracidium of *Postharmostomum* sp., with tufts of cilia; (c) miracidium of *Leucochloridiomorpha* sp., with ciliated bars; (d–f) coracidia of pseudophyllidean cestodes. (With permission, modified from Schell, 1985.)

trematodes, on the other hand, actively seek their molluscan hosts. This requires these larval stages to have certain sensory capabilities. Some miracidia exhibit a strong, if unsophisticated, chemotaxis, being attracted mostly by certain fatty acids or amino acids released from snail hosts. Some miracidia also have eye spots, indicating the potential for phototaxis. Still others respond to geotactic forces that ensure attraction to molluscan hosts which may live on bottom substrata. Whether miracidia or coracidia, however, the transmission window is almost always quite narrow from a temporal standpoint. Once hatched from an egg, neither miracidia nor coracidia survive for more than 36 hours. During their free-living existence, they do not feed, relying instead on stored nutrient reserves in the form of glycogen.

Most species of trematodes also have a second motile stage, the cercaria. In those that have free-swimming cercariae, the locomotory organ is the tail. The tail morphology (Figure 5.8) is varied and always quite effective in taking the parasite to the next host in its life cycle. Strategies associated with host-locating behaviour by cercariae are also variable. Some cercariae are equipped with eye spots that elicit phototactic responses. Geotactic, rheotactic and thermotactic receptors are also

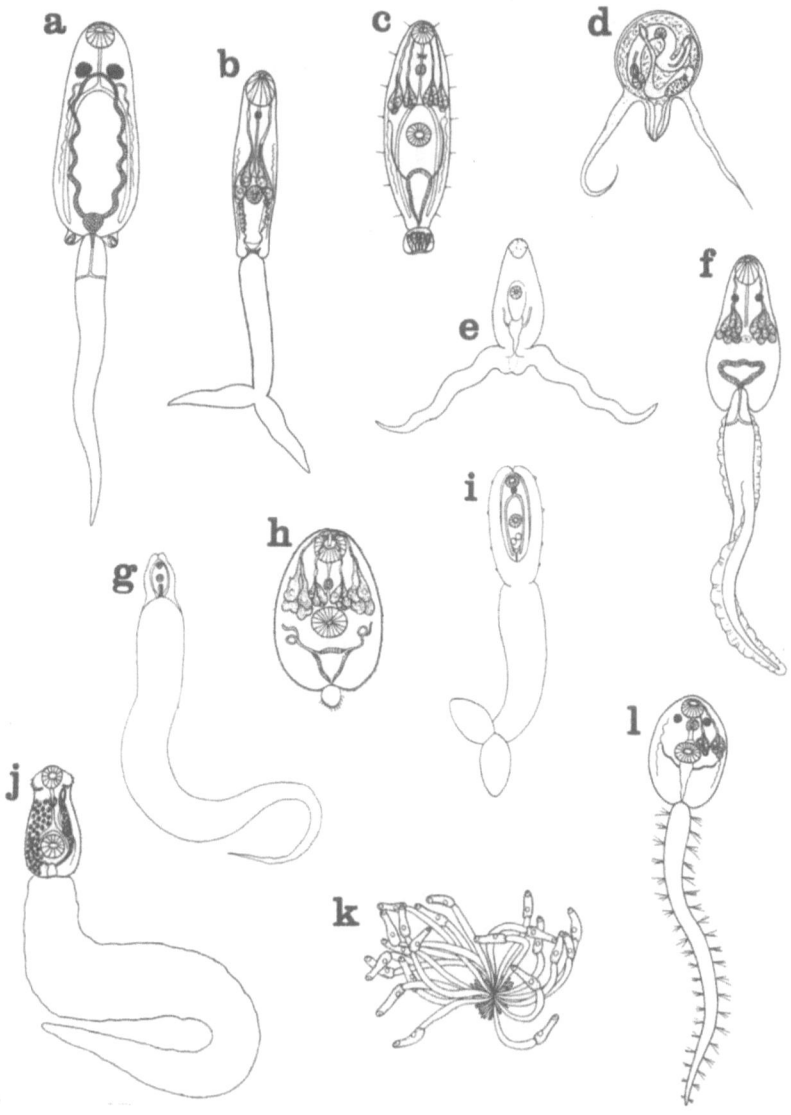

Figure 5.8 Tail morphology of cercariae from a representative group of digenetic trematodes: (a) leptocercous cercaria; (b) furcocercous cercaria; (c) cotylocercous cercaria; (d) cystophorous cercaria; (e) ox-head (gasterostome) cercaria; (f) pleurolophocercous cercaria; (g) cystocercous cercaria; (h) chaetomicrocercous cercaria; (i) furcocystocercous cercaria; (j) magnacauda cercaria; (k) zygocercous cercaria; (l) trichocercous cercaria. (With permission, modified from Schell, 1985.)

present among certain cercariae. In general, the swimming behaviour of cercariae will only place them in the vicinity of the next host. In a second step, direct chemical stimuli from the target host may encourage the proximal contact. An elegant study by Donges (1964) describes in detail the behaviour of cercariae of the strigeid trematode, *Posthodiplostomulum cuticola*, that uses fishes as the second intermediate host. For most of the time, cercariae of this species hang motionless in the water column, swimming only in short bursts in response to appropriate stimuli. The cercariae are quite sensitive to shadows that might be created by a fish passing over, or to their side; they also respond to turbulence such as would be created by the movement of a fish's fin or by the passage of water over a gill surface. They are not continuously responsive to repeated shadows that might be created by reeds blown by wind or by clouds passing over. In this way, glycogen reserves in the cercariae are preserved and their lifespans are extended.

There are other strategies that ensure successful transmission by larval trematodes, some of which are rather bizarre. The azygiid flukes, *Proterometra* spp., use fishes as definitive hosts and *Goniobasis* spp. as their first intermediate hosts. Cercariae released from these snails are enormous in size, reaching 1.5–2.0 cm in total length; only one or two may be released from an infected snail per day. Probably 99% of the tissue in one of these cercariae comprises the tail that is roughly Y-shaped; the body of the cercaria is embedded within the tissue of the tail. The two floppy ends of the cercaria's tail are used as muscular flaps to propel them upward in the water column. After flapping for a brief period of time, movement ceases and the parasite drifts downward. After a few seconds, the process is repeated. This will continue for as long as there is a glycogen reserve or until they are consumed. The swimming behaviour used by these cercariae mimics that of mosquito larvae and is employed to attract predaceous fishes. Because so few cercariae are produced by the snail, the probability of successfully completing transmission is considerably reduced. However, the mechanism used to attract fish predators is highly sophisticated and the overall success in transmission is high because of it. In another possible counter to any disadvantage that may be associated with reduced cercariae production, *Proterometra dickermani* has **neotenic** cercariae in which the uterus is filled with viable eggs when the parasite is released from the snail. Thus, when the cercaria reaches the intestine of its fish definitive host, egg production is already underway and there is no pre-patent period. In this way, infective propagules can be released immediately into the water from the piscine host.

As previously described, the cercariae of many hemiurid trematodes do not swim, but remain motionless until ingested by appropriate micro-crustaceans. Instead of using a locomotory device to decrease dispersal in space and thereby increase the chances of contacting the next host, these cercariae remain viable for months, thereby widening the temporal window in transmission. In *Halipegus occidualis*, there is evidence suggesting that the cercariae are hypobiotic, a way of reducing their metabolic rate and extending the parasite's lifespan (Goater, Browne and Esch, 1990).

Trematode miracidia usually come equipped with several cells which secrete enzymes that digest the snail's outside covering, allowing them to gain access to the internal tissues. Some cercariae also have penetration glands that permits them to penetrate the surface of their next host. When some cestode eggs or coracidia are consumed by an appropriate host, the larvae (**hexacanth embryos, or oncospheres**) use six tiny hooks to attach and manipulate the mucosal cells in order to gain entrance to the haemocoele of the host where they will continue development.

In aquatic systems, many cestodes have evolved locomotory processes. In contrast, all species of tapeworms in terrestrial systems developed a shell to protect the larval parasite (hexacanth embryo or oncospheres), from desiccation. These eggs, then, must be consumed by the next host in the life cycle. Several species of trematodes may employ terrestrial snails as intermediate hosts. In these situations, as with terrestrial cestodes, an egg with a desiccant-resistant covering is produced by the parasite. The problem for these terrestrial dwelling trematodes comes, however, in getting from the snail to the definitive host. This problem has been resolved in at least two ways. The cercariae of *Dicrocoelium dendriticum* are bound up in a secretion of slime released by the snail; the ants then consume these so-called 'slime balls'. Cercariae of *D. dendriticum* are thus transported in an aquatic medium created by the snail's salivary secretion. Among several other species of trematodes (*Leucochloridium* sp., *Neoleucochloridium* sp.), cercariae are not released from the snail. Instead, they are retained in sporocysts that develop in the snail's antennae. The parasite induces a melanization change in the host that causes the antennae to become a bright colour, resembling a caterpillar, for example. After losing its protective coloration, the snail thus becomes an attractive prey item for the avian definitive host.

Many nematodes in aquatic and marine systems have evolved uncomplicated, one-host life-cycle strategies. Others, such as the marine nematode, *Anisakis* spp., have life cycles that include two, three, or more hosts (although most of these are paratenic or transport hosts). Eggs of these nematodes will be eaten by a copepod or a euphasid crustacean where the larva that emerges employs a penetration stylet to enter the haemocoele. If the infected copepod is consumed by a fish, the larvae moult twice, penetrate the gut wall, and encyst in the mesenteries or the fleshy surface of organs such as the liver, remaining as a L_3 larvae. If the infected fish is eaten by another fish, the anisakid worms will penetrate the gut wall and repeat the migration to internal organs where they again encyst as L_3 larvae. The normal definitive hosts for *Anisakis* spp. are marine mammals. When infected fish or euphasids are eaten by these mammals, the parasites moult, become adults and attach to the stomach wall. Humans may also be infected with these worms if they eat raw fish, a practice already common in many parts of the world, but becoming more and more common in certain western countries. In this case, the worm will moult only once, becoming an L_4 larva, or pre-adult, that attaches to the stomach wall, producing a gastric granuloma and painful, ulcer-like symptoms. For the anisakid nematode, then, the entire life-cycle process is dependent on a series of predator–prey links.

For most nematode species in aquatic systems, transmission is passive. For those that use terrestrial systems, some are active and others are passive. However, even for nematodes that are actively transmitted, they have not evolved specialized locomotory organs as have many tapeworms and flukes in aquatic systems. Nematodes have a hard cuticle and this contributes to their ability to withstand desiccation. The filariform larvae of *Strongyloides*, *Necator* and *Ancylostoma* are all negatively geotropic and positively thermophilic. These larvae move to microelevations, e.g., blades of grass, and wait for a suitable host to pass. Other strongylid nematodes that have free-living larvae are acquired by herbivores as they graze and accidentally ingest infective larvae. Some nematodes, primarily the filarial worms, require an insect vector for transmission to occur. Among some of these species, motile microfilariae released from gravid females are carried in the bloodstream of the vertebrate definitive host if the vector is a blood-sucking anthropod; microfilariae of other species of filarial worms reside subcutaneously from where they will be acquired by biting arthropods (Bartlett and Anderson, 1989a, b).

Once a suitable host has been reached and entered, the parasite must then locate the proper site to develop to the next stage in its life cycle. Some migrations within a host are truly remarkable. In other species, migration is not required and they may remain inside the lumen of the stomach or intestine. If migration takes place, the parasite must be able to sense a cue, or a set of cues, and respond accordingly. This is one reason why some parasites become especially dangerous if they enter the wrong host; apparently, they are confused by the signals coming from the host and cannot respond properly, so their migration becomes aimless or erratic.

Establishment and development in the proper host will be affected by a number of factors. Among these are host and parasite age, host sex, host immunity, host stress, and competition, both inter- and intraspecific. All but stress have been previously considered within other contexts. Stress has been defined as 'the *effect* of any force that tends to extend any homeostatic or stabilizing process beyond its normal limit, at any level of biological organization'. (Esch, Gibbons and Bourque, 1975). The variety of factors known to induce stress in a potential host is extensive. These can range from extremes in temperature to mechanical irritation. In the classical scenario, the stressors will induce increased levels of circulating corticosteroids, but primarily cortisone. These hormones are anti-inflammatory in character and, if in high concentrations, will curtail the host's ability to effectively respond to invasive parasites.

As previously emphasized, the distribution of most parasites in their hosts is aggregated. Wakelin (1987) makes the assumption that the majority of hosts within a given population are exposed to parasites and that some hosts have higher densities because their immune protection is less efficient than the more lightly infected individuals. Implicit in this assumption is that the majority of hosts within a population are also more or less equally exposed, but this is probably not the case. It can be reasonably argued that the spatial and temporal heterogeneities of host and parasite infrapopulations will be completely sufficient for most parasite species to explain their aggregation in host populations. As pointed out by Crofton (1971b), overdispersion can be generated in a number of ways, and the host's immune responsiveness is but one of several contributing factors.

5.4 HOST BEHAVIOUR AND TRANSMISSION

It is quite apparent that many parasites have evolved mechanisms for altering host behaviour, even host appearance, to increase the likelihood of successful transmission (Holmes and Bethel, 1972; Barnard and Behnke, 1990; Hurd, 1990). Holmes and Bethel (1972) distinguish between parasites that make hosts more prone to ingestion and those that make infected hosts more visible than non-infected individuals. In the former category, *D. dendriticum* and its effects on the ant intermediate host when air temperature drops has been discussed. In the later category, *Leucochloridium* sp. and *Neoleuco-chloridium* sp. produce changes in the pigmentation of a snail's antennae that invites predation of the infected snail by the definitive host.

Moore (1983a, 1984a) described in detail the behavioural modifications of the isopod, *Armadillium vulgare*, induced by the acanthocephalan, *Plagiorhynchus cylindraceous* (Figure 5.9). These altered behaviours increase the chances of the parasite being transmitted to starlings, *Sturnus vulgaris*, the definitive host. In this system, infected, dark-coloured hosts were found more frequently on light-coloured backgrounds and in less humid locations, both of which would result in increased exposure to predation by starlings. Smith-Trail (1980) argued that the sort of

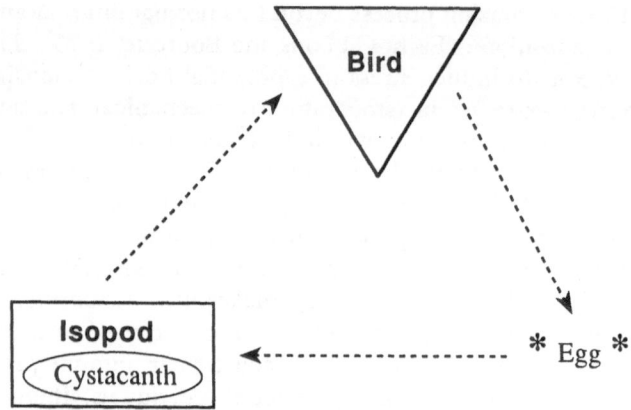

Figure 5.9 A schematic life cycle for the acanthocephalan, *Plagiorhynchus cylindraceous*. The symbols are explained in the caption for Fig. 1.1 (page 7).

behavioural changes that increased the chances of predation of an infested host were a form of 'host suicide' or self-sacrifice. The result would function to rid the parasite from the host population. Moore (1984a) disputed this kin-selection hypothesis, saying that the changes in host behaviour were of greater value to the parasite than to the host.

Based on the reviews by Holmes and Bethel (1972), Barnard and Behnke (1990) and Hurd (1990), it is clear the host behavioural modification by parasites is commonplace. Moreover, as noted by the latter author, the physiological and behavioural interactions between hosts and parasites 'serve to illustrate the complex manner by which these strategies are executed and are indicative of highly co-evolved associations'.

5.5 THEORETICAL CONSIDERATIONS

It was proposed that selection in the tropics favoured organisms with low fecundity and slow development while in temperate areas selection pressures gave competitive advantages to those species with higher fecundity and rapid development (Dobzhansky, 1950). MacArthur and Wilson (1967) coined the terms, *r*- and K- selection to describe these two forms of processes. Subsequently, Pianka (1970) represented selection processes as a continuum, with *r*-selected species at one end and K-selected species at the other. The *r*- end of the continuum was viewed as a quantitative extreme, 'a perfect ecologic vacuum, with no density effects and no competition'. At the other end, organisms have saturated the environment and density effects are greatest.

Jennings and Calow (1975) were among the first to examine parasites and parasitism within the framework of *r*- and K-selection. Their primary interest was in considering the relationship between the evolution of parasitism and parasite fecundity. It was reasoned that '*r*-strategists, with high fecundity, can be expected to have low calorific values because their resources are channeled into the production of the maximum number of progeny, while K-strategists will have high calorific values based on lipid reserves that buffer adults against possible reductions in food supply'. Free-living platyhelminths, it was noted, had higher levels of stored lipid and, simultaneously, larger calorific values and lower fecundity than parasites (Figure 5.10). A question regarding insufficient data for the Jennings and Calow (1975) propositions was raised by Boddington and Mettrick (1976) and discussed more extensively by Esch, Hazen and Aho (1977). However, the latter authors agreed with Jennings and Calow (1975) that it is probable, even likely, that a relationship exists between *r*- and K-selection, relative nutrient values, and evolutionary strategies among free-living and parasitic flatworms.

Esch, Hazen, and Aho (1977) discussed the *r*- and K-selection process and parasitism within a different context than that taken by Jennings and

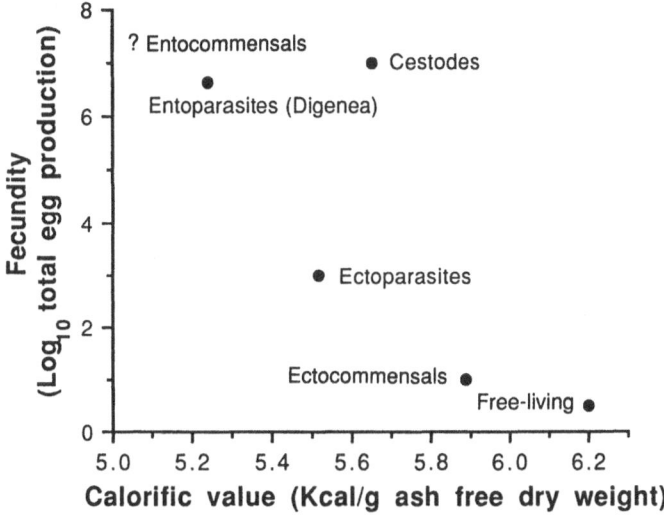

Figure 5.10 Relationship between calorific value (Kcal/g ash-free dry weight) and fecundity ($\log_{10}$ total egg production) of flatworms with different life styles. The value for cestodes is 'whole worms' including gravid proglottids with albuminous eggs as well as immature proglottids. Data for ectocommensals were estimated based on their low calorific value (5.080 ± 0.216 Kcal/g ash free dry weight) in conjunction with the negative trend shown in the figure. (With permission, from Jennings and Calow, 1975.)

Calow (1975). As mentioned above, Pianka (1970) identified a series of correlates within which he could compare r- and K-strategies. These included climate, mortality, survivorship, population size, intra- and inter-specific competition, and longevity. Esch, Hazen and Aho (1977) considered parasitism within the framework of each correlate. It was concluded that, in general, 'parasitic animals exhibit many characteristics of an r-strategist'. But, they also reiterated Pianka's (1970) position that there was no such thing as a perfect ecological vacuum; thus, within a given compound community, some parasites may reflect genetic tendencies toward one type of strategy and others toward another.

Between 1970 and 1981, no less than 48 papers considered various aspects of the r- and K-selection concept that had been applied to taxa ranging from corals to starfish and from parasites to mammals (for review, see Parry, 1981). Despite its wide application, the r- and K- concept in its original form has been questioned and several objections have been raised. For example, Parry (1981) stated, 'considerable confusion has resulted from the terms r- and K-selection being used as a label or as an implied explanation, without the distinction between these usages being made clear (Stearns, 1976)'. He argued that the dimensions of life-history patterns are multi-faceted and not just restricted to r- and K-strategies although these may also function as significant selection pressures. Some of the additional constraints on life histories that were identified included seasonality, habitat stability, the need to disperse, and predation. For the life histories of parasitic organisms, another dimension can be added, this one associated with the host itself. The host's immune capacity and the peculiarities of a host's gut physiology and morphology, for example, are important constraints in the evolution of the life histories of many parasites. Another set of factors would include all of the chemical cues used by a parasite in the internal migration within a host. In other words, parasite life histories are affected by many of the same forces that affect free-living organisms, but parasites must also cope with the added influence of a living environment(s).

MacArthur and Wilson's (1967) ideas regarding r- and K-selection were mostly based on 'the effects of population density on general demographic traits, especially on fitness' (Sibly and Calow, 1985). These latter authors noted that classifications of selection pressures based solely on density effects were unsatisfactory because they, like Parry (1981), recognized that selection pressures are multi-dimensional. Sibly and Calow (1985) proposed a new, two-dimensional model, based on two variables. The first was 'an index of age-specific survivorship (S)'. The second was 'an index of growth rate of offspring (G)'. Both variables are affected by intrinsic and extrinsic factors and both are indices that indicate the totality of selection pressures in a given habitat. After deriving the S and G terms of the model, they formulated and tested a set of life-cycle predictions by

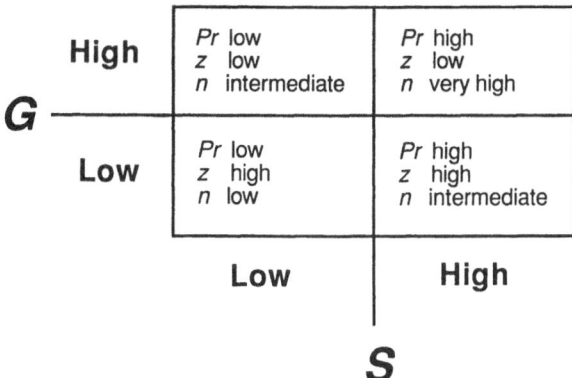

Figure 5.11 Life-cycle predictions in relation to a dichotomous classification of habitats by *G*, an index of growth rate of offspring, and *S* an index of survivorship. *Pr*, total investment in reproduction during a particular breeding attempt; *z*, investment per egg; *n*, number of eggs. (With permission, from Sibly and Calow, 1986.)

establishing a simple matrix that could be described by the high and low extremes of the *G* and *S* axes (Figure 5.11). They proposed that under conditions of high growth rates such as for parasites, differences in survivorship probabilities are likely to be important in the evolution of life cycles. They indicated that endoparasites have good conditions for growth, but that the chances of juvenile survival are poor; this is the opposite for free-living flatworms that have poor growth conditions, but higher juvenile survival.

Southwood (1988) appears to be much more benign in his treatment of the concept of *r*- and K-selection. Southwood (1977) had previously hypothesized that life-history strategies were developed using the habitat as a templet. In the more recent paper, he examined this hypothesis as well as several others in greater detail (Greenslade, 1972; MacArthur, 1972; Sibly and Calow, 1985; Grime, 1986; Hildrew and Townsend, 1987). Southwood's (1988) view was that habitat templets could be defined by two axes, one describing disturbance and another for adversity. Disturbance was considered in terms of the speed with which a habitat changes as a function of the needs of an organism. Thus, temporary and permanent habitats can be compared. The adversity axis, on the other hand, can be measured in terms of productivity, growth rates, stress, etc. Southwood (1988) further proposed that the optimal strategy for conferring fitness on a species would include five tactics (Figure 5.12). These were:

1. the ability to tolerate adversity or stress, including diapause;
2. the development of an ability to contain pressures from predation, parasitism and competition;

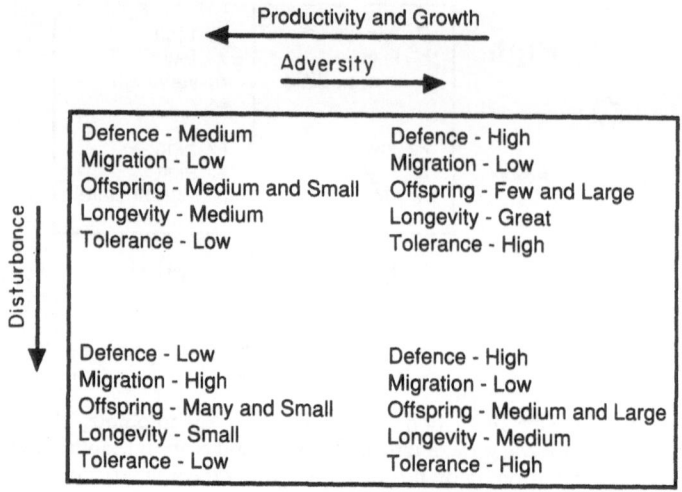

Figure 5.12 The relative importance of different tactics in particular habitats as predicted by various templet schemes. (With permission, from Southwood, 1988.)

3. the capacity to harvest food and invest it in somatic development;
4. the ability to migrate or change one's habitat; and
5. the ability to maximize reproductive effort with respect to the number of young in proportion to the weight of the offspring.

Based on this analysis he predicted that with increasing adversity and habitat disturbance, the life-history strategy of a species would tend toward greater defence, a capacity to migrate, medium- to large-sized offspring, moderate longevity and increased ability to handle environmental harshness.

How do parasites actually fit into these schemes? It can be argued that they do not fit very well. There are three problems with which parasites, in general, had to contend during their evolution. These included the development of successful transmission and survival strategies, and the generation of high fecundity. The fecundity problem was resolved in part, through the evolution of hermaphroditism by some species or the ability to reproduce asexually at one or more points in the life cycle, or both, in others. The residual component (see section 1.4.1) in the life cycle of many parasites has also been an important collateral step. Second, the host has been viewed traditionally as a potentially hostile, albeit stable and nutrient-rich, environment. The hostile aspect of the interaction resides with the host's ability to mount an immune response. On the other hand, the success of parasitism as a life style suggests that hosts may not be all that inhospitable. If a given parasite gains access to the correct host, then its chances of survival may be quite high even though some, as previously noted (e.g. Wakelin, 1987), believe that aggregated parasite infrapopulations occur because of uneven immune capabilities within the host population. Parasites may also be able to cope well with the host as a habitat because hosts and parasites (at least many of them) have been in a long, evolutionary 'arms race' (Dawkins and Krebs, 1979). Thus, many of the immunological ploys evolved by the host have been effectively countered by the parasite. Third, the evolution of many parasite life-history strategies has paralleled the life-history strategies of their hosts. Thus, the transmission of many parasites is almost completely dependent upon:

1. a behavioural 'transgression' by the host which places it in a vulnerable spatial or temporal position; in some cases the so-called 'transgression' is actually induced by a parasite;
2. predictable predator–prey interactions;
3. a deterministic pattern in the spatial or temporal distributions of infective propagules; or
4. some combination of the above.

The point being emphasized is that transmission dynamics usually are not 'accidental' from the perspective of time and space. It can be concluded that while adversity and disturbance may well be important factors in the evolution of life cycles by free-living organisms, the life cycles of parasitic organisms, have been much more likely to track the behaviour and trophic interactions of their hosts.

6   *Infracommunity dynamics*

6.1 INTRODUCTION

The work of Holmes (1961, 1962a, b) has been discussed in relation to understanding the nature of competitive interactions among parasites. These studies are also seminal from the standpoint of their bearing on parasite community dynamics since it was from these baseline efforts that quantitative, parasite community ecology emerged. To better explain the structure and dynamics of parasite communities, a hierarchical classification scheme (Fig. 2.1, page 29) was introduced by Bush and Holmes (1986a). The parasite **infracommunity** was defined as all of the parasite infrapopulations within a single host (Bush and Holmes, 1986a). The classification concept was subsequently extended by Holmes and Price (1986). They described all of the infracommunities within a host population as a **component community**. All of the parasite infracommunities within an ecosystem were considered as a **compound community**.

To establish a link between parasite colonization processes and the structure of parasite communities, Esch *et al.* (1988) introduced the **allogenic** and **autogenic** species concepts. This terminology was developed to assist in interpreting the nature of those mechanisms involved in parasite transmission at all three levels of community organization. Two kinds of parasites were identified, each based on the manner in which the host is used in transmission. Allogenic parasites employ fish or other aquatic vertebrates as intermediate hosts and then mature sexually in birds or mammals. Autogenic species include those in which the entire life cycle of the parasite is completed within an aquatic ecosystem. Since the allogenic-autogenic species concepts are applicable to each hierarchical level, further consideration will be deferred until the next Chapter.

6.2 THE EVOLUTION OF PARASITE COMMUNITIES

Price (1987) reviewed the processes of parasite community development through time within the framework of four different models and posed a set of questions for the diagnosis of characters that distinguish between the various models (Figure 6.1.). These questions alluded to current competition in the communities, the occurrence of vacant niches, the ability of new species to colonize a community, local extinction that would imply the potential for dynamic equilibria, and evidence for co-speciation in the phylogenies of parasites and their hosts.

The **co-speciation** model of Brooks (1980) and Mitter and Brooks (1983) proposes that parasite communities are presently non-interactive and that they have developed over the course of evolutionary association between the host and the parasite. Within the context of this model, co-evolution has occurred for such a long period of time that present competition as a means of structuring the community does not occur. In

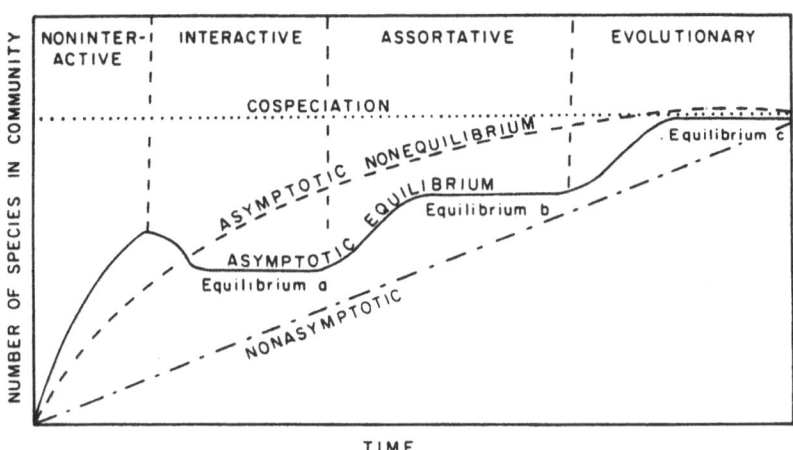

Figure 6.1 Summary of four models of community development, i.e., co-speciation, asymptotic non-equilibrium, asymptotic equilibrium and non-asymptotic equilibrium, through time. The categories, non-interactive, interactive, assortative and evolutionary, and the equilibria a, b, c, refer to the Asymptotic Equilibrium Model. Note that all models can produce the same number of species in the community although the mechanisms differ. (With permission, from Price, 1987.)

this sense such communities may be, in part, reflections of the 'ghost of competition past' (Connell, 1980). Aho (1990) suggested that the absence of 'unique' helminth faunas among many species of amphibians and reptiles seems to challenge Brooks' (1980) co-speciation hypothesis. In these groups of vertebrate animals, large numbers of phylogenetically unrelated parasites seems to co-occur, suggesting the process of 'host capture' (Chabaud, 1981; Goater, Esch and Bush, 1987) in structuring what appear to be non-interactive and isolationist infracommunities. Aho (1990) asserted that 'host–habitat selection for a helminth fauna overrides evolutionary trends in amphibian and reptile systems' and for this reason the co-speciation model for the development of helminth infracommunities in amphibians and reptiles does not apply. On the other hand, while it has probably not been an important organizing force for infracommunities in amphibians and reptiles, the co-speciation process unquestionably has been important for many host–parasite systems.

In the **non-equilibrium model** (Connor and McCoy, 1979; Lawton and Strong, 1981), a stable equilibrium is never reached because the community remains unsaturated and vacant niches always exist. Based on the analyses of Kennedy, Bush and Aho (1986), Esch *et al.* (1988), and Kennedy (1990), the enteric faunas in freshwater fishes are depauperate, non-interactive, and non-equilibrial in character. Kennedy (1990) indicated that vacant niches exist and that chance colonization events are the most likely operational factors in the development of communities in these organisms, just as 'host capture' was for amphibians and reptiles (Aho, 1990).

The **asymptotic equilibrium model**, based on the Theory of Island Biogeography (MacArthur and Wilson, 1967), holds the position that communities are maintained by balancing the rates of colonization and extinction. Both stochastic events and biotic forces, such as competitive interactions, cause extinctions. Wilson (1969) proposed that these types of communities evolve in a predictable sequence, beginning with a non-interactive phase, followed by interactive, assortive, and evolutionary phases. In the non-interactive stage, the island (host) is colonized; initially, population sizes are small and there are no opportunities for interaction. The parasite infracommunities of many vertebrate hosts appear to begin in this fashion, irrespective of whether they end up at the interactive or the isolationist end of the continuum. Subsequently, in the interactive stage, new species of potential competitors colonize and there are interactions that are accompanied by extinctions. It is difficult to classify communities of helminths that have what appear to be interacting species into any phase beyond this point because of the lack of long-term data as well as information regarding the biogeographical distributions of most parasite species. On the other hand, it would seem that the helminth communities (Bush and Holmes, 1986a, b) in lesser scaup, *Aythya affinis*, are either in the interactive phase or in one of the next two stages of community development. In the assortive phase, the species are being arranged in such a manner that the result is a more efficient co-existence. In other words, space and nutrient resources are apportioned to various species according to their competitive abilities and requirements. The final step in the development of the community is the evolutionary phase. During this period, the species are co-evolving; they are, in effect, 'fine tuning' the community. In this phase of development, there are no vacant niches and any recruitment of new species will mean displacement of existing ones. At this point, the community is saturated and in equilibrium.

The final evolutionary construct is the **non-asymptotic** model which proposes that new species are acquired in a linear fashion through time and that the communities remain unsaturated (Southwood, 1961). In this model, there are no constraints on colonization imposed by biotic factors, the assumption being that parasites are so specialized that they do not interact. This may apply to a few host–parasite systems but it is unlikely for most, even those that are considered non-interactive.

Arguments regarding the evolutionary classification of helminth communities are varied but, as Price (1987) pointed out, 'very different kinds of parasite communities exist and ...different processes operate in each, ...but only further study will permit an evaluation of the role of evolution in general'. This admonition is crucial. Understanding the nature of evolutionary processes in free-living systems is difficult at best, but developing a satisfactory scheme for parasite communities at any hierarchical level is several orders of magnitude more difficult. The problem resides in three areas. First, a long-term record is non-existent for any parasite infracommunity. Long-term studies on host–parasite systems are rare and, when they have occurred, they have been focussed at the metapopulation level. Second, more information on the biogeographical nature of parasite distributions is necessary; at present, it is inadequate to address questions of an evolutionary nature. Finally, it is virtually impossible to manipulate parasite infracommunities experimentally. One obvious exception was the Holmes (1961, 1962a, b) effort, but this is *the* exception when it comes to parasite systems in vertebrate animals. Some attempts have been made using snail–trematode systems, but it is doubtful that much broadly based information will emerge from these investigations. Studies on the evolutionary ecology of parasite communities are not at an impasse, but movement in this arena will be slow unless new approaches and novel methodologies are developed.

6.3 INFRACOMMUNITY STRUCTURE

6.3.1 Introduction

Competition is defined in terms of resource use and requirements by two or more organisms of the same species or by individuals of two or more species. In other words, it can be **intraspecific** (the former case) or it can be **interspecific** (the latter case). Competition will take place whenever two organisms or species contend for a resource that is in limited supply. The resource may be space, nutrients, or it could be a physicochemical character in the habitat.

If the interaction of two populations is indirect and both have equal access to the nutrient, then there is **exploitation** competition. **Interference** competition, on the other hand, occurs when two or more organisms or species are vying for the same resource and there is direct confrontation that reduces access to the resource for one or both contenders.

The **niche** is an n-dimensional hypervolume with multiple axes for all variables contributing to the fitness of a given species (Hutchinson, 1957). These include the more obvious abiotic factors such as nutrients in the gut of a host for an enteric parasite, or the moisture, warmth, and shade necessary for the development and survival of free-living larval stages. Other, more subtle, factors would include those that affect host specificity, the host's immune capacity, or the ability of a parasite to counter the host's immune response. The maximum range of all variables affecting a species describe the **fundamental niche**. However, the fundamental niche for a given species is seldom achieved because of the influence of a range of interacting biotic and abiotic factors and, instead, the organism occupies the **realized niche**. Price (1980, 1984a, 1987) makes a strong argument that vacant, or empty, niches occur within many parasite infra-, component and compound communities.

There are three possible outcomes for competitive interaction at the species level. The first is **selective site segregation**. Communities dominated by species exhibiting selective site segregation are non-interactive, implying the absence of present-time competition. The development of these kinds of communities must be considered within an evolutionary time-frame. The communities within many freshwater fishes, amphibians, and reptiles are probably organized in this fashion.

The second outcome is **interactive site segregation**. In these cases, competition is considered to be an important organizing force. The fundamental niches of competing species will be reduced or restricted in some manner. Bush and Holmes (1986a, b) considered the communities within lesser scaup and several other avian species to be highly interactive and structured, at least in part, through competitive interactions.

Gause (1934) originated the principle of **competitive exclusion**. Briefly stated, it infers that two species with very similar or identical requirements cannot co-exist simultaneously in the same space. Holmes (1973) has indicated that exclusion is a rather common phenomenon among enteric helminths in vertebrate animals. Indeed, his experimental manipulations of *Hymenolepis diminuta* and *Moniliformis dubius* in rats illustrate the principle clearly (Holmes, 1961, 1962a). Chappell (1969) described competitive interaction between the cestode, *Proteocephalus filicollis*, and the acanthocephalan, *Neoechinorhynchus rutili*, in a natural population of three-spined sticklebacks, *Gasterosteus aculeatus*. Interestingly, neither species was dominant in this system as both were forced from their preferred sites of infection.

The enteric infracommunity of a given host may include **core species**, or **satellite species** or both (for a complete discussion of the core-satellite species concepts, see Caswell (1978) and Hanski (1982)). According to Bush and Holmes (1986a, b), core species are regionally common and locally numerous; they would be present, for example, in more than 50% of the hosts sampled from a given host population. Satellite species, on the other hand, are not present in a large proportion of infracommunities and their infrapopulation sizes are generally small; a given species may be a satellite in one host species and a core species in another.

6.3.2 Structure in definitive hosts

6.3.2.1 Fish infracommunities

While investigations on helminth infracommunities from freshwater and some marine fishes are fairly extensive, the structure of helminth infracommunities in these hosts is not well understood. One of the problems, noted by Holmes (1990), is that while many of the studies on the faunas of fishes are quite appropriate for the questions being asked, the statistical analyses usually are based on data summed across infracommunities. This produces patterns that are irrelevant to interpreting infracommunity dynamics (see also Kennedy, 1985; Bush and Holmes, 1986a, b; Stock and Holmes, 1988; Bush, 1990). The proper approach in examining infracommunity structure is to treat each host separately and their infracommunities as replicates.

Data on enteric parasite infracommunities from a range of vertebrate taxa, but primarily fishes and birds, were analysed by Kennedy, Bush and Aho (1986). They observed that birds possessed faunas that were both rich and diverse, while at the other end of a proposed community continuum, freshwater fishes were comparatively depauperate and non-interactive. They attributed the latter observation to:

1. the narrowness in diet that decreases the chances for parasite recruitment;
2. ectothermy that leads to lower energy demands and less food consumption;
3. a less complex digestive system that decreases the diversity of potential niches; and
4. reduced vagility that restricts opportunities for exposure to a wider variety of parasites.

However, Kennedy, Bush and Aho (1986) predicted that communities of marine fishes might be different from those in freshwater fishes because the former group exhibited greater vagility and because there is a greater array of invertebrate species with, therefore, more potential intermediate hosts in the oceans.

Holmes (1990) examined the parasite communities in the rockfish, *Sebastes nebulosus*, and then compared his observations with the literature available on other marine fishes as a way of testing the Kennedy, Bush and Aho (1986) predictions (Fig. 6.2). Based on Holmes' (1990) study, it is

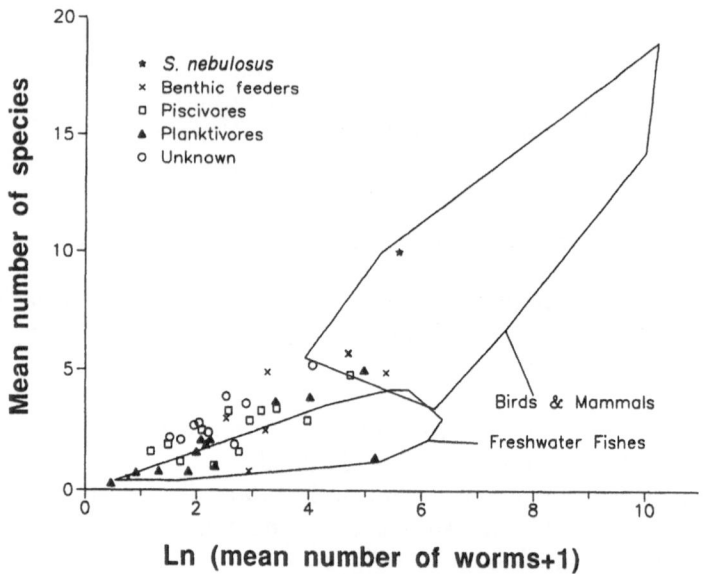

Figure 6.2 Community richness of gastrointestinal helminths of marine fishes (symbols) in comparison with those in freshwater fishes (lower left polygon) and birds and mammals (upper right polygon). (With permission, from Holmes, 1990.)

clear that both the diversity and richness of enteric helminth communities in rockfishes are relatively high. Indeed, they are similar to those in birds and mammals. He attributed the similarities to high host vagility and to the observation that enteric helminths of these marine teleosts are not very host specific. However, even with the rich and diverse infracommunities, rockfishes appear to have many empty niches. Rockfishes are apparently unusual in that many marine fishes do not harbour such relatively rich and diverse infracommunities. He suggested that determinants for species-poor, enteric infracommunities in many marine and most freshwater fishes were probably ectothermy and the low rate of energy flow through individual fish. With less exposure to potential intermediate hosts, fewer parasites will be acquired and the infracommunities will be less diverse and rich.

The work of Rohde (1976a, b, 1978a, b, 1979, 1980a, 1982, 1986) on the ecology of marine parasites, particularly the ectoparasitic mono-geneans, brings a somewhat different perspective to the understanding of community dynamics than that of Holmes (1990) for the enteric hel-minths. Rohde (1982) argued that infracommunities of gill monogeneans are not structured so much by competition as they are by reproductive barriers. He proposed that several hundred thousand niches were available on the gills and other surfaces for occupation by ectoparasites, but that only about 5000 species have been described to date. As a result, even though parasites may compete for resources, the role of competition in structuring ectoparasite infracommunities must be insignificant since there are so many empty niches. Rohde (1979, 1980a) proposed a **mating hypothesis** for niche restriction among ectoparasitic monogenetic trematodes of marine fishes. It argues that the restricted niche width is necessary to ensure reproductive contact among individuals within a small infrapopulation. The narrow niches also would result in a high degree of both site and host specificity among individuals in these small infrapopulations (Holmes and Price, 1986). Moreover, it suggests that 'species which have other means of establishing intraspecific contact, such as good locomotory ability or ability to establish large populations, have less restricted habitats' (Rohde, 1982).

As for marine fishes, the infracommunity structure in freshwater fishes has not been well studied (Kennedy, 1990). The reasons are basically the same for both groups of organisms. Information regarding aspects of parasite abundance and species numbers is quite extensive, but the man-ner in which these data were collected is inadequate to answer certain of the fundamental questions regarding determinants of infracommunity structure. In an effort to shed some light on this question, Kennedy (1990) undertook an extensive and thorough examination of the enteric parasite infracommunities within the European eel, *Anguilla anguilla*, throughout Great Britain and Ireland; some 39 localities were sampled and 842 eels were necropsied. A total of 19 species of helminths was recorded.

The parasite infracommunities were effectively depauperate, with a maximum of three species (Fig. 6.3) occurring in a single eel. The generalist acanthocephalans usually dominated the infracommunities in a given locality, but even the rarest of the parasites dominated in at least one locality. The dominance patterns to some extent reflected the fact that most infracommunities consisted of one species, implying that empty niches were apparently common. 'The overall impression, then, is one of independent assortment by each helminth species; if a species is present in a locality, it may occur in and dominate some eels and, if it is common, it will dominate the majority' (Kennedy, 1990).

He noted that eels are vagile and consume a wide array of prey items. Moreover, since they are catadromous and tolerant of a range of physico-chemical conditions, their parasites, which are also fairly tolerant, can be, and are, distributed over a wide range of habitats. He was, however, struck by the apparent lack of pattern and the very low diversity within the infracommunities in eels. As an explanation for the low diversity, he proposed that there may be a fixed number of niches (three) in the alimentary canal of eels and, no matter how many eel parasites might be present within a given habitat, the maximum number that can be recruited is three. Alternatively, he suggested that colonization events within a given

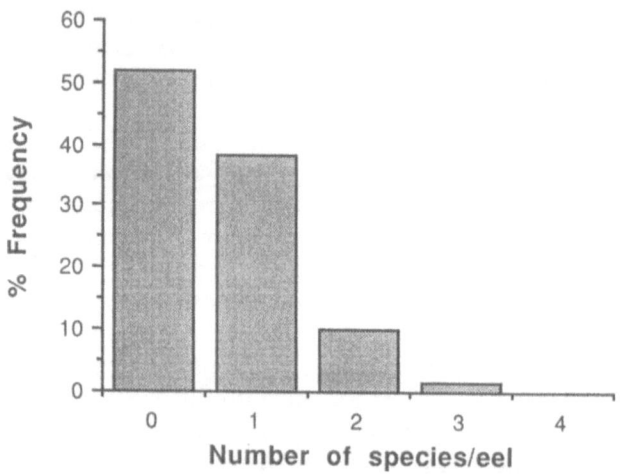

Figure 6.3 Frequency distribution of the number of helminth species per individual eel, *Anguilla anguilla*, taken from throughout Great Britain and Ireland. (With permission, from Kennedy, 1990.)

habitat, or that transmission and infection opportunities, could be the primary determinants of low diversity in the eel infracommunities.

Parasites within the helminth infracommunities of eels fall into three groups. There is a suite of specialists that is specific to the eel, presumably having evolved with it. There is also a group of generalists that can infect eels, but other fish species as well, and then there is a group of what Kennedy (1990) calls accidentals. He points out that eel helminth infracommunities have certain features in common with other groups of fishes. Salmonids, for example, have a suite of specialists, but can also acquire generalist acanthocephalans (Kennedy, 1978a, b). Moreover, he asserted that salmonid infracommunities seem to be organized by stochastic events much like those in eels. Therefore, competition as an organizing force for enteric infracommunities within freshwater fishes is probably not a major factor. Kennedy (1990) concluded that the eel infracommunities and probably most of those in freshwater fishes were stochastic assemblages, not structured communities, and described them as isolationist, not interactive, in character.

6.3.2.2 Amphibian–reptile infracommunities

The most thorough and recent review of parasite infracommunity dynamics in amphibians and reptiles is that of Aho (1990). This invaluable contribution was, however, largely restricted in its analysis of infrapopulation community organization because of the lack of appropriate data. He determined that helminth infracommunities in amphibians and reptiles are affected most often by three factors: geographical location or habitat quality, temporal considerations, and host demography. Several studies provided evidence for these conclusions. The helminth fauna from the painted turtle, *Chrysemys picta,* in a hypereutrophic lake in southwestern lower Michigan (USA) and in the slider turtle, *Trachemys scripta,* from a variety of habitats at the Savannah River Site in south-central South Carolina (USA) were examined (Esch and Gibbons, 1967; Esch, Gibbons and Bourque, 1979a, b). The Michigan turtle population was dominated by digenetic trematodes while the South Carolina population was dominated by acanthocephalans. It was speculated that the abundance of trematode species in Michigan was related to the high species diversity of the regional molluscan fauna as a consequence of the hardness (high calcium levels) of the water. Their scarcity in South Carolina was linked to water softness (low calcium levels) and low species diversity within the southeastern snail fauna. The range for several of the South Carolina acanthocephalans extends into southern Michigan, so this cannot explain their absence in the Michigan lake. However, as noted by Esch, Gibbons and Bourque, (1979a, b), local variability in habitat may

influence diversity of parasite infrapopulations. Where habitats were harsh
or uneven in quality in the South Carolina study, diversity was lower than
in normal aquatic situations, probably because of the adverse effects on
the distribution of intermediate hosts. Parasite community structure in
reptiles can also be affected temporally by changes in diet (Esch, Gibbons
and Bourque, 1979a, b) and by demographic conditions associated with
host age and size among amphibians (Goater, Esch and Bush, 1987).

The linear and radial distributions of helminth parasites within the
alimentary canals of the slider turtle, *Trachemys scripta*, from the Savannah
River Site, South Carolina, were examined by Jacobson (1987). She
reported that the parasites exhibited a strong site fidelity as might have
been predicted based on the work of Crompton and Nesheim (1976). An
interesting aspect of this study was the sympatric distributions of three
species of *Neoechinorhynchus*, robust acanthocephalans that occupy the
anterior part of the intestine. It was concluded that while there was a
moderately high species diversity and richness, the infracommunities in
these turtles were non-interactive.

In an early study, Schad (1963) investigated the mucosal–lumenal
(radial) and longitudinal distributions of *Tachygonetria* spp., nematodes
that parasitize the colon of the European tortoise, *Testudo graeca*. The
longitudinal distributions of the various species overlapped, but radial distributions were completely segregated, indicating strong site fidelity by the
various species. The site segregation was attributed to variations in morphology of mouth parts in the various species, suggesting differences in
feeding strategies within this group of nematodes.

Aho (1990) concluded his review by observing that parasite infracommunities in amphibians and reptiles are highly variable, but that they
are isolationist and non-interactive for the most part. He argued that
local/regional richness of amphibian and reptilian infracommunities may
be defined by local biotic and abiotic conditions, 'but the number of
species found reflects such regional processes as geographic dispersal and
the historical accumulation of parasites in different host animals'. The
infracommunities of helminth parasites in four species of desmognathine
salamanders (Goater, Esch and Bush, 1987) in the Great Smokey
Mountains of North Carolina (USA) support these conclusions. These
infracommunities were decidedly depauperate and isolationist in character. Parasite prevalence, intensity, and diversity were related in part to
host diet and habitat preferences among the different host species. These
workers suggested that ectothermy and low vagility, in conjunction with
the salamanders being generalist insectivores and having simple alimentary canals, collectively produced depauperate infracommunities.
These observations on salamanders were paralleled by studies on the
enteric infracommunities of newts, *Notophthalmus viridescens*, in Michigan

(Muzzall, 1991). The newt infracommunities also were depauperate and isolationist in character. Moreover, as with the parasite faunas in desmognathine salamanders (Goater, Esch and Bush, 1987), the infracommunities in newts tended to increase with host size. According to Muzzall (1991), this tends to conform to the **island size hypothesis** (Holmes and Price, 1986) that predicts richer helminth faunas in larger hosts.

6.3.2.3 Avian infracommunities

Based on the large numbers of species reported in ordinary surveys, it might be predicted that birds would have interactive infracommunities and, therefore, that competition would be an important structuring force. This question was addressed by Kennedy, Bush and Aho (1986) when they undertook their systematic comparison of bird and fish helminth infracommunities (Fig. 6.4). The main thrust was to develop an explanation as to why bird and fish communities differ. The differences were attributed to several factors. First, the complexity of avian diets in combination with endothermy are conducive to increased parasite diversity. Simply stated, the more food that is consumed, the higher the probability of parasite recruitment. Second, birds are highly vagile. With the exception of raptoral species that have depauperate infracommunities, highly mobile avian hosts are exposed to a greater diversity of potential intermediate hosts. Third, with the exception of seed-eating birds, most avian species have relatively broad diets and this also results in exposure to a wider range of potential intermediate hosts. Fourth, they predicted that selective feeding on prey that would serve as intermediate hosts for an array of parasites should also increase parasite diversity in avian hosts. Finally, the complexity of the avian alimentary canal increases the number of niches available for colonization. Based on this analysis, Kennedy, Bush and Aho (1986) concluded that these conditions would produce both higher species richness and abundance in avian infracommunities. Further, such conditions would also lead to intra-, or interspecific interactions, or both. An examination of studies that have used appropriate collecting protocols in combination with statistical analyses suggest, for the most part, that these predictions are correct (Bush and Holmes, 1986a, b; Stock and Holmes, 1987a, b, 1988; Goater and Bush, 1988; Edwards and Bush, 1989; Bush, 1990; Moore and Simberloff, 1990).

The size of the parasite infracommunities in lesser scaup, *Aythya affinis*, is enormous, with nearly 1×10^6 parasites present in only 45 hosts examined (Bush and Holmes, 1986a, b). The most common species of helminths were broadly distributed in over half of the small intestine. The core species 'were more evenly distributed along the intestine than

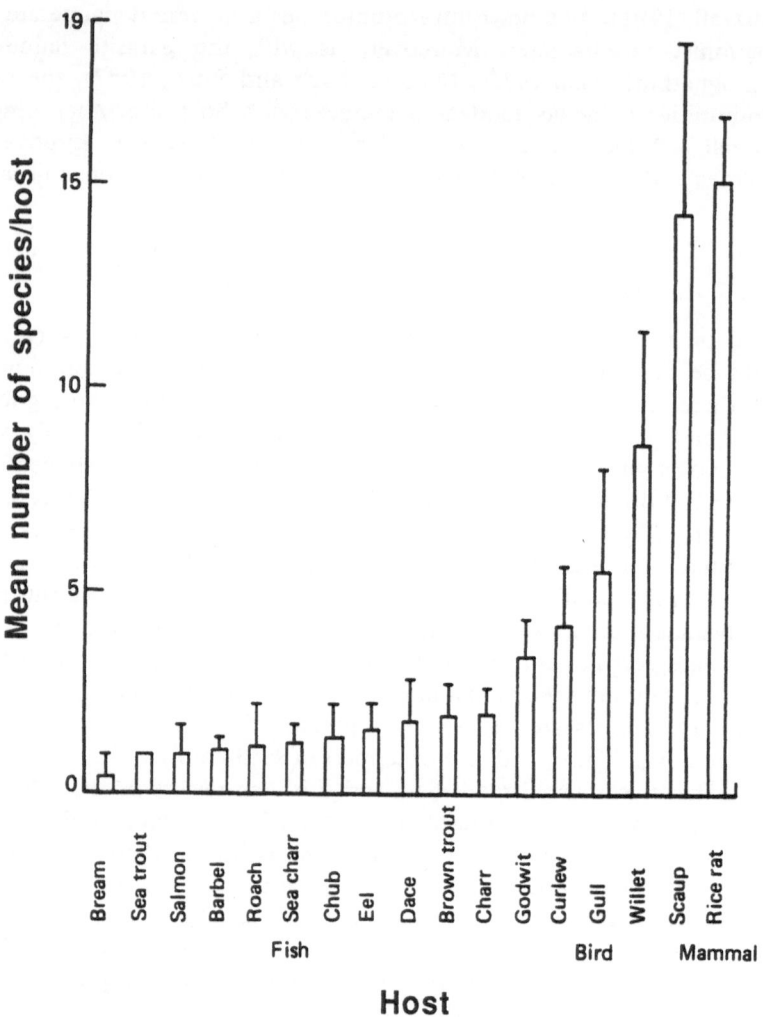

Figure 6.4 Comparative mean species richness in intestinal helminth communities. Data for fish represent the most diverse site for each species. Bars denote 1 SD (With permission, from Kennedy, Bush and Aho, 1986.)

expected by chance', with the secondary species present in the anterior and posterior ends of the gut. Satellite species were randomly distributed along the intestine. The infracommunities were divided into **guilds** (*sensu* Root (1967); a guild describes functionally similar species within a community, or species that share common resources). The guilds included small absorbers that were located immediately adjacent to the mucosa, large absorbers that were mid-lumenal, and the trematodes.

Antagonistic interactions between species occurred both within and between guilds, with the basic interactive component of the infracommunity being comprised of core and secondary species of absorbers, primarily congeners. The occurrence of trematodes and satellite species was essentially stochastic with respect to their contributions to the infracommunity structure.

The infracommunity structure was examined in four species of grebes (Podicipedidae) from several lakes in Alberta, Canada (Stock and Holmes, 1987b; 1988). Results showed that the various species of helminth parasites occupied predictable longitudinal positions in the small intestine. However, because the various helminths were also capable of occupying extensive longitudinal ranges of the intestine in different host species, they suggested that the parasites had a wide tolerance for physiological conditions within the intestine. Within individual birds, the realized niches were much smaller than their fundamental niches, suggesting the existence of an interactive community. Another observation of interest was the negative impact of the cestode, *Dioecocestus asper*, on infracommunity structure in the red-necked grebe, *Podiceps grisehena*. This cestode appeared to dominate infracommunities through interference competition rather than the more typical exploitative method. It was found to especially impact the distributions and species richness of small absorbers. Its effects were pronounced for satellite species, appearing to prevent their establishment. Core species on the other hand, were able to establish in the presence of *D. asper*, but their normal distributions were altered.

The infracommunities in the long-billed curlew, *Numenius americanus*, are not species rich or large (Goater and Bush, 1988). They are, however, interactive. The interactive nature of the infracommunity structure appears to be affected, at least in part, by the nature of parasite recruitment when the birds arrive on their summer breeding grounds. Thus, initially the infracommunities were depauperate, but they began to develop immediately. One cestode they acquired was *Dictymetra numenii*; it is a core and specialist species that first occupies and then dominates the anterior end of the small intestine. Another core species that the curlews recruited was the congener, *D. nymphae* and, in the presence of *D. numenii*, the former moved posteriorly. If *D. numenii* was not abundant, the tendency was for *D. nymphae* to occupy a more anterior position. Three other congeners occurred in the curlews, but these species were not prevalent or abundant; they were accordingly considered satellite species, along with two other cestodes, an acanthocephalan, and a trematode. They emphasized, in conclusion, that parasite infracommunities need not be complex in order to be interactive.

The study by Edwards and Bush (1989) on avocets, *Recurvirostra americana*, emphasized how the nature of habitats in which the birds feed influences the organization of parasite infracommunities. Avocets were collected from both ephemeral and permanent bodies of water in Alberta and Manitoba, Canada. The avocet infracommunities from ephemeral bodies of water primarily consisted of avocet specialists and a few host generalists. In these infracommunities, there was no evidence of competitive interaction. In contrast, the infracommunities in avocets taken from permanent bodies of water were very different. These infracommunities included the typical avocet parasites, but in addition there were many species normally associated with ducks, particularly lesser scaup. They occupied linear gaps along the intestine, but also overlapped the distribution of avocet specialists. They exhibited strong interaction among themselves and with the avocet specialists. Infracommunities in avocets from ephemeral bodies of water were thus non-interactive and had vacant niches while those from permanent bodies of water were saturated and interactive. The communities in avocets that appeared to be 'co-evolution-structured' were compared with the 'invasion-structured' communities described by Rummel and Roughgarden (1985); the former had fewer species, were loosely packed, less stable and were more likely to be invaded, while the latter were completely the opposite. In the normal habitat, avocet infracommunities were co-evolution-structured, but in an ephemeral habitat with a different set of intermediate and definitive hosts, avocets possessed invasion-structured infracommunities. The invasive component of the co-evolution-structured infracommunity was illustrative of Roughgarden's concept of 'supply-side' ecology (Lewin, 1986). As Edwards and Bush (1989) have shown in avocets, the supply of invasive or infective forms of helminth parasites is a significant factor in structuring their infracommunities. Application of the supply-side concept can be made to other host–parasite systems as well (see Bush, 1990; Holmes, 1990).

Willets, *Catoptrophorus semipalmatus*, from freshwater and marine habitats were examined with a view to comparing the guild structure of parasites from these two, quite different, habitats (Bush, 1990). He predicted that species packing should be similar across the geographical range of the host and that the same, or similar, number of parasite guilds should also occur. It was found, however, that transmission patterns of trematodes and cestodes in freshwater and marine systems were dissimilar. Those species that occur in marine environments appeared to 'spread the risk' by maturing in a wide range of definitive hosts while those in freshwater systems specialized in one, or a few, host species. By spreading the risk, Bush (1990) was implying that selection favoured conditions in which generalist species could more successfully evolve than specialists. He observed 'an essentially constant level of species packing irrespective of major environmental differences'. On the other hand, while species richness was constant, guild structure changed from freshwater to salt-water infracommunities. In freshwater systems there were three guilds, one that included large cestodes and acanthocephalans (the lumenal absorbers), a second that included the small cestodes (mucosal absorbers) and a third that included the mucosal trematodes. The lumenal absorbers dominated freshwater infracommunities. The same three guilds were present in saltwater infracommunities, but these were dominated by mucosal trematodes. He concluded that the willet communities were saturated, but still open to invasion.

Infracommunities within the bobwhite quail, *Colinus virginianus*, were examined (Moore and Simberloff, 1990) within the conceptual frame-works provided by the Holmes and Price (1986) interactive–isolationist model and the ectotherm–endotherm approach used by Kennedy, Bush and Aho (1986). A total of 12 species of intestinal helminths was recovered, but only six species were common. If the criteria used by Bush and Holmes (1986a, b) are applied for the identification of core species, only four could be classified in this manner. This included one cestode and three nematode species. When either one of two caecal nematodes was present in high densities, one of them regularly shifted its position, suggesting negative interaction; two of the cestode species also indicated a negative association. Moore and Simberloff (1990) conjectured that faunas in bobwhite quail superficially resembled isolationist infra-communities since there were only a few species and, when present, these were not abundant. On the other hand, they suggested that the probability of colonization was high and that interactions were occurring. They were not satisfied with the manner in which their infracommunities could be classified according to the models proposed by Holmes and Price (1986) or Kennedy, Bush and Aho (1986), arguing instead that these repres-entations will require modification as additional host–parasite systems are studied.

6.3.2.4 Mammalian infracommunities

As for most other vertebrate taxa, infracommunity investigations in mammals (see Pence, 1990) are mostly inadequate and basically for the same reason; the data have been summed prior to statistical analysis.

Hobbs (1980) undertook a thorough study of two species of pikas, *Ochotona* spp., from locations in the Yukon Territory and Alberta, Canada. No evidence for competitive exclusion was observed, but several positive, paired associations were identified. In these cases, the life cycles of the parasites were apparently similar so that if one species was recruited, the probability of recruiting the second was also high. In only one case was interactive site segregation noted. The infracommunities in pikas were described as being quite old in evolutionary terms, consisting mostly of specialist species. A cestode, *Schizorchis caballeroi*, was the only generalist and possibly the newest member of the infracommunity.

One of the most complete efforts on mammalian infracommunities was that of Lotz and Font (1985) who examined the big brown bat, *Eptesicus fuscus*, from two locations in Wisconsin and Minnesota, USA. A total of 21 species of helminths was identified from 82 hosts in Wisconsin and 13 species from 47 hosts in Minnesota. While relatively species rich, the infracommunities were not interactive and competition was not a structuring force. It was proposed that infrapopulation carrying capacities in the bats were quite large and that the host's immune response kept worm densities well below carrying capacity so that competition could not occur. They reasoned that if two species of helminth could compete for a limited resource, then the host and the parasite may do so as well. In this way, through co-evolution of the host and parasite, the host had developed a mechanism for keeping parasite infrapopulations at low levels, below the threshold where exploitation competition could occur.

Infracommunities in white-tailed deer, black bear, and coyotes from the southern and southwestern USA were examined by Pence (1990). Based on his analysis, it was concluded that both deer and bear had isolationist and non-interactive enteric infracommunities. Coyotes, on the other hand, had species-rich, and probably interactive, infracommunities. The latter suggestion could not be confirmed, however, because he did not examine the linear distribution of the parasites in the gut. On the other hand, studies of coyotes from other localities (Hirsch and Gier, 1974; Pence and Meinzer, 1979; Custer and Pence, 1981) indicated the importance of species interactions even though data in these investigations were summed across individual hosts and were thus inappropriate for comparative purposes.

Montgomery and Montgomery (1990) recently reported the results of a comprehensive, long-term study on the enteric infracommunity dynamics in wood mice, *Apodemus sylvaticus*, from several geographical locations in Northern Ireland. The long-term character and the geographical diversity in this investigation make it a most useful contribution. It was suggested that species composition in the helminth infracommunities of *A. sylvaticus* may be stable over a wide geographical range in Northern Ireland, even though there was some variation. As was noted, 'the parasite community associated with *A. sylvaticus* may be regarded as globally stable with zones of local instability both in time and space', a situation that was characterized as being analogous to certain insect populations described by Taylor and Taylor (1977). The infracommunities in *A. sylvaticus* also were described as non-interactive because there were 'neither strong nor consistent positive or negative interactions between pairs of helminth species'. Finally, species composition and abundance in *A. sylvaticus* were not due to interspecific interactions, but were attributed to the population dynamics of each species acting independently of the other. In other words, it seemed that spatial and temporal resource partitioning were of great significance in organizing the infracommunities in the wood mice populations.

6.3.3 Structure in intermediate hosts

Almost nothing is known about the structure of parasite infracommunities in intermediate hosts. Even less is known about the nature of parasite interaction, if it occurs, in intermediate hosts. The exception to this generalization applies to antagonisms described in molluscan hosts, with special reference to digenetic trematodes. Intramolluscan interactions among trematodes have been observed for many years (Cort, McMullen and Brackett, 1937; Cable, 1972; Kuris, 1974, 1990; Lauckner, 1980; Sousa, 1990). From these studies, two dominant themes have emerged. The first is that multiple species (double and triple) infections in most snails occur at a rate that is significantly less than would be expected by chance alone. This well-documented observation evoked a long series of investigations designed to interpret the nature of interactions within snails that might preclude the establishment of double infections (for a review, see Lim and Heyneman, 1972). Second, studies indicate that antagonistic interactions involving larval trematodes in snails include both competition and predation.

Most of the recent information regarding intramolluscan antagonism in natural snail populations comes from studies on two host–parasite systems (Kuris, 1990; Sousa, 1990; Fernández and Esch, 1991a). These investigators have examined the highly complex interactions within the parasite fauna of the horn snail, *Cerithidea californica*, in two separate locations on the coast of California (USA), and in the pulmonate snail, *Helisoma anceps*, in the Piedmont area of North Carolina (USA).

Kuris' (1990) analysis was based mainly on collections made by Martin (1955, 1972) in Upper Newport Bay, California; 12 995 snails (all older than 2 years) were collected and checked for shedding cercariae. Seventeen trematode species were identified. The focus of his effort was a comparison of expected and observed frequencies of double infections among trematodes with different combinations of intramolluscan larval stages (redia–redia, redia–sporocyst, sporocyst–sporocyst). As was noted, certain developmental processes associated with trematodes in snails can be linked to either competition or predation because of the finite nutrient and spatial resources available to the developing larvae. Intramolluscan development begins with either the penetration of a free-swimming miracidium or the ingestion of an egg. By far the majority of species use the first method. After gaining access to the snail, the larva migrates to a specific organ or tissue where it develops into a sporocyst; morphologically, the sporocyst is little more than a shapeless sac. In some species, sporocysts then give rise to rediae, all of which have a mouth and a primitive, but incomplete, gut; this larval stage is able to ingest host tissue as well as prey on the larvae of other trematode species. (An interesting footnote here is that there is no evidence of cannibalism among the redial stages. This means that trematode rediae have evolved mechanisms of self-recognition which must be rather sophisticated). Some species do not produce rediae, remaining in the snail as sporocysts. Site specificity within the snail is high, at least at the beginning of the infection. Later, when space or nutrients become limited, rediae, or sporocysts, or both, may spill over into adjacent tissues.

The various combinations of two-species trematode infections that occurred in *C. californica* were identified and compared with the type of developmental stage that occurred in each species (Kuris, 1990). The aim of this analysis was to determine if either competition or predation was a structuring force at the infracommunity (infraguild) level. In order for antagonistic interaction to be a significant determinant in organizing the infraguilds, Kuris contended that three criteria must be met. Thus, parasite prevalence should be high, a dominance hierarchy must exist, and interference competition should be more important than exploitation competition. All three criteria were satisfied for the trematodes in the horn snail, making their infraguilds potentially interactive. The process of structuring the infracommunities occurred through intense interspecific interaction such that death or complete suppression of inferior competitors was the outcome. Kuris (1990) concluded that the infraguilds were strongly influenced by antagonistic interactions, so much that he was able to construct an elaborate dominance hierarchy for many of the species in *C. californica*. On the other hand, among most other trematode assemblages, competition is unlikely to affect infraguild interactions because parasite prevalences are generally too low. When antagonism does occur, it causes site displacement, total exclusion, or crowding effects such as size reduction, delayed maturity or reduced fecundity of subordinate species (Kuris, 1990). Similar guild structures were reported by Lie, Basch and Umathevy (1966) in the freshwater snail, *Lymnaea rubiginosa*. In that assemblage, however, dominance was established through priority of occupancy, conferring a strong temporal component to the processes that organize trematode infraguilds.

Sousa (1990) described the results of a 7-year study on the infraguild structures in *C. californica* at two sites (different from Kuris' (1990)) on the California coast. From a total of 4462 snails examined, he identified

15 species of trematodes. Sousa (1990 and 1991) made three important observations with respect to negative antagonistic interactions for the infraguilds in his investigation. First, double infections were rare, occurring with much less frequency than would be expected by chance alone. Second, direct evidence was presented for the existence of hierarchical antagonism among species occurring together. Third, he recorded hierarchical species replacement over time. There were, therefore, competitive interactions that affected infraguilds in his snail populations, confirming some of the observations and conclusions of Kuris (1990).

Although negative interactions may clearly prevent the co-existence of species at the infracommunity level, the extent of exclusion of subordinate species also may be insignificant, even among the most subordinate species as was shown in a recent mark–recapture study by Fernández and Esch (1991a). This investigation demonstrated that interspecific antagonism could occur in *H. anceps*, but that the observed frequency of multiple infections under field conditions was quite low. Instead, the infrequent occurrence of intramolluscan antagonism was more directly related to spatial and temporal factors that influenced the distribution of infective stages for the snails. The upshot of the transmission process, as it affected the trematode infraguild in *H. anceps*, was that most species simply did not overlap, so that there was little opportunity for negative interaction.

Throughout this discussion, the parasites within a given snail have been referred to as infraguilds. This is based on the terminology used by both Kuris (1990) and Sousa (1990). However, it may be incorrect and should be noted at this point. For example, some species of larval trematodes occupy only the hepatopancreas, others only the gonad, while others may occupy the hepatopancreas initially, subsequently spilling over into the gonads. Since a guild includes only functionally similar species, or species that share a common resource, *sensu* Root (1967), it would seem that species occupying different organs should be assigned to different guilds. Moreover, trematodes species with only sporocysts, or sporocysts and rediae, should not be considered as members of the same guild because they are functionally dissimilar. Based on these arguments, two conclusions can be drawn. First, the concept of the guild as it applies to the intramolluscan stages of digenetic trematodes clearly requires further refinement. Second, the infraguild structure in molluscs is definitely much more complex than would appear from any superficial consideration of the system.

While this Chapter is devoted primarily to infracommunity structure, it is appropriate that the discussion of parasite assemblages be extended here to the component community because of the strong implications for infracommunity interactions in molluscs. Sousa (1990) proposed two alternate hypotheses regarding the determinants of structure at the host population, or component community, level. The first was that inter-specific, hierarchical antagonism in molluscan infracommunities was involved in structuring the component guild. Second, he suggested that spatial and temporal factors associated with parasite transmission, in combination with the duration of host exposure to infective agents, were the factors structuring guilds at the component community level.

Sousa (1990) argued that the 'number of uninfected susceptible hosts never becomes sufficiently limiting, given the rates of snail and parasite recruitment into the system, to drive parasite diversity downward'. This would occur if the parasite guild were dominated by species at the top of the dominance hierarchy. While interspecific competitive interactions may have occurred, any reductions in species numbers were 'more than com-pensated for by increases in both the number and equitability of other parasite species in other host populations'. The dominant parasite species varied markedly from site to site within his sampling area and the prevalences of various species of trematodes changed considerably from one year to the next. His view of the component community in *C. californica* was that it possessed unlimited resources and that it was structured primarily by variability in the spatial and temporal transmission processes employed by the different species of parasites. At the com-ponent community level then, transmission processes rather than inter-specific interactions determine the composition and abundance of species.

Fernández and Esch (1991b) came to similar conclusions when they examined the component community in *H. anceps*. The structuring processes were attributed to four factors. First, there was temporal heterogeneity in abundance of infective stages, mostly miracidia, in the pond. The heterogeneity was not the result of stochastic events; it was related to the timing of visitation to the pond by migratory definitive hosts. Second, the various species of trematodes in the community differentially responded to the presence and abundance of different size classes of snail hosts during the recruiting season. Thus, some species preferred small snails and others preferred larger ones, etc. Third, there was variable mortality of snails infected with the different species of trematodes. Some snails would become infected, shed cercariae for a brief period, and then die; others would become infected, shed cercariae and then not be able to survive the winter. The outcome of this kind of differential mortality was that the diversity of the component community was predictably affected. Finally, the parasite community in *H. anceps* had to be 'reset' each year because of the annual life span of the host. Resetting in this case was regarded as a perturbation that involved the loss and replacement of the entire snail population and its component parasite community on an annual basis.

The distribution and dynamics of the infra- and component parasite community was also investigated in the same *H. anceps* population on a microgeographical scale (Williams and Esch, 1991). Thus, individual snails were removed from the pond and each time as much information as possible was recorded about the precise location of their capture, e.g., water depth, distance from shore, substratum type, etc. It was found that snails infected with a given parasite species were usually associated with specific substrata, water depths, or stretches of shoreline dominated by certain biotic or abiotic features (Figs. 2.5, 2.6, page 35). These observations were subsequently verified by Fernández and Esch (1991a) and extended to identify fixed sequences of parasite species recruitment throughout the spring, summer and autumn months. These studies serve to emphasize the processes of temporal and spatial niche partitioning in the infra- and component communities in *H. anceps*. In effect, the parasites were partitioning the snail as a resource base from the standpoint of time, microgeographical space, and site of infection within the snail. The parasites were not using antagonistic or stochastic tactics to structure their community, but rather the far more subtle processes predicted by Sousa (1990). As clearly stated by Price (1984a), 'the resource base on which communities are built must be understood in detail if the communities themselves are to be understood'. It would appear that trematode communities in molluscan hosts are not structured just through 'the simple addition of the properties or characteristics of the species members' (Fernández and Esch, 1991b). However, a predictive capability for such communities can be acquired when the responses of individual species to variable resource bases are determined and then combined with a probabilistic perspective on the patches that may be colonized (Price, 1984b; Fernández and Esch, 1991b).

6.4 THE SCREEN/FILTER CONCEPT

At the conclusion of Holmes' (1990) paper on the parasite community biology of marine fishes, he presented a cogent perspective on his views regarding the determinants of parasite community structure (Fig. 6.5). It was, in part, derived from previous reviews in which he attempted to deal with many of the same issues (Holmes, 1987, 1988). At the broadest scale, he proposed that the quality of a regional host and parasite fauna (the compound community) will be the reflection of a combination of historic and zoogeographical events. A great variety of factors will impact on the biology of host and parasite communities at this level. The nature of compound communities could be, for example, affected by colonization or local extinction of either hosts or parasites. Natural and anthropogenic

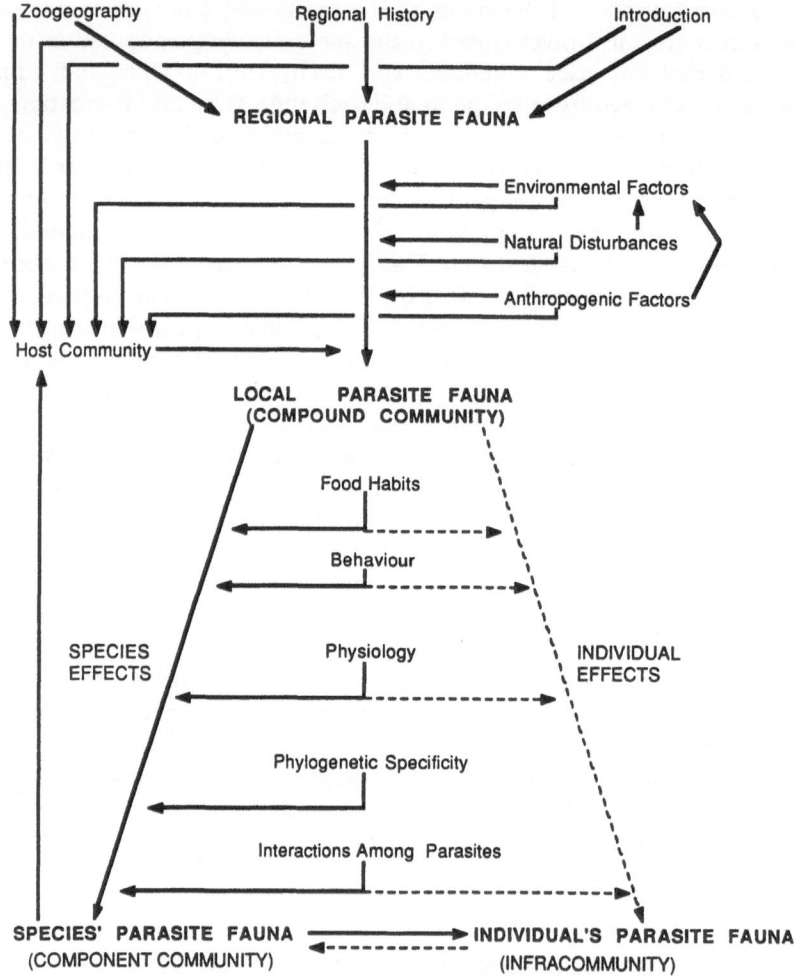

Figure 6.5 Determinants of parasite community structure. (With permission, from Holmes, 1990.)

forces may also directly affect these communities, or perhaps modify the environment that could then impact the community structure. Consider, for example, the anthropogenic impact of cultural eutrophication on *Crepidostomum cooperi* (section 3.1) or the effects of selenium pollution on *Bothriocephalus acheilognathi* (section 3.6.5). In both situations, pollution affected the population biology of the host and the parasite such that the parasite community structure was altered at all three hierarchical levels.

Holmes (1990) proposed that there is a pool of potential parasites within the compound community to which members of the host community are exposed, an application of Roughgarden's supply-side concept (Lewin, 1986). Then, at several different levels, a series of filters acts to influence the successful transmission of the parasites. Some of these filters are external to the host and may be either biotic or abiotic in character. Others are associated with the host directly; the host's immune capacity, for example, would operate at this level. The final filter rests with the potential for parasite–parasite interaction at the infracommunity level. At the one extreme, in isolationist parasite communities, certain kinds of screens prevent either the exposure or the establishment of parasites, and interspecific interactions would not occur with regularity. At the other extreme, in interactive infracommunities, screens that block exposure or establishment are largely ineffective, making interactions common. Based on this perspective, Holmes (1990) projected that factors affecting the component communities in marine teleosts are deterministic in some cases and stochastic in others.

The screen/filter paradigm is not new, having been developed from Roughgarden's (Lewin, 1986) supply-side concept. Its application to host–parasite systems is more recent (Holmes, 1990). As noted, the latter author considered the screen/filter concept only within the context of component communities in marine teleosts. He did not extend the idea to other host taxa. It is probable, however, that the factors affecting component communities throughout vertebrate and invertebrate taxa are both deterministic and stochastic as he proposed for component communities in marine fishes. On the other hand, intuitively it seems that in those cases where there is a clear co-evolutionary component in a given host– parasite relationship, deterministic factors should play a greater role than stochastic ones in structuring the community.

7 Component and compound communities

7.1 INTRODUCTION

As previously outlined, component and compound parasite communities should be considered within the context of specific hierarchical scales. Thus, a **component** community consists of all the infracommunities of a species within a given ecosystem, while a **compound** community is composed of all the component communities within the same ecosystem. A discussion of these community concepts has been combined in this chapter for two reasons. First, it was a matter of convenience and clarity as far as the authors were concerned. Second, because so few studies have actually been conducted at the compound community level, it was believed that a chapter devoted solely to this level of organization would not stand alone very well.

7.2 CORE–SATELLITE/GENERALIST–SPECIALIST SPECIES CONCEPTS

The idea of core and satellite species is not a new one, having been developed a number of years ago by Caswell (1978) and Hanski (1982). **Core** parasite species are those that occur with a high frequency within a given host species. Generally, core parasite species will also occur in large numbers within a component community. In contrast, **satellite** species are found with less frequency and in smaller numbers (Holmes and Price, 1986).

Generalist species will be found in a broad range of unrelated host taxa. **Specialists,** on the other hand, are more restricted, usually to a single host species, or closely related host species. In many instances, generalist species will be unable to mature sexually in certain host species even though they may commonly occur in those hosts. Specialists will always be able to mature sexually in their normal hosts. In summary, the core–satellite species concept is based on the frequency and density of parasites within a range of host taxa while the generalist–specialist concept reflects the nature of host specificity.

One of the first, if not the first, applications of the core–satellite concept to parasites was made by Bush and Holmes (1986a, b) in their study of the enteric helminth communities in lesser scaup ducks, *Aythya affinis*. Within the highly complex assemblage of 52 species, eight were found with sufficient regularity to be designated as core species. Interestingly, the core species in scaup were associated with two different suites of parasites, one using the amphipod, *Hyalella azteca*, and a second that employed another amphipod, *Gammarus lacustris*, as intermediate hosts. A similar pattern in the occurrence of certain core species was observed among parasite communities in four species of grebes (Stock and Holmes, 1987b) and for certain core species of component communities in two species of insectivorous bats (Lotz and Font, 1985); there was thus a clear link between the presence of certain core species and the preference of particular prey species in the host diet.

Holmes (1990) used the extent to which core species are shared as a way to assess the predictability of component communities in marine fishes. All species having a median of two or more 'core species per component community had at least two individual helminth species that attained core status in at least 50% of those component communities'. Thus, the sharing of core species accounted for the high level of similarities in component communities among these marine species. Habitat, however, can have a significant impact on the occurrence of core and satellite species in the same host species. Three different cestodes were found to be both core and specialist species in willets, *Catoptrophorus semipalmatus*, from freshwater habitats (Bush, 1990). In contrast, all three species were absent in willets taken from marine habitats, and they were not replaced by new specialist species. This observation is similar to those of Holmes (1990) and Kennedy (1990) regarding variation in transmission dynamics of parasites and the phenomenon of host specificity in marine versus freshwater fishes.

Goater and Bush (1988) also raised the question of host specificity in considering the nature of core–satellite species in long-billed curlews, *Numenius americanus*. It was stated that 'while high diversity may not be required for interactive communities, the presence of sympatric congeners might be crucial. The magnitude of their importance may be enhanced if they are host specialists and they use the same intermediate host species'. Holmes (1990) indicated that host specificity played but a minor role in affecting the nature of enteric helminth communities in marine fishes although he also noted that the generalization may not apply to more tropical fishes or elasmobranchs.

Host specificity also appears to be of little consequence in determining the nature of enteric helminth communities in amphibians or reptiles (Aho, 1990). Aho cited the presence of only a few specialists and the dominance of infracommunities by generalists. However, he noted that a comparatively large number of helminth taxa are core species in several host species from distinct populations. Similarly, Bush (1990) pointed out that trematodes from birds in marine habitats seemed to 'opt for spreading the risk among a large number of host species' and, accordingly, there were fewer specialists and more generalists than among trematodes from freshwater hosts. In contrast, many freshwater helminths, and especially cestodes, appear to have specialized in a single host or a small group of host species. Bush (1990) concluded that host specialists were more typical of birds in freshwater systems and host generalists among birds from marine systems, but was unable to provide insight as to the nature of selection pressures that might produce such a pattern.

7.3 DETERMINANTS OF COMPONENT COMMUNITIES

The determinants of component community structure are varied for both the host taxon and the type of parasite, e.g., ectoparasite versus endoparasite. The range of determinants includes both ecological and evolutionary factors. Within an ecological context, determinants would include the vast mix of physical and biotic characters that shape the niche of the species and influence the dynamics of local populations. Historic and zoogeographic factors will affect the availability of host populations (Holmes, 1990). Ontogenetic or other developmental phenomena also can affect the structure of a component community. These may operate through the modification of host behaviour, or physiologically by changing the physical conditions in the gut. If the nature of those forces that affect the metapopulation dynamics of a parasite within a host population is examined, it will become apparent (not surprisingly) that they are very similar to those that affect infracommunities. Since the complexity of these interacting variables is so extensive, only a few will be considered here.

7.3.1 Fishes

The component parasite community in eels, *Anguilla anguilla*, from the British Isles was considered as an assemblage of parasites created by stochastic colonization events which, in turn, were influenced by environmental and habitat conditions (Kennedy, 1990). When the component communities of other species of freshwater fishes were examined and compared, his conclusions were unequivocal. It was thus contended that the parasite infracommunities in freshwater fishes are not usually species rich and that they appear to possess many empty niches. Kennedy, Bush and Aho (1986) had previously attributed these characteristics to reduced vagility, ectothermy, etc., among piscine hosts.

Since freshwater fishes must contend with land barriers, they tend to be more isolated and generally do not move great distances; as a group, therefore, freshwater fishes are relatively less vagile than their marine counterparts. Accordingly, most freshwater fishes are not only restricted in the extent to which they are able to disperse the infective agents of parasites, they are simultaneously restricted in the range of potential parasites to which they are exposed. Ectothermy further reduces the necessity to consume large quantities of food and, in turn, reduces the risk of exposure to infective agents of parasites. Having made this point, however, it is important to note that if the checklists of parasites for North American freshwater fishes are examined, the numbers of species in some hosts are quite large (Hoffman, 1967). This at least suggests the potential for interactive site segregation as well as the possibility of complex component communities in freshwater fishes. Therefore, care must be exercised in making broad generalization regarding the nature of component communities in freshwater systems before substantially more data are available.

Component communities, at least in some marine fishes, appear to be quite dissimilar from those in freshwater fishes (Holmes, 1990). Polyanski (1961a) identified several factors that he believed were important in influencing the richness and spatial patterns of component communities in marine fishes. First, the range of potential intermediate hosts is greater in marine systems than in freshwater and could, therefore, affect the diversity of the component community. Second, many marine fishes have longer life spans than their freshwater counterparts; the former are, therefore, likely to be exposed to more infective agents over a lifetime. Third, with greater vagility comes more opportunity for exposure to infective agents of parasites and marine fishes are much more vagile than most freshwater fishes. Host population density is also critical since, with increasing host densities, the rates of parasite transmission will be greater; many marine fishes have population densities that are quite high and, therefore, conducive to higher parasite transmission rates. Finally, host size may be an important factor; generally, larger hosts will have greater species diversities, although this is not always the case.

The component communities of the rockfish, *Sebastes nebulosus*, off the coast of British Columbia, Canada, were examined by Holmes (1990); he also reviewed the data from several studies regarding component communities in other marine fishes. Rockfish component communities were relatively rich, much more so than the component parasite communities of freshwater fishes. This observation was attributed to a couple of factors. First, either marine fishes are more vagile or the intermediate hosts for their parasites are highly vagile, or both. In any event, increased vagility of either the intermediate or definitive host would lead to greater exposure to more, and different, parasite infective stages. Second, marine helminth parasites also appear to exhibit a relatively low degree of host specificity; this means that many of the species infecting rockfishes are generalists. Holmes (1990) concluded that 'factors acting at the host species level are important determinants for some helminths, and stochastic factors are important for others'. It also was argued that the long-term success in using parasites as 'biological tags' is evidence (Margolis, 1963; MacKenzie, 1983, 1987) for the greater importance of deterministic rather than stochastic factors in shaping the component communities in marine fishes.

7.3.2 Amphibians and reptiles

While amphibians and reptiles consume a wide range of prey items and should, therefore, have diverse and rich component communities, they are in fact the least diverse and most depauperate of all the vertebrate groups (Aho, 1990). The lack of parasite species diversity can, in great measure, be attributed to their ectothermy, their possession of a morphologically simple intestine with reduced niche numbers, and their low vagility. There is, on the other hand, a great deal of regional variation in species richness among different host groups. Predator–prey interactions were not considered as significant factors influencing the helminth distribution and abundance patterns in most amphibians and reptiles (Aho, 1990). For the most part, therefore, there is no indication that their infra- and component communities are structured by interactive factors. Instead, they seem to be the product of evolutionary and stochastic events, some of which are not unlike those that appear to influence the component community structure in freshwater fishes.

7.3.3 Birds

Four determinants of component community structure have been examined with some frequency in birds (Bush, 1990). The first includes an examination of patterns in the zoogeographic variation of component communities, emphasizing primarily the nature of change during host migration (Buscher, 1965, 1966; De Jong, 1976; Hood and Welch, 1980; Tallman, Corkum and Tallman, 1985; Wallace and Pence, 1986). The second area of focus is the manner in which component communities change with host age (Bakke, 1972a; Forrester *et al.*, 1984; Moore *et al.*, 1987), although as Bush (1990) notes, age and seasonal phenomena are continuous variables and not independent of each other; when changes do occur, they are primarily the result of diet switching and, therefore, of variation in exposure to infective agents. The third determinant to receive considerable attention is host gender and its effect on component community structure (Cornwell and Cowen, 1963; Threlfall, 1968; Moore *et al.*, 1987); in these cases, alterations may be reflective of differences in diet due to gender-related behaviour, feeding intensity, vagility, etc. Finally, there are several studies that have assessed the influence of seasonal changes on component communities (Bakke, 1972b; Hair, 1975; Moore and Simberloff, 1990). Bush (1990) made the point, however, that there is a common thread that runs throughout all of these studies and this relates to host diet which is, in turn, primarily a function of habitat.

Based on these and other observations from a review of pertinent literature, Bush (1990) asked rhetorically, 'to what extent can helminth communities in avian hosts be invaded by additional individuals or species?' In order to answer this question, the infra- and component communities of willets, *Catoptrophorus semipalmatus*, were examined in great detail in birds taken from their breeding grounds (freshwater habitat) and their wintering grounds (marine habitat). The choice of willets for the study was a priori based on their abundance plus the fact that all their enteric helminths are acquired via ingestion. Willets also undergo intercontinental migrations, which increases the potential for exposure to a variety of prey species, and they possess rich parasite faunas with tight species packing.

Neither age nor gender played a significant role in the parasite component community structure of willets (Bush, 1990). Regionally, whether from marine or freshwater habitats, willets possessed rich parasite communities. There was, however, a pronounced difference in the nature of parasite guilds in birds from the two locations. At marine sites, the primary food sources were molluscs and crustaceans while at freshwater sites, willets fed mostly on amphipods and aquatic insects. The birds have two guilds that were common at both sites (mucosal absorbers and a nematode guild). Communities in willets from freshwater sites were dominated by absorbers and had reduced numbers of trematodes. In contrast, birds from marine habitats were dominated by trematodes and lumenal absorbers occurred in low numbers. Further, the data were interpreted to mean that the communities in each habitat type were saturated and that the switch in guild type was simply a function of the external environment. The scenario for community development envisaged was based primarily on the 'supplying' of helminths from the local habitat. This was followed, then, by the application of appropriate screens and filters (Holmes, 1987, 1990) that 'will determine what subset of those helminths available will actually colonize any given species' (Bush, 1990).

7.3.4 Mammals

The major determinants of component community in mammals include host age, gender, habitat, and season (Kisielewska, 1970; Haukisalmi, Henttonen and Tenora, 1987; Pence, 1990). Each of these determinants seems to be of importance in one system or another. Pence (1990) investigated factors that affected helminth communities in coyotes, *Canis latrans*, from the Brushlands of southern Texas. Significant, age-related, and seasonal effects on the mean number of parasite species and on the ranked density values of total individual parasites were observed. These results were attributed, at least in part, to the greater number of species and individual parasites in juvenile coyotes collected during the fall months. The increasing diversities and numbers with age also result from helminth parasites in coyotes being retained for long periods and not being lost annually as they frequently are in migratory birds, for example.

In investigating the component communities of white-tailed deer, *Odocoileus virginianus*, black bear, *Ursus americanus*, and coyotes from a variety of habitats in North America, Pence (1990) concluded that the major factor in generating species diversity was the variety of habitats used by the different host species, a situation not unlike that described by Bush (1990) for component parasite communities in willets from freshwater and marine habitats. Pence (1990) reported that coyotes in more xeric southern Texas (excluding the Gulf coast) tend to have species of parasites that use terrestrial arthropods as intermediate hosts and that this dominance pattern is reduced in populations from more northerly latitudes where they are replaced by parasite species that use vertebrate intermediate hosts.

7.3.5 Summary

There are two problems in viewing determinants of component and compound helminth communities from the standpoint of the host's taxonomic status, e.g., fish, amphibian, reptile, bird or mammal. First, there is extensive variability within each group. For example, some birds are aquatic and some are terrestrial and most evidence presently suggests that

Table 7.1 Impact of host-related factors on various characteristics of parasite infra-, component and compound communities

Host-related factors	Characteristics of the Helminth community							
	Number of individuals	Number of species	Niche size	Core species	Specialist species	Generalist species	Site specificity	Interactive/ isolationist
Host abundance								
intermediate host	higher host density = higher number of parasites	higher host number = more species of parasites	variable	higher host number = more core species	variable	higher host number = higher parasite number	not predictable	variable
definitive host	higher density = higher number of parasites	higher host number = more species of parasites	variable	increase in number	variable	higher host number = higher parasite number	not predictable	variable
Biotic interaction								
competition	often reduces numbers	often reduces numbers	narrower niche width	not predictable	not predictable	not predictable	increase in specificity	interactive
immunity	may reduce numbers	may reduce numbers	not predictable	not predictable	not predictable	not predictable	not predictable	may reduce interactions
Endothermy	potentially exposed to greater numbers	potentially exposed to more species	not predictable	usually more core species	not predictable	usually more generalists	not predictable	not predictable
Complexity of gut	variable	more complexity usually increases number of species	variable	more complexity >core species	more complex >specialists	more complex >generalists	not predictable	not predictable
Diet								
herbivory	low	low	not predictable	low	more	few	not predictable	typically isolationist
omnivory	highly variable	highly variable	not predictable	highly variable	variable	variable	not predictable	variable
carnivory	highly variable	highly variable	not predictable	highly variable	variable	variable	not predictable	variable

Table 7.1 (continued)

Characteristics of the Helminth community

Host-related factors	Number of individuals	Number of species	Niche size	Core species	Specialist species	Generalist species	Site specificity	Interactive/ isolationist
Breadth of diet	not important	if wide, then usually with large number of species	not predictable	if wide, then usually more core species	variable	if wide, then usually more generalists	not predictable	wide diet = potential interaction
Life style								
aquatic	variable, but potentially high	highly variable	not predictable	highly variable	variable	variable	not predictable	variable
terrestrial	variable, but often low	variable, but often low	not predictable	highly variable	variable	variable	not predictable	variable
Host vagility								
low	typically low	low	not predictable	typically few	not predictable	typically few	not predictable	isolationist
high	typically high	typically high	not predictable	typically many	not predictable	typically many	not predictable	often interactive

aquatic birds have richer parasite faunas than terrestrial species. This sort of argument could be extended to other vertebrate taxa, including mammals, amphibians, and reptiles. Second, there are too few studies on which to make accurate judgements with respect to the exact influence of the determinants. Thus, most of the published work on bird communities is based on freshwater and marine species and little has been done with terrestrial species. This could easily produce biased conclusions.

An attempt has been made to summarize what is known about various aspects of parasite community ecology. The summary (Table 7.1) was created as a set of predictions regarding the nature of determinants and how they individually impact the infra-, component, and compound parasite community structure. For example, high densities of intermediate hosts usually indicate large numbers of individual parasites because of enhanced transmission probabilities, while low densities suggest the opposite. Intra- or interspecific competition would have the tendency to reduce parasite densities. The effect of endothermy would be to enhance transmission probability because of increased feeding intensity. Gut complexity may, or may not, be an important factor in determining the number of individual parasites present, but it may affect the number of niches available for colonization. The nature of the host diet is generally regarded as an important determinant, although it is not always a good indicator of the number of parasites that might be present.

Each of these determinant factors should be considered within the context of the number of species present, the occurrence of core, specialist, and generalist species, site specificity, and the interactive/isolationist continuum. As indicated above, however, data for such factors are limited. On the other hand, as more and more information is gathered, an effort should be made at some point to attempt multivariate analyses to see if determinants such as life style, habitat, diet, vagility, etc., will produce a more realistic clustering of various helminth groups and communities.

7.4 HABITAT VARIABILITY (SUCCESSION) AND PARASITISM

Before assessing the relationships between the variability of aquatic habitats and compound parasite communities, it is first necessary to describe certain of the physical and biological characteristics of a typical lentic habitat in the north temperate zone. Lentic habitats were chosen for analysis because most of the research that has been done at the compound parasite community level has been conducted in lakes rather than streams, and most of it has been accomplished in the northern temperate areas of the world.

If oxygen concentration and temperature in mid-summer are measured from the surface to the bottom at 1 m intervals in a hypothetical (and pristine) lake in the northern part of the north temperate zone, the data will appear as seen in Figure 7.1(a). If the graph is carefully analysed, a number of important features will become apparent. For example, the temperatures at the surface and bottom are substantially different. At the top, they are in the range 26–28°C, while at the bottom the temperature approximates 4°C. There is also a very sharp temperature gradient between 9 and 12 m below the lake's surface. Because of this sharp temperature gradient, the lake is said to be **thermally stratified**. The upper part of the lake is called the **epilimnion**, the middle part is the **metalimnion**, and the bottom is the **hypolimnion**. The metalimnion also is sometimes referred to as the **thermocline** because of the marked temperature gradient that occurs there.

The temperature gradient also creates a very strong density gradient. One of the unique qualities of water is that it is most dense when its temperature is 3.96°C. As water temperature increases or decreases around this point, it becomes less dense. The density gradient created in the thermocline is exceedingly powerful, so much that water in the epilimnion and in the hypolimnion does not mix as long as the lake remains thermally stratified. In effect, there are two lakes. However, as the autumn months pass in the north temperate zone, less and less infrared radiation from the sun strikes the surface of the lake. Simultaneously, heat radiates away from the surface of the lake at a rate that is much more rapid than it is gained and temperatures in the epilimnion decline. Eventually the density gradient in the thermocline becomes less pronounced and then, at some point in the autumn, the gradient disappears completely and water from the epilimnion and the hypolimnion mix. This is referred to as **autumn turnover** or **mixis**. At this point, the temperatures at the top and bottom are isothermal (Figure 7.1(b)). Then, when the lake ices over in winter, it becomes stratified again (Figure 7.1(c)). In the spring, the ice melts and disappears, leaving an isothermal temperature profile for a brief period (Figure 7.1(d)). As the infrared radiation increases in the spring, water temperatures begin to rise and the lake again stratifies thermally. Lakes that undergo two periods of stratification and two periods of mixis are **dimictic**. Many lakes in more southern areas of the north temperate zone do not stratify in the winter and their vertical thermal profiles remain isothermal. These are **monomictic** lakes, turning over but once each year.

The impact of temperature on other physical characteristics and on the biology of lake systems is profound. Oxygen profiles, for example, are also shown in Fig. 7.1. In the summer, epilimnetic water with temperatures of 28°C, will carry about 8 mg/l of oxygen, while the hypolimnetic water will hold approximately 12 mg/l of oxygen. This is simply because water will dissolve more oxygen (or any gas) at lower temperatures and at higher

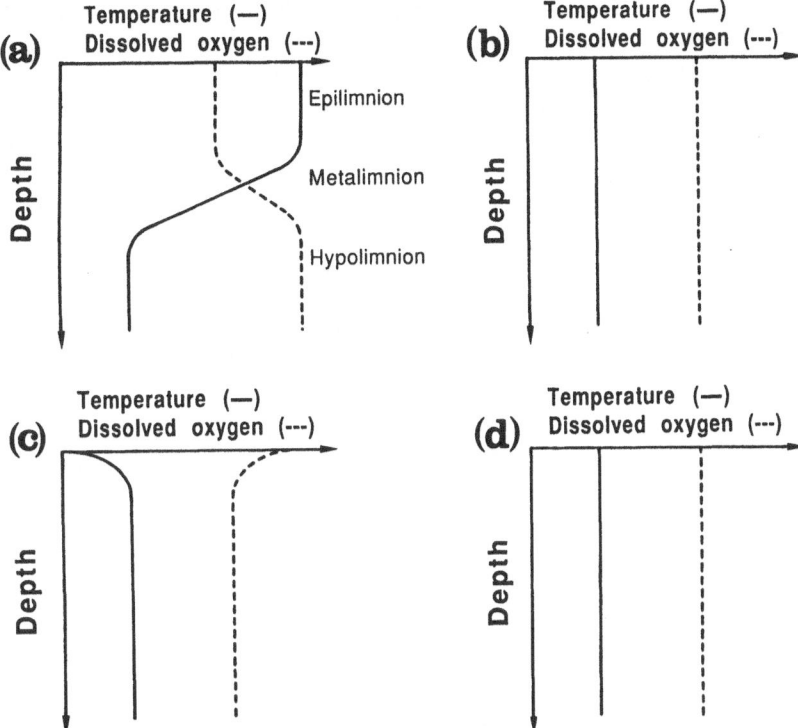

Figure 7.1 Oxygen and temperature profiles in a northern temperate, oligotrophic lake throughout the year: (a) middle of summer; (b) autumn, just after mixis; (c) middle of winter; (d) spring, just after mixis.

pressures, both of which are characteristic features of the hypolimnion in the hypothetical lake. Most important to much of the biological component of the lake, the hypolimnion carries oxygen throughout the period of thermal stratification.

The aquatic system just described would be **oligotrophic** in character. In addition to the typical physical parameters just described, the biological character of an oligotrophic ecosystem is also distinct. For example, such a lake will be relatively unproductive in terms of the amount of carbon fixed through photosynthetic processes. The oligotrophic lake will possess a widely diverse flora and fauna, although the population densities of those species that are present will not be very high. Finally, since the lake will also have oxygen in the hypolimnion throughout the summer months, it can, and probably will, support a cold-water fishery.

Ecological succession can be described as the natural process by which a community changes and matures through time; the process is non-cyclic and is effected by the organisms within the community itself. Ecological succession will alter the hypothetical oligotrophic lake in a radical manner. Succession in a freshwater system such as this is called **eutrophication**. Under natural conditions it will take place within an evolutionary time scale, that is to say, during the course of perhaps hundreds or, more likely, thousands of years. On the other hand, eutrophication can be greatly speeded up through the influence of man. Indeed, many of the physical and biological changes that would normally occur in the long term can be made to take place within the course of a few years. The manifest physical changes that occur in the ecosystem during eutrophication will dramatically affect the structure of the free-living community. As in the case of any form of succession, these physical changes are the result of biological processes that generate eutrophication itself. The compound parasite community within an ecosystem under-going eutrophication is no less affected than the free-living one. The essence of these changes and the forces affecting them have been the focus of a number of studies over the years and are worth considering at this point because they provide some insight into the structure and dynamics of compound parasite communities.

In order to examine these forces, it is first necessary to understand the nature of eutrophication itself. Primary production in an aquatic eco-system is related to a large number of factors. However, nutrient loading, mostly in the form of phosphates, is the limiting factor in triggering and sustaining the eutrophication process. In **cultural eutrophication** (eutrophication created by the activities of man), the sources of phosphate coming into an aquatic ecosystem may vary widely, but can include faulty septic systems, fertilizer run-off, pulp mill operations, concentrated live-stock feeding facilities, raw or partially treated sewage, industrial waste, etc.

The significance of phosphate as a limiting factor is related to the role it plays in DNA and ATP synthesis by plant cells. When phosphates are added to an aquatic system, primary production by phytoplankton and other vegetation increases. Phytoplankton turnover is rapid and, when individual cells die, they sink down to the substrata of the hypolimnion where they are decomposed by bacteria. The decomposition process is aerobic, meaning that oxygen in the hypolimnion is required for it to occur. Oxygen that is consumed through the aerobic degradation of dead phytoplankton in the hypolimnion cannot be replaced until the period of autumn turnover because during the time of thermal stratification, the water in the hypolimnion and epilimnion does not mix. As nutrient loading progresses, the speed at which oxygen disappears from hypo-limnetic water also increases until, at some point during the summer months, it can be completely eliminated (Figure 7.2).

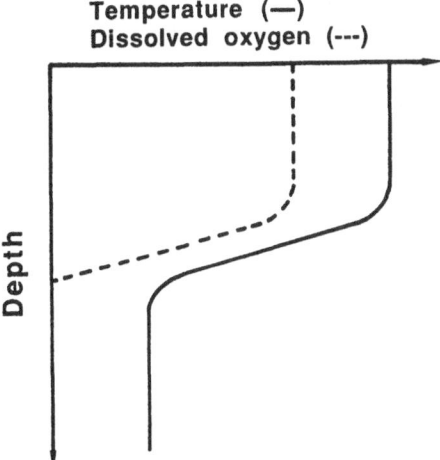

Figure 7.2 Oxygen and temperature profiles in a northern temperate, eutrophic lake during the late summer when the hypolimnion is completely anoxic.

The impact of hypolimnetic anoxia is enormous with regard to composition of the flora and fauna in a lake. Organisms that live in the hypolimnion must either adapt to anaerobic conditions, migrate into warmer water and adjust metabolically, or become locally extinct. The outcome for benthic and planktonic invertebrates and fishes in an anoxic hypolimnion is usually local extinction. The effects of hypolimnetic anoxia will extend to the epilimnetic community as well because many organisms within the latter zone prey extensively on various benthic insects that live for periods of time in hypolimnetic substrata and have emergent stages in their life cycles, e.g., mayflies, midges, etc. In the epilimnion, the phytoplankton community will also be changed; some oligotrophic species will be out-competed and eliminated by eutrophic species that thrive when the phosphate concentration is increased. In certain situations, blue-green algae will become abundant and may create physical conditions that are incompatible for survival of some species of zooplankton and even fishes. When the composition of the phytoplankton community changes, so will that of the zooplankton community.

Another group of organisms that is quite sensitive to eutrophication includes the molluscs, not only those that normally occur in hypolimnetic substrata, but also those that live in the more shallow epilimnetic substrata. It is worth noting that the physical changes associated with the eutrophication process can be reversed if nutrient loading within the system is reduced; in other words, the desirable pristine character of an oligotrophic lake can be restored. However, the emphasis here is on reversal of the adverse physical changes, because the original biotic character of the lake cannot be restored. This is primarily because many species become locally extinct and their re-introduction, whether by design or accident, is not always successful.

The question for many parasitologists working in lentic habitats has been, to what extent does the eutrophication process affect compound parasite community structure and, if it does, what is the process by which such changes are manifested? The earliest investigations to address this problem were those by Wisniewski (1955, 1958) who examined eutrophic Lake Druzno and mesotrophic Lake Goldapivo in Poland. The effort included the examination of 170 species of potential intermediate and definitive hosts; parasite flow was established as being towards the dominant predators in the system which were fish-eating birds and mammals, primarily the former. Thus, nearly 75% of all the parasites in the lakes were found to mature sexually in birds and only 17% in fishes. In these systems, therefore, allogenic species were found to be the most abundant.

A slightly different approach to the same question was taken by Esch (1971) who investigated the parasite faunas in species of centrarchid fishes and in molluscs from oligotrophic Gull Lake and eutrophic Wintergreen Lake in southwestern lower Michigan (USA). In general, his observations were in agreement with those of Wisniewski (1955, 1958). In effect, the parasite fauna in the oligotrophic system was dominated by autogenic species, while in the eutrophic system allogenic species were more abundant. Esch (1971) speculated that eutrophic systems were 'open' and subject to more opportunities for aquatic–terrestrial interaction such as may occur between predatory birds and their piscine prey. Oligotrophic systems, on the other hand, were described as more 'closed', with less aquatic–terrestrial interaction and less opportunity for the transmission of parasites through piscine intermediate hosts to avian definitive hosts. The hypothesis was, therefore, essentially linked to the nature of predator–prey relationships; according to Esch (1971), this was the primary determinant of compound parasite community structure and dynamics in oligotrophic versus eutrophic ecosystems.

Chubb (1963) and subsequently Kennedy (1978a) compared the compound parasite communities in several fish species in lakes from a number of locations in England and Wales. The composition of the fauna was viewed more in terms of individualistic patterns for the various lakes rather than from the standpoint of the trophic conditions of a given body of water. Kennedy (1978a) emphasized that local conditions, stocking programmes, fish migration, etc., are unique for each lake and, therefore, are principally involved in structuring the compound parasite communities in brown trout, *Salmo trutta*, within the various systems. On the other hand, Kennedy and Burrough (1978) examined the compound parasite communities from brown trout and perch, *Perca fluviatilis*, in a number of the same lakes and were able to discern patterns that matched the trophic status of the lake, not unlike that of Wisniewski (1955, 1958) and Esch (1971). However, they did observe an exception to the pattern and used it to reinforce Kennedy's (1978a) position regarding the individuality of lakes and the composition of compound parasite communities. Halvorsen (1971) and Wootten (1973) rejected the notion that parasite faunas are representative of a lake's trophic status or limnological condition. Instead, both investigators asserted that the composition of parasite communities is reasonably constant within a given host species from one lake to another.

In a slightly different approach, a major portion of the compound parasite community structure from 10 different species of fishes was examined (Leong and Holmes, 1981) in oligotrophic Cold Lake, Alberta (Canada). They concluded that the compound parasite community's composition is characterized by parasites of the dominant host species and that the community should be viewed within the context of the number of individual hosts and parasites instead of the number of species. When considered within this framework, salmonids and salmonid parasites were dominant. A significant premise in their position was related to the so-called **exchange hypothesis**. Thus, 'in Cold Lake, an exchange of parasites between hosts played an important part in enriching the parasite community in each species of host and in determining the similarities in the parasite communities between host species'. A similar pattern was observed for the rockfish component communities off the coast of British Columbia, Canada (Holmes, 1990).

7.5 THE BIOCOENOSIS AND PARASITE FLOW

According to Noble *et al.* (1989), a biocoenosis is defined as 'a community of living organisms whose lives are integrated by requirements imposed by a circumscribed habitat and by mutual interactions'; it is also sometimes considered in terms of a **species network**. The biocoenosis is certainly an appropriate concept for free-living species since it has clear implications for food webs, predator–prey interactions, and energy flow at the ecosystem level. The concept can be adapted for parasite flow when the compound community is viewed in terms of a **parasitocoenosis**. As was just discussed, such an approach was adopted by Wisniewski (1958) for Lake Druzno in Poland; the goal was to determine if there was any sort of 'directional' movement of parasites through hosts within the ecosystem. After sampling many thousands of potential vertebrate and invertebrate hosts, it was concluded that parasites were moving primarily from fishes, amphibians, and certain invertebrates, to birds. In other words, parasites were following food chains dominated by the top predators within the system. The dominance pattern was linked to the eutrophic character of the lake. Many eutrophic systems, but certainly not all, have extensive littoral zones that are favourite habitats for the foraging of fish-eating birds. Such conditions are generally conducive to transmission patterns that involve shore- and wading birds. The theme regarding the component community and the importance of habitat variability as a determinant of structure thus also becomes important in the flow of parasites within the compound community.

Holmes (1990) emphasized the importance of food webs in the structuring of compound communities in marine teleosts, but also pointed out the real paucity of research in this area. One of the features of importance in structuring compound communities of marine reef fishes, particularly rockfishes, was the preponderance of generalist species in the component communities. It was surmised that each of the more common fish species presumably would bring its own cadre of parasites to the reef and that these would be exchanged with many of the other species of hosts occupying the site. As a consequence of such exchanges, the species richness in a given host community would be greatly enhanced.

A similar explanation was provided by Neraasen and Holmes (1975) for the flow of parasites through four species of geese in Canada. Basically, they proposed that the geese, on arrival at their breeding ground, released infective agents of parasites into the habitat so that over time a pool of infective agents would accumulate. The pool provided a source of infective agents for members of the same or different species of potential hosts. Then, according to this scenario, an individual 'host will contribute to, and draw from, the infective pool in accordance with its abundance and suitability as a host'. These observations led to the conclusion that factors such as the density of potential hosts, the migration routes used by the different host species, diet, and degree of host specificity shown by the different species of parasites, all contributed to the nature of the compound parasite community in the area.

7.6 THE ALLOGENIC–AUTOGENIC SPECIES CONCEPTS

The allogenic and autogenic species concepts, defined in an earlier section of this chapter, were introduced by Esch *et al.* (1988) as part of an effort to examine and explain the helminth colonization patterns and parasite community structure among freshwater fishes in Great Britain. One of the important observations made by these authors was that colonization abilities by allogenic species were significantly greater than those of autogenic species. The basis for such an assertion was simply related to the fact that allogenic parasites mature sexually in birds and mammals that usually have the capacity to migrate widely. Autogenic species are confined to fishes which are generally restricted in the distance they are able to migrate. Autogenic species would tend to have highly patchy distributions and be more unpredictable in occurrence than allogenic species, primarily because of the more limited colonization abilities in the former group.

Parasites in three groups of British fishes were characterized according to their allogenic/autogenic dichotomy (Esch *et al.*, 1988). The first included the salmonids, that were dominated by autogenic species. These species also accounted for a substantial part of the similarity within and between the various study sites. By contrast, cyprinids were dominated by allogenic species; this also accounted for much of the similarity within and between collecting areas. Anguillids were intermediate, with neither allogenic or autogenic species showing clear dominance. They concluded by stating, 'more importantly, our conclusions relating to the different contributions of autogenic and allogenic species to community structure in the three groups of fish hosts are novel and unexpected, and are not simply logical consequences of the dispersal abilities of the helminths themselves. An understanding of colonization strategies, therefore, including the separation of autogenic and allogenic species and recognition of the different roles of both transient/resident and euryhaline/stenohaline host species, provides important clues for evaluating the stochastic nature of parasite community structure'.

8 *Biogeographical aspects*

8.1 INTRODUCTION

Brown and Gibson (1983) defined biogeography as 'the science that attempts to describe and understand the innumerable patterns in the distribution of species and larger taxonomic groups'. These patterns of distribution consistently have been shown to be non-random and require explanations in terms of process(es), after which the events that led to the present distributions can be reconstructed. Biogeographical studies necessitate a significant amount of diverse information from disciplines such as systematics, ecology, evolution, palaeontology and geology.

Biogeographical research has been conducted using both historical and ecological approaches. Historical biogeography attempts to understand the sequences of origin, dispersal, and extinction of taxa and explain how geological events such as continental drift, glaciations, emergence and submergence of land masses, etc., have influenced the present patterns of distribution. Ecological biogeography, on the other hand, focuses mostly on extant species and attempts to account for distribution patterns in terms of interactions between organisms and their physical and biotic environment now, and in the recent past. In general, the thrust is to identify the processes that limit the distribution of a species and that maintain species diversity; in doing so, explanations are generated for phenomena such as gradients of species richness and colonization patterns (Myers and Giller, 1988).

The beginning of biogeography as a discipline can be traced back 200 years to the great explorer and naturalist, Alexander von Humbolt. His studies, beginning in 1805, were the first to conceptualize and quantify the primary role of climate in the distribution and morphology of plants around the world. The study of animal distribution lagged behind phytogeography several years. The first significant zoogeographical effort was made by W.L. Sclater in 1858; at that time, he proposed a number of faunal regions based on the distribution of birds. Interestingly, the regions he created are very similar to the present day classification of biogeographical realms, i.e., Nearctic, Palearctic, Oriental, Australian, Ethiopian, and Neotropical. At least 20 more papers were published before the turn of the century, marking the beginning of a prolific era in the biogeographical investigation of free-living organisms.

Biogeographical studies on parasites began in 1891 with the work of H. von Ihering, who recognized the significance of parasitic organisms in the zoogeography of host species. He noted the presence of closely related species of *Temnocephala* (commensalistic turbellarians) on species of freshwater crustaceans on both sides of the Andes and in New Zealand. Later, Metcalf (1929, 1940) examined the opalinid protozoans in leptodactilid frogs. He observed that the same genus of ciliates (*Zelleriella*) was present in frogs from South America and Australia, but was absent in frogs and toads from the Northern Hemisphere. Metcalf reasoned that leptodactilid frogs had not migrated into Eurasia because none of the Eurasian amphibians had *Zelleriella* as part of their parasite faunas. On the other hand, non-leptodactilid frogs in Brazil, such as *Bufo* and *Hyla*, were parasitized by *Zelleriella*. He speculated the parasites had been acquired as a result of host capture from sympatric leptodactilids. Metcalf postulated a close relationship between South American and Australian leptodactilid frogs based on the similarity of their opalinid protozoans. He also reviewed some of the previous studies on avian lice which led him to suggest close relationships between certain species of European and North American birds, as well as between the rhea and the ostrich (two species then thought to be unrelated). This approach in establishing 'phylogenetic' relationships between host species, based on the similarity of their parasites, was referred to as the 'von Ihering method' of biogeography.

The next significant set of contributions to the biogeography of parasitic organisms was made by Manter (1940, 1955, 1963). His interests were in the geographical distribution and affinities of digenetic trematodes along the Atlantic and Pacific coasts of Central America, of trematodes in marine fishes on a global scale, and on the zoogeographical affinities of trematodes in freshwater fishes of South America. Since then, several important communications have been published, with a virtually exclusive emphasis on marine parasites. Polyanski (1961b), for example, provided a significant contribution to the zoogeography of parasites in marine fishes of the Soviet Union. Szidat (1955), Kabata and Ho (1981), Fernández and Durán (1985) and Fernández (1985) used the distribution of parasites in hakes, *Merluccius* spp. (bentho-demersal marine fishes) to suggest their possible place of origin and dispersal routes. Lebedev (1969) distinguished 10 marine zoogeographical regions from a two-dimensional geographical perspective, based on the distribution of 216 genera of monogeneans, and 420 genera of digenetic trematodes. Unfortunately, because the degree of knowledge was highly variable for different regions of the ocean and some of them still remain understudied, i.e. the Pacific and Atlantic coast of South America, extensive areas could not be included in his classification. Finally, Noble (1973), Campbell, Haedrich and Munroe (1980) and Campbell (1983, 1990) provided important information regarding the distribution of parasites in deep sea fishes.

Thus far, Rohde (1976a, 1978a, b, 1980b, 1982, 1984, 1989) has contributed the most to our understanding of the ecological biogeography of marine parasites. His work on marine monogenetic and digenetic trematodes has been the nucleus for expansion in this arena. Similar approaches in terrestrial or freshwater ecosystems are quite preliminary and, except for a few studies (Price and Clancy, 1983; Gregory, 1990), comprehensive efforts in these areas have not been made.

8.2 FACTORS AFFECTING THE GEOGRAPHICAL DISTRIBUTION OF PARASITES

As indicated, the geographical distribution of parasites and their hosts is not random. The processes involved in shaping these patterns on a global scale include:

1. the division and separation of biota by natural geographical barriers;
2. the fragmentation of land masses that affected their original geographical distribution;
3. the ability to move and increase their distributional range actively or passively, regardless of vicariant events; and
4. the evolutionary time frame within which these events have occurred.

Dispersal success will then be influenced by more specific abiotic and biotic factors that affect the ability to cope and survive under new environmental conditions once a barrier has been breached. Abiotic factors will have various effects depending on the nature of the new habitat that an organism colonizes. For many free-living organisms, competition, predation, and parasitism have been recognized as important biotic determinants for animal distributions; competition and predation also are significant for many parasitic organisms.

Depending on the degree of specificity for a given species, parasite distributions are also affected by host distribution. On the other hand, as will be discussed later in this Chapter and in Chapter 9, parasites may also influence the geographical distribution of their hosts in some cases. In these situations, the exclusion of one or more host species via parasite-mediated competition may occur and thereby increase the geographical range of another host species.

As scale dimensions decrease, the variables affecting host and parasite distributions become more specific. Thus, pH, dissolved oxygen, water chemistry (the presence of calcium, for example), flow rates, trophic condition, substrata types, can affect animal distributions in freshwater ecosystems; the ephemeral character of a body of water also may play an important role. In terrestrial environments, the soil type and composition, moisture content, humidity, vegetation, and altitude are determinants of biogeography. Finally, in marine systems, temperature, salinity, depth, physiography of the substrata, and currents (mainly as distinctive bodies of water), are all important in affecting distributions.

The host itself is one of the most significant components in influencing the acquisition and establishment of parasites. Host sex and reproductive stage, age, feeding habits, vagility, and immune capabilities, can impact parasite distributions. Parasite–parasite interactions may also affect distribution via competition; thus, antagonistic intramolluscan inter-actions involving the larval stages of digenetic trematodes may dictate the structure of parasite infracommunities in some cases. Other parasite–parasite interactions may be indirect, such as when mediated by the host immune system through enhanced immunity to secondary invaders, or by non-reciprocal cross immunity.

8.2.1 Freshwater systems

Some of the factors affecting parasite distribution in freshwater systems have been described by Bailey and Margolis (1987) in their study of the parasite fauna of juvenile sockeye salmon, *Oncorhynchus nerka*, in a series of lakes along the northwest coast of North America. They reported that the parasite fauna was similar among lakes with similar trophic status and within the same biogeoclimatic zones. Their analysis also revealed that parasites with similar modes of transmission within a geographical range tended to cluster. This was particularly true for *Philonema agubernaculum*, *Diphyllobothrium* sp., *Eubothrium* sp. and *Proteocephalus* sp., helminth parasites that are transmitted by common planktonic intermediate hosts such as *Cyclops* spp. They observed a substantial variation in the pattern of infection over time in relatively mesotrophic lakes and less so in oligotrophic lakes. This finding supports the allogenic–autogenic species concept (Esch *et al.*, 1988) and its relationship to the trophic status of a given lake (discussed in Section 7.5). Thus, in oligotrophic systems most of the parasite species were autogenic, completing their life cycles within the confines of the system; this allowed for a degree of 'predictability' from year to year in terms of species distributions. Mesotrophic lakes, on the other hand, had an increased number of allogenic species whose life cycles are completed in birds and mammals that are not confined to the lakes. In these systems, the presence of many larval parasites in fishes will depend on the migration patterns and residency times of their 'ephemeral' definitive hosts. Bailey and Margolis (1987) concluded that 'evidently, geography influences the characteristics of the parasite fauna, but the trophic status of the lake and many biotic variables clearly have strong influences on the parasite faunas studied'.

A similar pattern was noted by Chubb (1970) in his study of the parasite fauna in British freshwater fishes, which emphasized aspects of parasite diversity in terms of host distribution and lake trophic status. He observed that the distribution patterns of acanthocephalans, for example, were determined primarily by the presence and absence of isopod inter-mediate hosts. He reported that salmonid fishes and their parasites were dominant in oligotrophic lakes while cyprinids and their parasites dominated eutrophic systems. Mesotrophic lakes possessed mixed host species populations and had a greater range of both host and parasite species (Fig. 8.1).

The significance of water flow and drainage systems was investigated by Arai and Mudry (1983) within the context of potential parasite movement across the Continental Divide in North America as a consequence of a possible water diversion in northeastern British Columbia, Canada. They

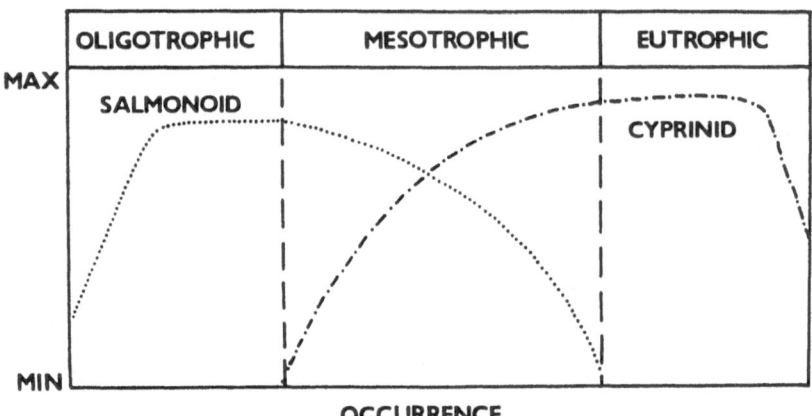

Figure 8.1 Occurrence of salmonid and cyprinid fishes in relation to lake types. Salmonid fishes are dominant in oligotrophic waters, cyprinids in eutrophic lakes. Mesotrophic lakes contain mixed populations and have the greatest range of species. (With permission, from Chubb, 1970.)

found 88 parasite species in 20 species of fishes from both the Arctic and the Pacific drainage basins. Twenty-six parasite species had disjunct distributions; 17 occurred in hosts from both drainages systems. This suggested an association of the parasites within a given drainage system and indicated a restricted dispersal across the Continental Divide even though nine species had direct life cycles which, presumably, should facilitate wider distributions.

A determinant typically overlooked when dealing with trematode distributions in freshwater systems is the level of calcium (hardness) in the water. This is an important limiting factor for the establishment and maintenance of molluscan communities. The diversity and abundance of molluscs, in turn, constrains the occurrence and distribution of most trematode species.

Ectoparasites of freshwater fishes (protozoa, monogeneans, and copepods) are directly affected by salinity levels. Polyanski (1961a) reported that some protozoans, e.g., myxozoans and ciliophorans, disappeared completely as salinity increased. The best examples for the effect of salinity on parasite distributions are provided, however, by anadromous and catadromous fishes such as *Oncorhynchus* and *Anguilla*; the former fishes lose their freshwater ectoparasites soon after they reach marine waters and the latter after entering freshwater. Loss of ectoparasites from these hosts is rapidly followed by loss of their enteric helminths. There are, however, some exceptions to this pattern. Among the copepods, for example, some species in the genus *Ergasilus* show a wide range of salinity tolerance. Thus, Johnson and Rogers (1973) grouped the copepod species present in several Gulf of Mexico drainage systems into five categories; these included coastal, estuarine, coastal and inland, estuarine and inland, and inland species. Those with greater tolerance for salinity were the coastal and inland species (*E. versicolor* and *E. clupeidarum*) and the estuarine and inland species (*E. arthrosis*), although their degree of host specificity was not extremely high. Interestingly, these copepods are normally associated with hosts that have a high tolerance to salinity changes, e.g., *Mugil cephalus*, *Alosa* spp. and *Derosoma* spp.

8.2.2 Terrestrial systems

A significant number of 'terrestrial' parasites are dependent on freshwater systems. Many cestodes, and most trematodes and some acanthocephalans, rely almost exclusively on aquatic or semi-aquatic intermediate hosts in completing at least one step of their life cycles. In these cases, the factors affecting their distribution and dispersal in terrestrial definitive hosts will be constrained by processes operating within aquatic systems.

Most cestodes in exclusively terrestrial animals such as lizards, some turtles, birds and mammals rely on predator–prey interaction in at least one stage in their life cycles. As a result, patterns of parasite distribution on a relatively local scale are determined by the presence of both intermediate and definitive hosts; these, in turn, are affected by local climatic and physiographical conditions. Ultimately, these patterns will be influenced by biotic factors related to the host such as sex, age, feeding strategy, vagility, and foraging areas (Dogiel, Petrushevski and Polyanski, 1961).

Distributions of parasites with direct life cycles, e.g., some nematodes and arthropods such as lice and fleas, depend to a great extent on special environmental conditions, mainly temperature and humidity. The latter factor is extremely significant for nematodes with free-living stages since humidity conditions, temperature, and substrata types directly affect the survivorship and transmission of rhabditiform and filariform larvae.

8.2.3 Marine systems

The abiotic factors affecting the distribution of marine parasites are similar to those in freshwater systems. They are, however, determined mostly by **temperature–salinity (T–S) profiles** and their association with specific masses of water.

A now classic example of the effect of temperature on parasite distribution in the ocean is that of the hemiurid trematodes, *Lepidapedon sachion*, *L. elongatus* and *Derogenes varicus*. Manter (1934) found these trematodes at depths ranging from 270 to 550 m off the coast of Florida (USA). However, in more northern latitudes, these species are abundantly distributed in much shallower waters. *Derogenes varicus* is truly cosmopolitan, with an almost complete lack of host specificity. Polyanski (1961a) reported it from 21 fish species in the Barents Sea at a wide range of depths. Love and Moser (1983) listed it from 49 hosts along the west coast of North America, also from a variety of depths. Hewitt and Hine (1972) recorded it in 10 fish species from New Zealand, while Suriano and Sutton (1980) and Fernández (1985) found it in bentho-demersal fishes (50–150 m) from the continental shelf off Argentina and Chile in South America. This type of bipolar distribution has been reported many times for free-living benthic species that follow isothermal bands of water across the ocean floor (Brown and Gibson, 1983). Cold water (4°C) is much closer to the surface at the poles than in the tropics. These masses of polar water move from pole to pole across the Atlantic and Pacific Oceans following their submergence at the Antarctic Convergence (75°S), an oceanic frontal system delimiting the Antarctic from the sub-Antarctic. It is believed, then, that the distribution of *D. varicus* in tropical waters is determined by its association with masses of cold water that move through these areas at greater depths.

The significance of the Antarctic Convergence from a strictly oceanographic view has been widely acknowledged and so has been its impact on the distribution of marine, or marine-dependent, organisms. Its importance in parasite distribution is also becoming more apparent. Thus, Hoberg (1986) studied certain ecological and biogeographical characteristics of the acanthocephalan, *Corynosoma* spp., in Antarctic seabirds. He examined various aspects of the host–parasite relationships and the geographical distribution of different species in piscine paratenic hosts and in avian and mammalian definitive hosts. It was concluded that host–parasite co-evolution probably had an important influence on the species composition of acanthocephalans in different hosts within historical times. However, the current restricted distribution of *Corynosoma* spp. in their hosts, suggested to him that oceanographic factors such as the Antarctic Convergence could be limiting the range of some of the parasites by affecting the distribution of the amphipod first intermediate hosts and piscine paratenic hosts that are confined to specific masses of water in the southern oceans.

The potential effect of depth on parasite distribution is not yet clear because of the confounding influences of other correlated variables such as temperature, pressure, and relative abundance of food items. Temperature, however, is likely to be a significant factor. Unfortunately, no studies have been published, with the exception of that of Campbell (1983) on the parasites of *Coryphaenoides armatus*, that deal with the distribution of parasites in a given host species at different depths while simultaneously accounting for changes in other parameters. Campbell *et al.* (1980) indicated that the decrease in overall parasite diversity and abundance with depth is probably the result of changes in abundance and diversity of the free-living fauna and, thus, of potential intermediate hosts. This suggests that the processes influencing the distribution of free-living organisms with depth might also affect parasites with indirect life cycles.

8.3 PATTERNS OF DISTRIBUTION

The goal of historical biogeography is to reconstruct processes involved in the origin, dispersal and extinction of taxa and biotas. Vicariant events and dispersal patterns that affect the host and the parasite, host specificity, host capture, co-evolution, and co-accommodation, are some of the elements commonly considered as significant in these processes.

The relative importance of each one of these factors varies according to the parasite group and the host(s) involved. The validity of generalizations in this area are questionable unless their limitations are well defined. Because parasites and their hosts can provide basic information about their mutual history (the 'von Ihering method'), biogeographical analysis can be conducted on a given parasite taxon throughout its range, or it can also be pursued by examining the nature and distribution of parasites of a given host taxon. To propose biogeographical classifications based exclusively on parasite distribution (without considering the host) is probably not the best approach, given the strong influence of host specificity, co-evolution, and historical association.

Many terrestrial and freshwater species are highly restricted in their distribution because of the powerful barriers in these environments. Land barriers between bodies of water and *vice versa*, mountain ranges, and the patchy distribution of favourable habitats will limit the dispersal of species. Depending on the nature of the hosts, of course, many of these barriers can be breached. Present and historical land discontinuities such as the current structure of the continents and their historical genesis (i.e., Pangaea, Laurasia, Gondwana) and the appearance and disappearance of land bridges between them (i.e., the Bering Strait, Isthmus of Panama), are also critical factors affecting the distribution of both parasites and their hosts.

In the marine environment, patterns of distribution are affected by somewhat different physiographical factors. First, while the ocean is continuous and in constant movement, it certainly is not homogeneous. There are very distinctive masses of water whose integrity is constant, and that normally have a distinctive fauna at least in terms of their pelagic and planktonic species (McGowan, 1971; Pickard, 1975). Second, it is a three-dimensional realm and organisms are influenced by geographic location and depth. Third, factors that in the terrestrial environment constitute bridges or barriers, indicate just the opposite for marine biotas. For example, a bridge was formed between North and South America (the Isthmus of Panama) 6 million years ago and facilitated the dispersal of many terrestrial and freshwater organisms between the two continents; it also, however, caused isolation of the Pacific and Atlantic marine biotas across the tropical belt. In contrast, the submergence of Beringia significantly reduced exchanges of fauna between the Palearctic and Nearctic terrestrial realms, while it created a connection and allowed for the exchange of biotas between the north Pacific and Arctic Oceans.

8.3.1 Some patterns and problems of parasite distribution in terrestrial and freshwater habitats

There have not been many comprehensive attempts to examine zoo-geographical aspects of terrestrial parasites within the context of their host's distribution. Moreover, most of the studies are rather limited in terms of the 'known' host distributions. Because most authors have recognized the parallel nature of host and parasite distributions, many of the studies have included reference to host–parasite specificity and co-evolution, and have used parasites as 'biological markers' or indicators of possible host origins.

As already mentioned, von Ihering (1891, 1902) and Metcalf (1929, 1940) were the first to assign palaeobiological significance to the distribution of parasites and their hosts (*Temnocephala* in crustaceans land opalinids in frogs). Johnston (1912) reported that trematodes present in Australian frogs had their nearest morphological relatives in trematodes of anurans from South America, Africa, Asia and Europe, not among trematodes living in other Australian vertebrates. At the time, highly complex and contrived explanations, some involving hypothetical land bridges of different types, were necessary to explain the nature of these rather 'spotty' distributions. However, with the development of modern geology and geophysics, the notion of a dynamic earth crust with moving continental plates, and of the oceans expanding and shrinking at oceanic ridges and subduction zones, the early explanations were greatly simplified. In doing so, a reliable account of the history of the continental masses was generated (Fig. 8.2). (For a detailed explanation of plate tectonics and Continental Drift, see Dietz and Holden (1970)). It is now widely accepted that the break-up of the ancestral continent of Pangaea about 150 million years ago, led to geographical isolation (a vicariant event) and the resulting allopatric speciation of many organisms, including parasites.

Prudhoe and Bray (1982) provided a detailed account of the platyhelminth parasites in Amphibia worldwide. Although their emphasis was on systematics, they recognized that most genera of amphistome trematodes infecting recent amphibians, particularly frogs, have arisen through geographical isolation. Sey (1991) in his monograph on amphistomes agreed with Prudhoe and Bray (1982). Moreover, because he included most host groups, he was able to characterize the extent of dispersal by the parasites. Thus, amphistomes in non-mammalian hosts (mostly amphibians and reptiles) have intercontinental distributions that agree with the distribution of intermediate and definitive hosts. Moreover, these distributions can be best explained in terms of the main continental disjunctions caused by continental drift (vicariance). The distribution of amphistomes in mammalian hosts, on the other hand, is much more variable. Although the role of continental drift is still recognizable in their distribution patterns, their present distribution has been determined mostly by the greater dispersal capabilities of their hosts.

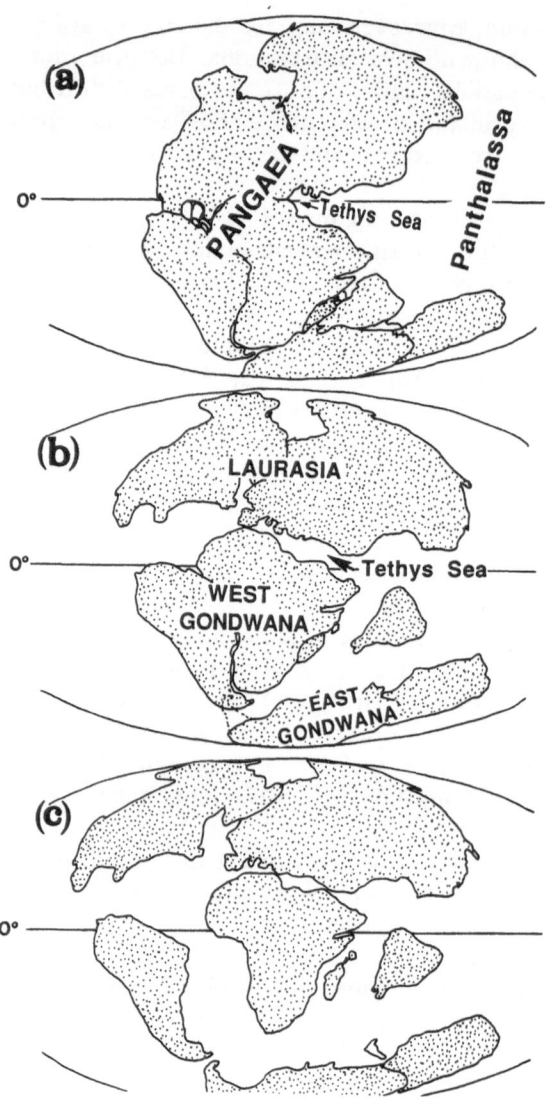

Figure 8.2 Configuration of the land masses and oceans at different past times based on actual views of plate tectonics and continental drift. (a) Supercontinent Pangaea in the Triassic, 200 million years ago. Panthalassa was the ancestral Pacific Ocean and the Tethys Sea the ancestral Mediterranean; (b) In the late Jurassic or early Cretaceous, after 65 million years of drift (135 million years ago) the Northern Pangaea became Laurasia; the Southern Pangaea, after becoming Gondwana, split into East Gondwana, West Gondwana and India. The Atlantic and Indian Oceans are already open; (c) At the end of the Cretaceous, after 135 million years of drift (65 million years ago), the Mediterranean is clearly recognizable, the Atlantic has become a major Ocean separating South America from Africa, while Australia still remains attached to Antarctica.

A rather different pattern, with vicariance and dispersal but no speciation after these events, is shown by the nematode *Rhabditis (Pelodera) orbitalis*. *Rhabditis orbitalis* is a microbotrophic nematode that, as adults and juveniles, lives in the nests of arvicolid and murid rodents throughout the Holarctic (North America, Europe and Asia). Third stage larvae, however, are parasitic and enter the orbital fluid of rodents where they remain as parasites for several days. Schulte and Poinar (1991) completed reciprocal crosses between European and American strains, corroborating the specific status of both populations. They argued that because *R. orbitalis* has an Holartic distribution, the association with the rodents must have evolved when the continents were still connected by the now submerged Bering Strait. They attributed the lack of divergence between American and Eurasian populations to the similarity in ecological conditions present on the two continents. However, it is also possible that the time that has passed since the disappearance of the land connection between Eurasia and America has not been sufficient to allow for genetic divergence.

The similarity in patterns of parasite distribution in freshwater and terrestrial parasites was demonstrated by Manter (1963) in his study of the geographical affinities of trematodes in South American fishes. Intercontinental relationships normally included related genera present in fishes in South America, Africa and India, all components of the Gondwana supercontinent. The different subfamilies of amphistome trematodes, for example, had the strongest pattern of Gondwanic distribution since they have an exclusive African–South American distribution. On the other hand, fishes from North America and Eurasia share at least eight genera and related species of digenetic trematodes not present in other areas, probably as a consequence of their past history as components of the Laurasian supercontinental fauna. Studies like those of Manter, completed 20 years ago, provided an important baseline for the development of zoogeographical studies on parasitic organisms. Unfortunately, not many investigations have followed.

Brooks (1977, 1979) combined biogeographical and phylogenetic approaches in examining the evolutionary aspects of host–parasite relationships from an historical perspective. In a cladistic analysis of plagiorchid trematodes from frogs, Brooks (1977) identified three main parasite lineages. Each was characterized by one genus whose species were distributed on a single, or at least adjacent, continent. Each genus had a different centre of origin and their dispersal tracks were similar, but not congruent. Species of *Glypthelmins*, for example, are found in Africa, Asia, and North, Central, and South America. The phylogenetic relationship among these species conforms with a pattern of vicariance of widely distributed ancestral taxa and the subsequent movement of the continents. On the other hand, the dispersal from South America to North America and from Africa to Asia, probably occurred after the continents reached their present configuration. The dispersal from South to North America probably started 65 million years ago (end of the Cretaceous) when the land bridge between the continents appeared for the first time. Brooks (1979) also examined the evolutionary relationships between crocodiles and their trematodes. Based again on phylogenetic and biogeographical analyses, he argued that the intercontinental relationships of the trematodes were coincident with the patterns of continental fragmentation since the Cretaceous; these relationships were also congruent with the biogeographic relationship of the crocodiles. The results provided a strong indication of co-evolutionary events, but also biological evidence for the proposed patterns and timing in the fragmentation of the supercontinents.

Based on studies such as these, it is important to emphasize the *relative* contribution of continental drift, geophysical events, host specificity or host–parasite associations, and local ecological conditions, in describing the observed patterns of distribution of the different groups.

8.3.2 Pattern of distribution in marine parasites.

Much information is available about patterns of distribution of marine parasite species; importantly, a substantial number of these have been studied on a more or less global scale in association with their hosts. The reasons for the difference between marine, and terrestrial or freshwater, systems are mostly related to economic factors. Seafood is a significant source of protein and an important component of the human diet almost worldwide. As a result, many studies have been undertaken to assess parasites as potential pathogens for both fishes and the human consumer. Similarly, the realization that some parasites could be used as 'biological markers' for the identification of breeding stocks, foraging areas, or migratory routes, also increased the funding for research directed toward the elucidation of these more 'practical' problems. From a purely academic perspective, however, these studies have provided valuable information about host–parasite co-evolution and the origins of some host groups. Because most of the work has been focussed on parasite dispersal in relation to patterns of host specificity and host distribution, the main thrust of that which follows will be on these factors.

The broad geographical distribution of the trematode, *Derogenes varicus*, in association with masses of cold water, has already been mentioned. There are, however, other features of some importance in its ubiquitous distribution, such as its almost non-existent host specificity for both definitive and intermediate hosts. Køie (1979) examined its life cycle and determined that cercariae from naturally infected molluscs (*Natica* spp.) were infective to calanoid and harpacticoid copepods, as well as to benthic

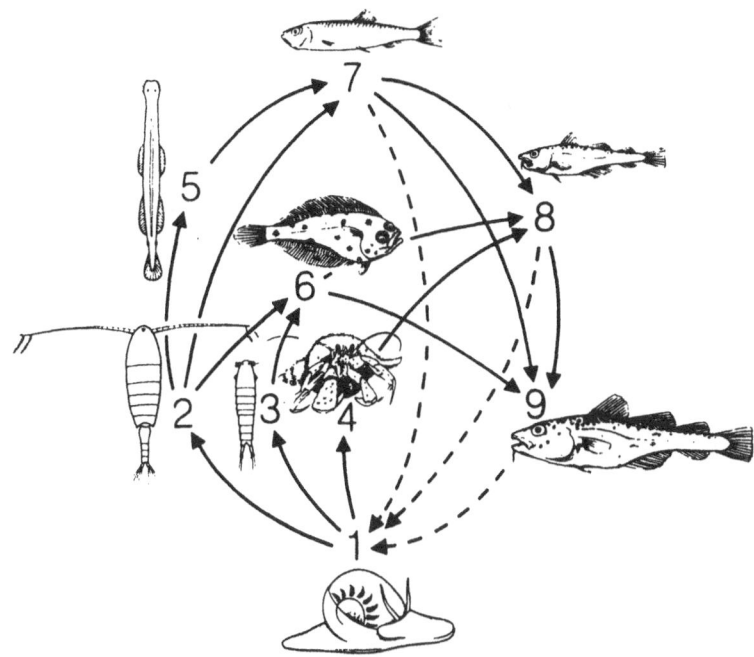

Figure 8.3 The life cycle of the trematode, *Derogenes varicus*, involving planktonic invertebrates: 1. *Natica* spp. (benthic mollusc); 2. calanoid copepod; 3. harpacticoid copepod; 4. hermit crab (benthic); 5. *Sagitta* spp.; 6. small fishes; 7. planktophagous fishes; 8. benthofagous and piscivorous fishes; 9. piscivorous fishes (large cod). (With permission, from Køie, 1979.)

hermit crabs where they develop into mesocercariae (Fig. 8.3). A diverse group of fishes with different feeding habits, both benthic and planktivorous, as well as chaetognaths (*Sagitta* spp.), become infected upon ingestion of hosts carrying mesocercariae. Definitive hosts include fish species with diverse feeding habits, ranging from planktivorous to piscivorous in both the pelagic and benthic regions. Moreover, adult worms also can be transferred from host to host upon predation of fishes infected with adult trematodes.

For parasites with complex life cycles, this type of host utilization seems to be the only possible way to achieve a global distribution. Another species exhibiting this phenomenon is the nematode, *Anisakis simplex*, whose life-cycle complexity was discussed in Chapter 5. It has a similar global pattern of distribution, with an almost complete lack of host specificity for intermediate hosts and innumerable paratenic hosts. It also has a rather broad spectrum of definitive hosts, that includes most baleen whales and marine dolphins.

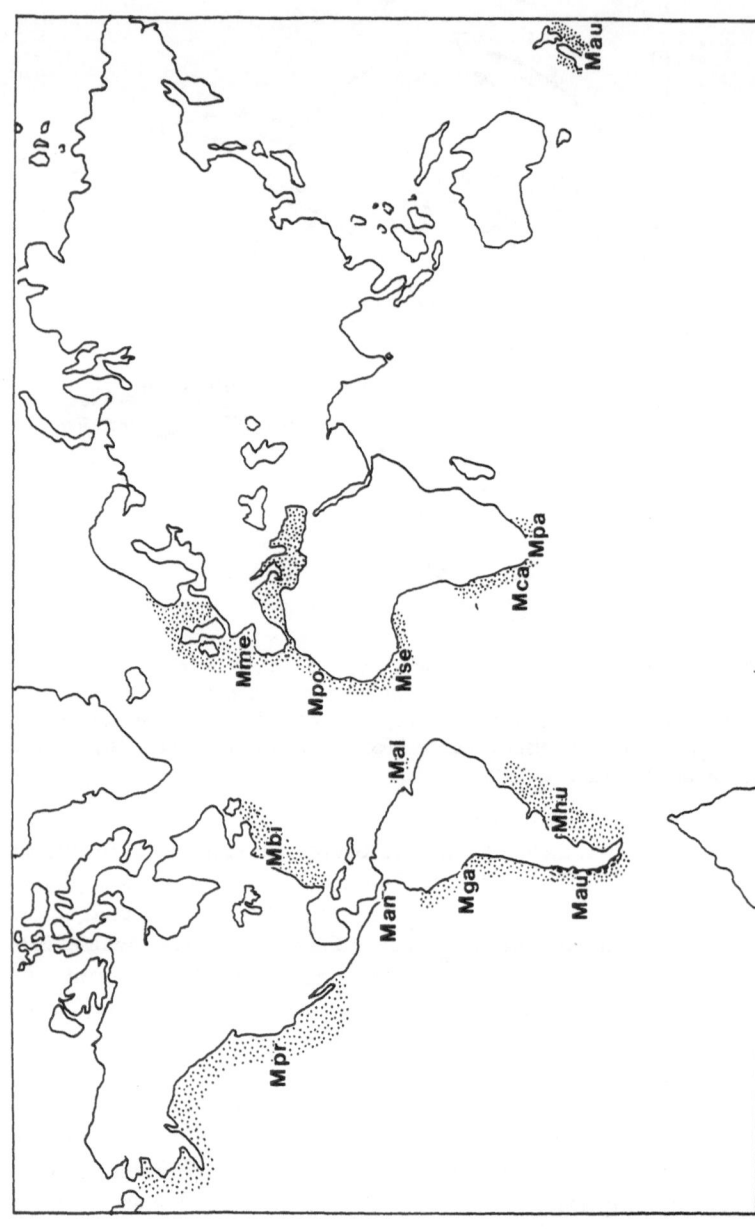

Figure 8.4 Geographical distribution of living species of *Merluccius* (Pisces, Gadiformes): Mal, *M. albidus*; Man, *M. angustimanus*; Mau, *M. australis*; Mbi, *M. bilinearis*; Mca, *M. capensis*; Mga, *M. gayi*; Mhu, *M. hubbsi*; Mme, *M. merluccius*; Mpa, *M. paradoxus*; Mpo, *M. polli*; Mpr, *M. productus*; Mse, *M. senegalensis*.

8.3.2.1 Parasites of Merluccius spp., Mugil cephalus and Clupea harengus

The genus *Merluccius* (Gadiformes, Merluccidae) includes 12 species of hake, each one with a discrete geographical distribution (Fig. 8.4) along the continental shelf of the Atlantic coast of Europe, Africa (including the Mediterranean) and America, as well as New Zealand and the Pacific coast of America (Inada, 1981). Approximately 70 parasite species (Table 8.1) have been reported from *Merluccius* spp., ranging from protozoans to isopods (Fernández, 1982).

The parasite fauna of the different species of *Merluccius* from the different localities can be separated into four categories. The first includes parasites specific for *Merluccius* spp., with a relatively low degree of diversification; as a result, they are present in the different species of *Merluccius* throughout its distribution. These species are **stenoxenic** (host specificity restricted to a given host genus or family). The cestode, *Clestobothrium crassiceps*, two species in the monogenean genus, *Anthocotyle*, and two, closely related species in the copepod genus, *Chondracanthus*, are included in this group. The second group consists of parasites specific for each species of *Merluccius*; these have a significant degree of diversification and closely related, congeneric parasite species are harboured by closely related hake species. These parasites are **oioxenic** (strict specificity, or restricted to a single host species). The *Aporocotyle* species complex (blood flukes), as well as the different species and morphs of the copepod, *Neobrachiella*, are in this category. A third group (**eurixenic**, broad host specificity) includes several, non-specific parasite species, with most having restricted geographical distributions; accordingly these are components of local faunas only. Included are various copepods, trematodes and larval cestodes. The fourth group has non-specific parasite species (also eurixenous), but with a rather cosmopolitan distribution; it is comprised of *Hysterothylacium aduncum*, *Derogenes varicus* and the plerocercoid of *Hepatoxilon trichiuri*.

A strikingly similar pattern of host–parasite association has been recorded for parasites of the striped mullet, *Mugil cephalus* (Perciformes, Mugilidae). *Mugil cephalus* is a cosmopolitan, euryhaline species present in tropical and temperate latitudes along continental coasts. Because the striped mullet is a commercially important species, its parasites have been documented worldwide resulting in an extensive data base. A total of 207 species has been reported (Fernández, 1986) from 15 major oceanic regions (Fig. 8.5). Three main groups of parasites have been described; these include parasites that are specific for *M. cephalus* (oioxenous), those that are specific for the striped mullet and other species in the same family (stenoxenous), and non-specific generalists (eurixenous). The degree of specificity is relatively high for monogeneans, digeneans and

Table 8.1 Parasite species present in the different species of *Merluccius*

	Merluccius paradoxus	*Merluccius senegalensis*	*Merluccius merluccius*	*Merluccius bilinearis*	*Merluccius albidus*	*Merluccius productus*	*Merluccius gayi*	*Merluccius hubbsi*	*Merluccius australis Chile*	*Merluccius australis New Zealand*
Monogenea										
Anthocotyle americanus			X	X			X		X	
Anthocotyle merluccii			X	X						
Dichidophora denticulata			X	X						
Trematoda										
Adinosoma robustum			X	X			X			
Aporocotyle argentinensis								X		
Aporocotyle australis										
Aporocotyle margolisi					X					
Aporocotyle spinosicanalis										
Aporocotyle wilhelmi						X				
Derogenes varicus			X	X	X		X	X	X	
Erythrophallus merluccii								X		
Gonocerca crassa			X	X						
Gonocerca phycidis			X	X						
Hemiurus levinseni			X	X	X					
Hemiurus ocreatus			X	X						
Lecithaster gibbosus			X	X	X					
Lepidapedon elongatum				X						
Podocotyle reflexa			X	X						
Sterrurus praellanus			X	X						
Tubulovesicula angusticauda									X	
Cestoda										
Abothrium gadi			X							
Bothriocephalus scorpii			X	X						
Callitetrarhynchus gracilis						X				
Clestobothrium crassiceps			X	X	X	X	X	X		
Grillotia dollfusi						X				

Table 8.1 (continued)

	Merluccius paradoxus	Merluccius senegalensis	Merluccius merluccius	Merluccius bilinearis	Merluccius albidus	Merluccius productus	Merluccius gayi	Merluccius hubbsi	Merluccius australis Chile	Merluccius australis New Zealand
Grillotia erinaceus			X	X				X		
Grillotia heptanchi								X		
Hepatoxilon trichiuri	X						X			
Lacystorhynchus tenuis						X				
Parabothrium bulbiferum										
Tentacularia coryphaena						X				
Tetrarhynchus cysticercus	X									
Acanthocephala										
Bolbosoma sp. (acantella)			X							
Corynosoma sp. (acantella)						X	X	X		
Echinorhyncus acus				X						
Echinorhyncus pumilio			X							
Nematoda										
Anisakis sp. (larvae)			X	X			X	X	X	
Capillaria gracilis				X	X					
Hysterothylacium aduncum			X	X	X			X		
Hysterothylacium sp. (larvae)								X	X	X
Terranova decipiens (larvae)						X				
Copepoda										
Acanthochondria phycidis						X				
Caligus debueni						X				
Chondracanthus merluccii	X	X	X	X				X		
Chondracanthus palpifer					X		X	X		
Clavella adunca			X	X						
Clavella stellata			X	X						
Clavella uncinata			X	X						
Cygnus gracilis										

Table 8.1 (continued)

	Merluccius paradoxus	Merluccius senegalensis	Merluccius merluccius	Merluccius bilinearis	Merluccius albidus	Merluccius productus	Merluccius gayi	Merluccius hubbsi	Merluccius australis Chile	Merluccius australis New Zealand
Epibrachiella impudica			X							
Haemobaphes cyclopterina			X							
Haemobaphes diceraus					X					
Lepeophtheirus dissimulatus					X					
Lernaeenicus tricerastes						X				
Lernaeocera branchialis				X						
Neobrachiella branchialis			X							
N. insidiosa f. insidiosa	X	X	X							
N. insidiosa f. lageniformis				X						
N. insidiosa f. pacifica					X	X		X	X	X
Neobrachiella merluccii		X	X			X				
Phrixocephalus carcellesi						X	X			
Trifur tortuosus							X	X		

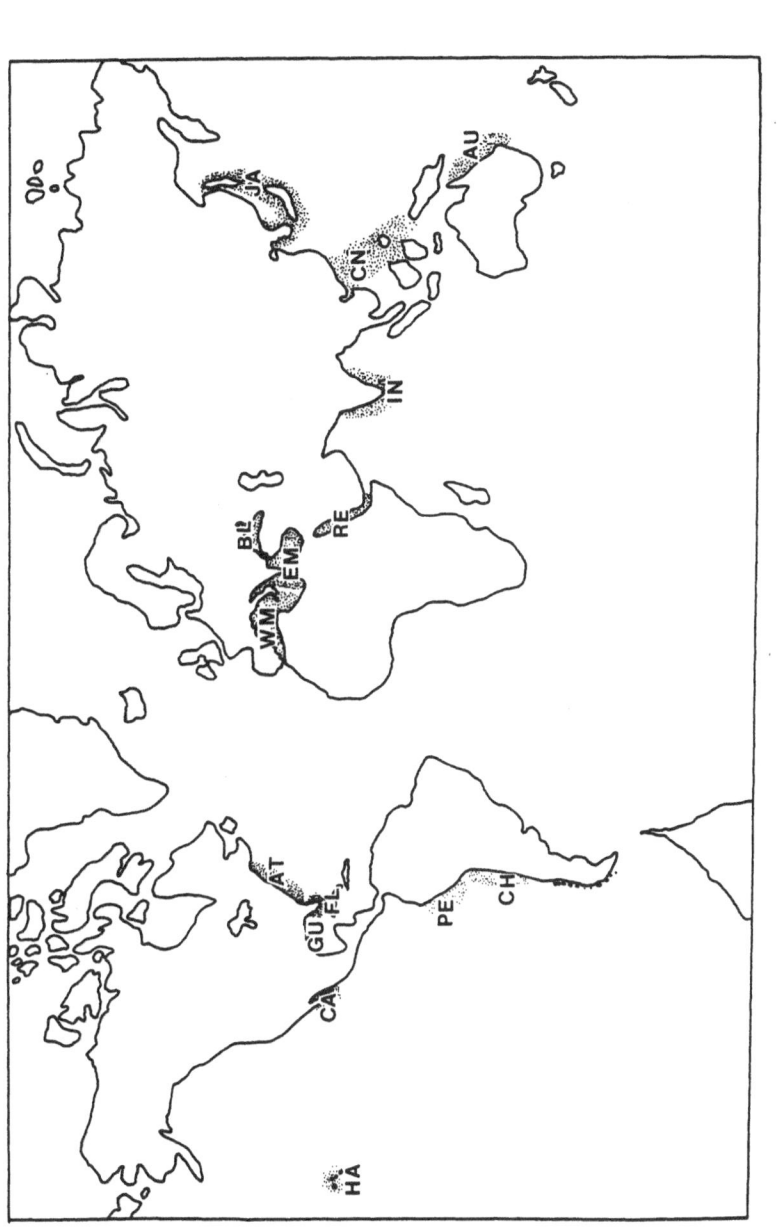

Figure 8.5 Geographical areas where the parasites of *Mugil cephalus* (Perciformes, Mugilidae) have been studied: AT, Atlantic Coast of USA; AU, Australia; BL, Black Sea; CA, California; CH, Chile; CN, China; EM, Eastern Mediterranean; FL, Peninsula of Florida; GU, Gulf of Mexico; HA, Hawaii; IN, India; JA, Japan; PE, Peru; RE, Red Sea; WM, Western Mediterranean.

Table 8.2   Degree of specificity of the different parasite taxa in *Mugil cephalus* throughout its distribution

	Number of parasite species			
	Only in M.cephalus (oioxenous)	In Mugilidae (stenoxenous)	Generalists (eurioxenous)	Total
Monogena	8	8	4	20
Trematoda (adults)	28	23	7	58
Trematoda (metacercariae)	0	0	0	40
Cestoda	0	0	3	3
Nematoda	4	0	3	7
Acanthocephala	5	0	2	7
Hirudinea	0	0	1	1
Copepoda	4	14	23	41
Isopoda	0	0	4	4
Branchiura	0	0	4	4
Protozoa	0	3	18	21

acanthocephalans (Table 8.2). Copepods and nematodes are almost equally divided between generalists and host-specific species. Larval trematodes and cestodes, branchiurans, isopods and protozoans are mostly generalist species.

If these parasites are considered in the context of geographical distribution, several parasite assemblages similar to those described for *Merluccius* spp. can be identified. There is a group of species with local distributions that are specific for *M. cephalus* or for members of the family. There is another group that also is specific for the striped mullet, or for the family Mugilidae, but these have a wide geographic distribution. Finally, there are several species that are local and non-specific, and several that are wide spread and non-specific. In areas where *M. cephalus* is the only mugilid present, most of its parasites are host-specific. However, in those areas where a significant number of mugilid species are sympatric, there is a high degree of exchange, and the parasites specific for the family Mugilidae are dominant. This is particularly true in the Mediterranean Sea, Red Sea and Black Sea, where the mugilids are diverse; the research efforts in these locations also have been intense and may, in part, account for the plethora of information regarding mugilid parasite distributions. Another relevant feature of this system is the taxonomic nature of the specific parasites throughout the host's distributional range. A number of local, highly specific, parasite species are congeneric and closely related to those in adjacent areas. Mono-geneans like *Ligophorus* spp. and digeneans within the genera *Dicrogaster*, *Saccocoelium* and *Saccocoelioides*, exhibit such a pattern. This indicates the probable isolation of different fish populations and speciation of the parasites. Since *M. cephalus* possesses many closely related and highly specific parasites with a restricted distribution, as well as parasites with broad geographical distributions, it suggests that the different parasite groups have different rates of evolution, or at least have been subjected to different selection pressures.

When the geographical distribution of parasites in *M. cephalus* is analysed using Jaccard's similarity index to compare the different areas, the clustering of the different regions is plainly related to the proximity of the areas studied (Fig. 8.6). Three main groups representing significant geographical and oceanographical discontinuities are defined (as shown by clusters I, II and III in Fig. 8.6): these include the American, Mediterranean and Indo-Pacific regions. The clustering between the different regions within the main groups suggests that the distribution and composition of parasite communities is greatly influenced by the contact between host populations. In this case, dispersal of the definitive host and phylogenetic relationships among the parasites seem to be much more significant for parasite distribution than vicariant events. A similar host–parasite pattern was observed by Prudhoe and Bray (1973) in deep benthic fishes with broad geographical distributions in the Northern Hemisphere.

Herring, *Clupea harengus* (Clupeiformes, Clupeidae), is an important pelagic species present in the northern temperate waters of the Pacific, Atlantic and Arctic Oceans. MacKenzie (1987) conducted an extensive survey of the herring and its parasites, emphasizing zoogeographical patterns. Although his characterization of the different parasites (80 species) was based on the relative importance of herring as a host, the same categories previously recognized for hakes and mullets could be also recognized. He identified parasites for which herring was the primary host, those for which herring was one of several equally important hosts, those for which herring was a host of minor importance and, lastly, accidental parasites. Those species for which herring was a primary host were herring-specific. The parasite species in the second group included those that were mostly found in other clupeid fishes (like *Engraulis* spp. and *Alosa* spp.) as well as several species found in non-clupeid fishes; these non-clupeid hosts, such as salmonids and carangids, have ecological habitats similar to the herring at some point in their life cycles and this facilitates the exchange. The parasite group considered as being of minor importance included opportunistic, non-specific parasites, mainly larval cestodes and nematodes. Finally, the accidental parasites, those for which

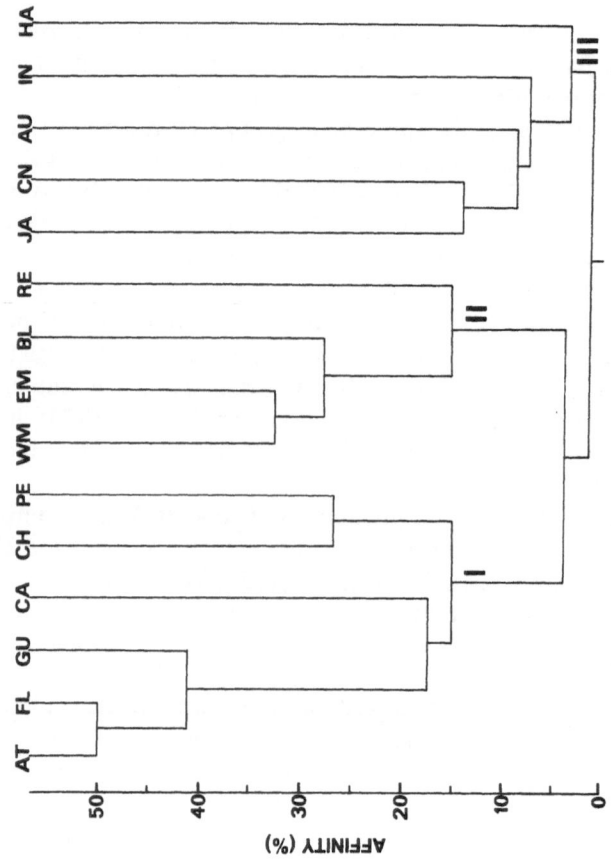

Figure 8.6 Clustering of different geographical regions based on the parasites of *Mugil cephalus*, using Jaccard's similarity index and the unweighted pair-group method with mean averages. Geographical regions are keyed as in Figure 8.5.

herring was an unsuitable host, were acquired either during migrations through regions of low salinity, or when herring moved to shallower waters to spawn. These parasites were for the most part gyrodactilid monogeneans, acquired probably by direct or close contact of herring and the primary host. Although the same parasite categories could be recognized in herring, non-specific local parasite faunas seem to be of greater importance in herring than in hakes or mullets.

8.4 ECOLOGICAL ASPECTS

The pure aims of biogeography can be redirected, not so much in terms of distributions (Brown and Gibson, 1983) which implies that the researcher already has an idea of the underlying processes, but toward a search for the causes of distributions, i.e., the 'how', 'when' and 'why' of distributions (Jablonski, Flessa and Valentine, 1985).

Various aspects of parasite distribution have been examined throughout this chapter in terms of vicariance and dispersal events, host specificity, and the impact of evolutionary history. All these elements contribute, in different ways, to the generation of patterns of species diversity across different ecological gradients and geographical settings. Many patterns of species diversity have been related to the distribution of free-living terrestrial and marine organisms. Some of these include gradients in species richness that are influenced by latitude, altitude, salinity, depth, light intensity, and the frequency or intensity of physical disturbances.

Pianka (1983) and Brown (1988) reviewed a number of current hypotheses directed at understanding these patterns and emphasized the idea that the different hypotheses are not mutually exclusive; normally, a combination of several provides the best explanation. The hypotheses included for consideration were: time since perturbation, climatic stability and predictability, productivity, habitat heterogeneity, interspecific interactions (competition and predation), ecological and evolutionary time, and rates of speciation and extinction. Similar patterns to those described for free-living organisms have been documented for parasites. Some authors have emphasized the Theory of Island Biogeography, where the host as a species, as a population, or as an individual, is treated as an island.

8.4.1 Latitudinal gradients of species diversity

The significant increase in diversity from the poles to the equator is one of the more clear-cut patterns in biogeography. Although the rate of increase in species richness towards the equator varies among taxonomic groups, the trend is consistent in the higher taxonomic groups. The data available

Table 8.3 Latitudinal gradients in numbers of monogenean and digenean genera of commercial marine fishes. (From Lebedev, 1969.)

Biogeographical zone	Number of species of commercial fish	Number of genera of: Monogenea	Digenea
Arctic	approx. 20	15	12
North-boreal	40-50	20	50
South-boreal	>50	50	160
Tropical	100-120	95	300
North-antiboreal	40-50	40	80
South-antiboreal plus Antarctic	approx. 20	3	10

Note: parasites of tropical fish are much less well known than those of fish from higher latitudes and, since 1969 when this table was composed, many new genera of Digenea and Monogenea have been described from tropical oceans.

for parasites are far from being comprehensive. Nonetheless, observations on several genera of Monogenea and Digenea in marine teleost fishes show the same trend. Rohde (1978b, 1982) summarized data from Lebedev (1969) (Table 8.3) and showed that the number of monogeneans and digeneans increases toward the tropics. He emphasized that this trend might be even stronger because the number of Monogenea are probably underestimated. Many more monogenean surveys have been conducted in the temperate than in the tropical regions. The increase in number of species toward the equator also is correlated to some extent with the increase in number of host species available. Moreover, if the number of monogenean species/fish species (relative richness) is considered (Rohde, 1980b), the trend of increase towards the Equator persists (Fig. 8.7). One of the few exceptions to this pattern is for parasites of marine mammals; in these hosts, species richness is greater in the cold waters of northern temperate zones (Table 8.4). A partial explanation for this observation is the greater diversity of marine mammals in the Arctic regions when compared with the tropics.

One hypothesis commonly used to explain the latitudinal pattern of species richness relates to changes in niche breadth and the intensity of inter specific interactions. It has been argued, for example, that species in tropical latitudes are less limited by abiotic factors and more by biotic factors, with inter-specific interactions being more important than intraspecific ones. Of the various parameters that can be used to define the niche of a parasite species, food requirements, host range and micro habitat seem to be the most important (Rohde, 1979).

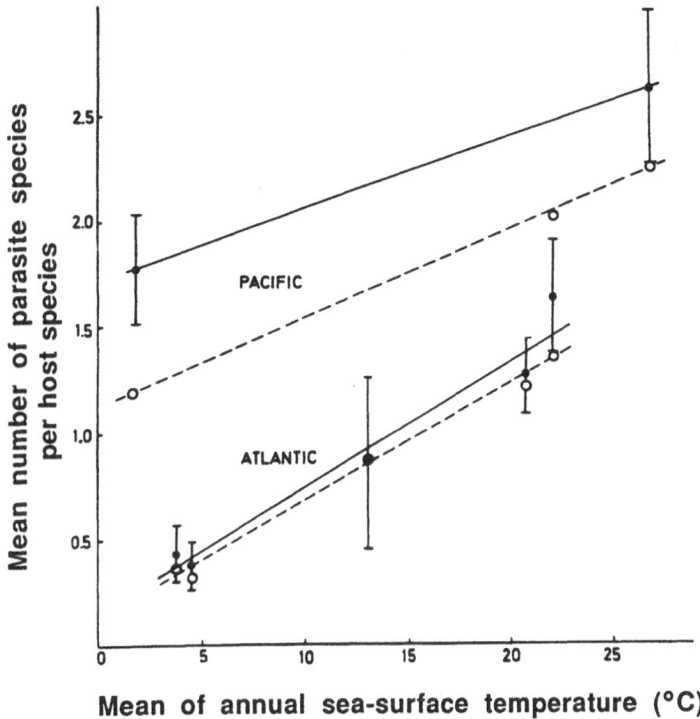

Figure 8.7 Relative species diversity (average number of parasites per host species) of monogenean gill parasites on teleost fish in the Pacific and Atlantic Oceans plotted against the approximate mean of annual sea surface temperature at various localities. Open circles: total number of monogenean species/total number of host species examined at each locality; closed circles: *as above*, but monogenean species occurring in *x* host species counted *x* times. (With permission, from Rohde, 1980b.)

Rohde (1978a, 1981, 1982) examined this question when he attempted to identify correlations between diversity gradients and possible gradients in the 'width' of the fundamental niches of parasites. Based on observations from several parasitological surveys, Rohde (1978a) indicated that the Digenea of marine fishes have more restricted host ranges in warm than in cold waters. Monogenea, on the other hand, have similar host ranges at all latitudes (Fig. 8.8). The only significant data set available to document microhabitat use (ectoparasites of marine fishes) showed a lack of correlation between the number of microhabitats available and the number of parasites present (Fig. 8.9). Similarly, the ratio of mono-pisthocotylean to polyopisthocotylean monogeneans calculated from many extensive surveys in different localities did not show any trend (Rohde,

Table 8.4 Helminths species of pinnipeds and cetaceans at different latitudes. Data from Dogiel (1964). (From Rohde, 1982.)

	Trematodes	Cestodes	Nematodes	Acanthocephalans	Total
Arctic	6	7	9	3	25
Boreal	30	24	31	9	94
Tropical	7	2	18	4	31
Antiboreal	0	9	21	8	38
Antarctic	1	7	5	1	14

1982). The rationale for the latter effort was that polyopisthocotylean monogeneans are obligatory blood feeders, whereas most mono-pisthocotyleans feed on mucus and epithelial cells and thus represent different feeding guilds. Based on these results, Rohde (1982) concluded that the fundamental niches of marine parasites were not significantly affected by the number of species in the community and that interspecific competition probably was not important in structuring the infra-communities. Therefore, the increase in species richness toward the equator could not be explained in terms of narrower niches allowing a tighter species packing. The evident conclusion was that many empty niches were available in hosts from temperate and cold regions.

Climatic stability and predictability do not differ significantly between cold and warm seas (Rohde, 1989). Although it is known that temperature is more constant in tropical and polar waters than in temperate

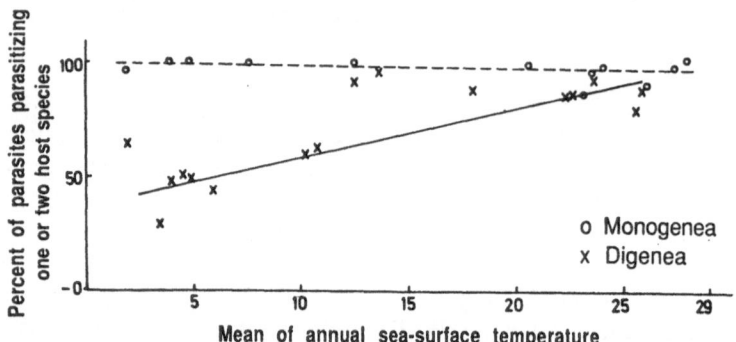

Figure 8.8 Host specificity of marine Monogenea and Digenea at different latitudes as indicated by the percent of species parasitizing one or two host species, plotted against means of annual sea surface temperature at various localities. The difference between both regressions is highly significant ($P < 0.001$). (With permission, from Rohde, 1978a.)

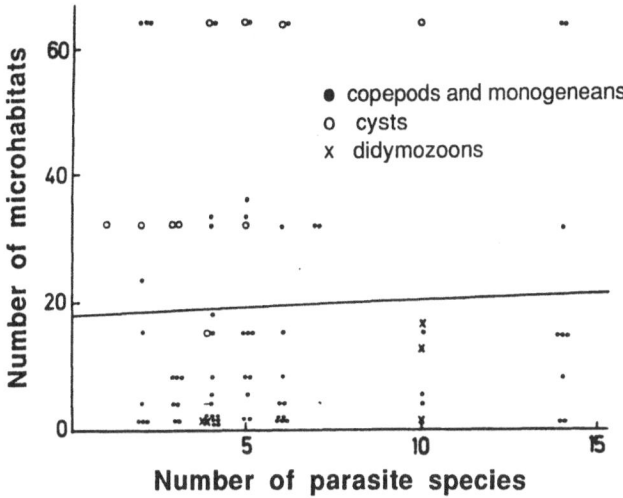

Figure 8.9 Number of microhabitats per species of parasite on the gills and in the mouth cavity of species of fish with various numbers of parasite species. (With permission, from Rohde, 1981.)

areas, he argued that the fluctuations are not significant and noted that temperature fluctuations in tropical waters can be dramatic during tidal cycles. Spatial heterogeneity (at least for monogeneans) is not significant in a latitudinal gradient, especially when the nature of the fish host is considered. Thus, counts of gill filaments and surface area are quite similar among fishes at different latitudes. Productivity is known to be similar in cold and warm waters and biotic interactions such as competition and predation also seem to be unimportant. Parasite competition, as discussed in terms of niche width, was not significant in the studied group (monogeneans), and predation, as seen in the cleaning symbioses, for example, provided insufficient evidence because little is known about this association at higher latitudes, or about its ecological effects.

Rohde (1989) considered two other hypotheses, one related to ecological time and the other to evolutionary time. He disregarded the importance of ecological time, arguing that temperate regions have a long, undisturbed history and probably have existed as long as tropical ones. Only the hypothesis related to evolutionary time provided an acceptable explanation for latitudinal patterns in species diversity. Although the actual time available for evolution to occur has been the same at all latitudes, a longer 'effective' time is assumed for the higher temperatures as a consequence of the accelerated physiological processes and shorter generation times of the organisms. Presumably, this would have led to an increase in mutation rates and provided more morphological and physiological 'grist' for selection. However, the potential interaction, or complementarity between the latter hypothesis and some of the previous ones presented by Pianka (1983) cannot be disregarded, especially if parasites other than gill monogeneans are considered.

Finally, another pattern emerges on examination of data in Tables 8.3 and 8.4. Thus, there is a much greater species richness in the Northern Hemisphere when compared to the Southern Hemisphere. This observation was also noted by Szidat (1961) for the parasites of marine fishes and mammals in the Atlantic and by Parukhin (cited by Rohde, 1982) for the nematode fauna of some 200 fish species worldwide.

8.4.2. Species diversity in the different oceans

A quantitative analysis of species richness (Rohde, 1986) between the Atlantic and Pacific Oceans, suggests that the relative diversity of monogeneans is higher in the latter body of water (Fig. 8.7). Parukhin (cited by Rohde (1982)) also reported that the number of nematode species was smaller in the southern Atlantic than in the southern Indo-Pacific. Rohde (1982), citing data from Lebedev (1969), noted that endemicity was greater in the Indo-Pacific than in the Atlantic. Rohde (1986) further compared data from different latitudinal regions within and between oceans. The analysis revealed that the differences between the two oceans were almost entirely due to a much greater number of species in the northern Pacific than in the northern Atlantic.

Rohde (1982, 1986) proposed that, again, an evolutionary-time hypothesis might best explain the differences in diversity between the two oceans. Based on geological evidence and the theory of Continental Drift, the Pacific Ocean has existed for the longest geological time, whereas the Atlantic began to form and expand 'only' 150 million years ago, with the break up of Pangaea. Although the pre-existent Pacific and the newly formed Atlantic have been in contact several times in their geological histories, the present degree of endemicity of their biotas suggests that the exchange between them has been rather limited (Ekman, 1953). In addition to the evolutionary explanation, Rohde (1986) provided an alternative based on ecological time. In this case, during the last glaciation events an ice sheet covered a significantly greater area of the continental shelf in the north Atlantic than in the Pacific, possibly leading to extinction of more Monogenea in the former than in the latter.

The idea of species–area relationship (as will be discussed in section 8.4.4) to explain the differences between the two oceans, was rejected by Rohde (1986) after studying the diversity of gyrodactilid monogeneans in the Pacific and Atlantic Oceans. Gyrodactylids are cold-water forms and, as a result, their distribution and origin are likely to be restricted to the northern, colder areas. Thus, they could not have migrated into these northern areas from warmer seas, making the north Pacific and north Atlantic Oceans equivalent in terms of parasite origins. These two regions are also similar in terms of surface area available when the complexity of the shoreline is included in a planimetric analysis. Thus, if the surface areas are similar and if colonization from the tropics is not a possibility, then the greater diversity of the Pacific Ocean is more likely related to evolutionary and ecological time.

8.4.3 Parasites and depth gradients

In aquatic environments, both marine and freshwater, diversity of free-living organisms generally decreases with increasing depth. The decrease in temperature and light, the increase in pressure, and the lack of seasonal change, are a few of the factors responsible for establishing and maintaining this gradient. In marine environments, depth gradients must be considered within the context of bottom topography, e.g., continental shelf, slope, rise, and abyss. The three-dimensional distribution of organisms (epipelagic, mesopelagic, bathypelagic) also should be addressed since the physicochemical factors affecting each of these zones are different.

8.4.3.1 Pelagic fishes

Pelagic fishes are distributed in the water column and are not associated with the substrata. They feed upon other pelagic organisms such as fishes, tunicates, crustaceans, chaetognaths and dead organisms or organic material falling through the water column. A few investigators have addressed the nature of parasitism in meso- and bathypelagic fishes, including some species that migrate vertically (Collard, 1970; Noble, 1973; Gartner and Zwerner, 1989). In general, these studies have shown that the distribution of biomass in pelagic waters and host longevity determines parasite diversity. Host biomass decreases from the epipelagic realm towards the bathypelagic, and so does parasite diversity. Organisms living in the mesopelagic region have adapted to maximize energy sources unavailable to bathypelagic fishes and that accounts, in part, for their greater parasite diversity. These adaptations include diel vertical migrations to overlying waters where food is more abundant. Other factors influencing parasite distribution and richness include the schooling behaviour of mid-water fishes that provide concentrations of prey for predators. The organically rich layer created by changes in water density at the permanent thermocline also is believed to attract populations of potential hosts (Campbell, 1983).

Based on the few studies mentioned above, mesopelagic fishes harbour many larval parasites common to pelagic prey that are peculiar to a given geographical region. Adult helminths (mostly monogeneans and acanthocephalans) are rare, and only a few trematode species with a relatively low specificity are present. Larval cestodes and mainly larval nematodes are dominant; this is probably due to the low specificity exhibited by these particular life-cycle stages. In contrast, bathypelagic fishes have the poorest parasite fauna partly because host density is very low. Their diet includes mostly crustaceans and some bottom-dwelling invertebrates that apparently are not suitable intermediate hosts for parasites in deeper water. Nonetheless, bathypelagic fishes have a few parasite species that are highly specific for them and that are, therefore, well adapted to the extreme environmental conditions (Campbell, 1990).

8.4.3.2 Benthic fishes

Bottom dwelling fishes, at any depth, have a greater parasite diversity than pelagic fishes at similar depths. Among benthic fishes, however, parasite diversity also decreases from the continental shelf into the abyss at various rates, depending on the feeding habits of the fish, the nature of the free-living species assemblages, and the nature of physical and biological oceanographic phenomena occurring in the upper layers. The studies by Manter (1934), Campbell, Haedrich and Munroe (1980) and Campbell (1983, 1990) summarize most of the information and provide some generalizations for the assemblage of parasites in deep, benthic fishes of the northwest Atlantic. For example, parasite densities of deep benthic fishes are directly related to host population densities. Host specificity of monogeneans and digeneans is high. Monogeneans exhibit oioxenic specificity (strict host specificity); most trematodes and cestodes are oioxenic or stenoxenic (narrow host specificity), while acanthocephalans and nematodes are mostly eurixenic (lacking host specificity). Overall, the deep-water parasite fauna is less diverse than in shallow waters. The best evidence supporting this assertion is shown by the parasites of the rattail, *Coryphaenoides armatus* (Fig. 8.10). The decrease in parasite diversity is paralleled by a decrease in macrofaunal abundance with depth and with changes in the consumption of certain prey, probably determined by their availability. Although not shown in Figure 8.10, Campbell (1983) indicated that the abundance of parasites without intermediate hosts, particularly monogeneans, does not change significantly until they reach the lower slope and abyssal plains (4000–5000 m deep), where they disappear.

Two major factors basically determine parasite diversity with increasing depth in both the benthic and the pelagic regions; one is the relative size of the deep-sea ecosystem and the other is the permanence of the host's

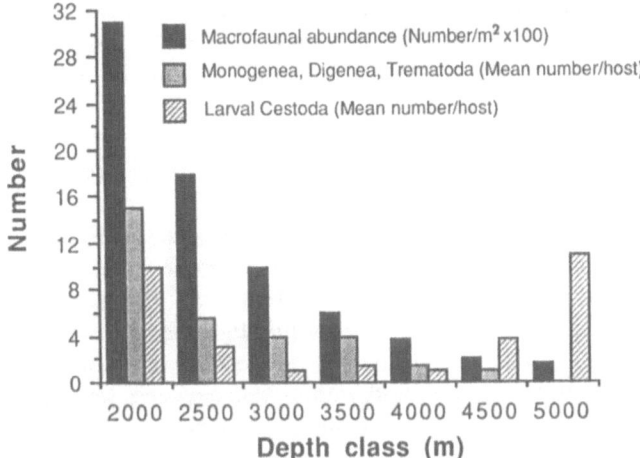

Figure 8.10 Rates of infection by helminth parasites in *Coryphaenoides armatus* and overall abundance of macrofaunal invertebrates, arranged by depth. The number of *C. armatus* examined in each depth class, from shallow to deep are 17, 86, 85, 11, 10, 3, 1. (With permission, from Campbell, Haedrich and Munroe, 1980.)

association with particular invertebrate communities. Parasite diversity is correlated with the diversity of the invertebrate fauna, providing suitable intermediate hosts for the completion of complex life cycles. Similarly, low fish density and the low level of interaction of the fish host population are probably major constraints for the establishment of parasites with direct life cycles. The differences between pelagic and benthic organisms at a given depth are attributed to differences in the food supply which, in turn, translates into parasite supply. Meso- and bathypelagic fishes are spread throughout a significant volume of water and their life spans are also relatively short, up to 2 years. The short life span means briefer periods of exposure and the recruitment of fewer parasites. Benthic fishes, on the other hand, have significantly longer life spans and live in a more restricted, two-dimensional zone with a concentrated and extensive food supply. Thus, their recruitment time is longer and the diversity and quantity of parasites to which they are exposed are greater.

8.4.4 Species–area relationship

It has long been noticed that the number of species increases systematically in samples from larger areas, but that the rate of increase in number of species decreases with progressively larger areas. In other words, once a plateau in species richness has been reached, the number of species then stays level despite an increase in the surface area sampled. This pattern has been developed to the extreme in MacArthur and Wilson's (1967) Theory of Island Biogeography; it emphasizes that habitat heterogeneity increases as surface area increases (Williamson, 1988). In this sense then, area should be considered more as an indicator than as an explanation for species number.

The species–area relationship has been applied to parasites on a number of different scales. Rohde (1989) examined these relationships at three different levels using gill parasites of marine fishes. The gills of fishes seem to represent an ideal system for examining species–area relationships because the gills of various fishes do not differ greatly in structure and because hosts can be studied both within, and between, latitudinal regions. Using fish length as an indicator of gill size, Rohde (1989) found that larger fishes with larger gills did not have more monogeneans than smaller fishes (Fig. 8.11). He also examined monogenean data from three species of *Scomber* (Perciformes; Scombridae) of similar size, but with different ranges of geographical distribution (scombrids include mackerel). He did not, however, obtain a general correlation between host range and parasite species richness. He then considered the copepod data for several scombrid genera and found a significant relationship between the number of copepod species and the number of geographical areas where the host genus occurred. Most importantly, however, there was a significant relationship between the number of copepod species and the number of host species in the genus. Indeed, this latter factor accounted for 93% of the variance in his

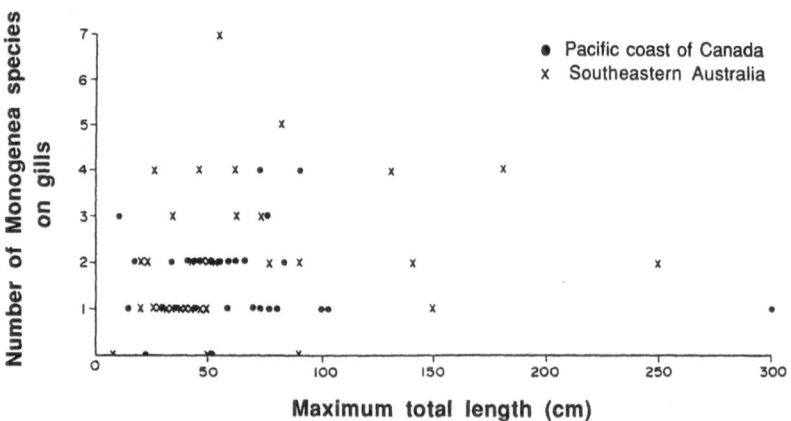

Figure 8.11 Number of gill monogeneans in marine teleost fish of different lengths (total length) at two localities. (With permission, from Rohde, 1989.)

regression analysis. Thus, diversity of the host group was the most important factor determining parasite richness. The data previously discussed on parasites of *Mugil cephalus* (Table 8.2) support, to some extent, Rohde's conclusions for the ectoparasitic component communities of *Scomber*. Thus, a higher proportion of the parasites in *M. cephalus* are species specific for Mugilidae, but will occur on related host species in the same area.

Rohde (1989) concluded that 'area as indicated by size of fish, gills, geographical range of hosts, and geographical area occupied by a fauna (as in the monogeneans of the north Atlantic and north Pacific), is not predominant in determining species richness. The effect of area, if it exists at all, is probably overridden by other factors such as evolutionary time, diversity of the group, etc.' Factors related to phylogeny and historical processes that affect the host can also override the effects of area.

Price and Clancy (1983) employed a similar approach in examining the relationship between parasite richness and host geographic range, size and feeding habits of freshwater fishes in Great Britain. Unlike Rohde (1989), however, Price and Clancy (1983) found a significant correlation between geographical range of the host and the total number of parasites per host. Geographical range accounted for 68% of the total variation and feeding habits for an additional 5%. A similar relationship, but on a different scale, was reported by Kennedy (1978a) for the number of parasite species in *Salmo trutta* in British lakes. Thus, 86% of the variance in parasite species was accounted for by the area of the lake. In contrast, a study of the parasites from Atlantic char, *Salvelinus alpinus*, did not indicate any correlation between parasite diversity and island size (Kennedy, 1978b).

The relationship between number of parasites and the geographical range of the host species, at this point, is clearly subject to debate. The correlation between these two variables could be the consequence of three different phenomena. The first is the existence of a causal relationship between the two variables as shown by Price and Clancy (1983). Thus, host species with larger ranges may acquire more parasite species because larger areas will include the ranges of more parasite species. Second, it is possible that the relationship is a function of a correlation between the number of parasite species and the host range with a third variable. For example, host species with larger ranges may have been more heavily sampled and so more parasites species have been identified (Kuris and Blaustein, 1977). Host species with larger ranges may be larger, providing more niches for more parasite species. It is also possible that host species with larger ranges may occur in higher population densities or may be more gregarious, thereby facilitating increased rates of parasite transmission (Dogiel, Petrushevski and Polyanski, 1961; Price, 1980).

A third possibility, one that has not been extensively investigated, is that biological variables may be confounded by the effects of phylogeny (Pagel and Harvey, 1988). Thus, a relationship between parasites and host range across all species may be a result of the host's phylogenetic history. In effect, species that share common ancestry would more likely be the subject of similar evolutionary constraints. They are also likely to be morphologically similar and occupy similar niches. It would seem, therefore, that their parasite infracommunities should also be similar. In these terms, Price and Clancy's (1983) results should be re-examined by controlling for the phylogenetic association of the hosts.

On the other hand, Gregory (1990) observed a significant relationship between the number of helminth parasite species per host and the geographical range for Holarctic waterfowl. The importance of his work rests with the identification and control of variables related to sample size, and with host variables such as body size, population density, social tendency and taxonomic association. From a methodological point of view, Gregory (1990) concluded that the critical assessment of parasite richness may depend upon the sample size of each host species, and, unless this is controlled, apparent relationships between parasites and biological variables (such as geographical range) may be spurious.

8.5 APPLIED ASPECTS OF BIOGEOGRAPHY

8.5.1 Parasites as 'biological markers'

As mentioned earlier in this Chapter, one of the reasons for the greater quantity of parasitological data from marine fishes when compared to freshwater or terrestrial organisms, is the economic value of the hosts. The efficient management of fisheries requires a complete understanding of the biology of the host. Sometimes, however, traditional fisheries biology methods and techniques are inadequate to answer questions regarding observed patterns of biogeographical distribution. In these cases, observations on parasites may provide the appropriate answers.

Parasites were first uses as markers some 35 years ago (Margolis, 1956) to distinguish between Asian and North American stocks of sockeye salmon, *Oncorhynchus nerka*, that occur in the north Pacific and adjacent areas. Based on the nature of differing parasite faunas, Hare and Burt (1976) were able to distinguish populations of the Atlantic salmon, *Salmo salar*, from various tributaries in the Miramichi River system in Canada. Recently, Margolis (1990) identified two trematodes as extremely good markers for populations of anadromous steelhead trout, *Oncorhynchus mykiss* (= *Salmo gairdneri*), originating in freshwaters of the Pacific coast of North America. Steelhead trout have a freshwater range that extends from California to Alaska, continuing to the east and west coasts of Kamchatka and the northern coast of the Okhotsk sea. Metacercariae of *Nanophyetus salmincola* and adults of the trematode, *Plagioporus showi*, only occur in the North American freshwater systems, where they are acquired by juvenile fishes prior to their seaward migration. These parasites are able to survive in the steelhead trout throughout all of its oceanic migration, an important part of the basic requirements outlined by Margolis (1990) for parasites used in stock identification of anadromous salmonids. He indicated that such markers also must have a restricted freshwater distribution in a specific geographical scale, that their life cycles should preclude the possibility of inter-host transfer at sea, and that they must have long life spans, including the ability to survive during the marine migration of the host. Additionally, a high prevalence of the parasites is desirable in order to increase the chances of obtaining infected hosts.

The same standards are employed in identifying markers for marine fishes. The restricted geographical distribution of the parasites also will be influenced by the presence or absence of appropriate intermediate hosts. Ectoparasites (monogeneans and copepods) are not useful because they are more susceptible to changes in various abiotic parameters such as temperature and salinity; moreover, they are easily transmitted horizontally from host to host, regardless of the original area of infection.

Larval stages usually make the best markers because of their inability to leave the host and because they are almost invariably sequestered in parts of the body not exposed to the vagaries of the external environment. In marine fishes, larval nematodes have been widely employed. Grabda (1974), for example, found that the degree of infection with *Anisakis* sp. varied in different populations of herring in the Baltic Sea. Platt (1975, 1976) studied larvae of the nematodes, *Anisakis* sp. and *Phocanema decipiens*, in several populations of cod, *Gadus morhua*, in the Arctic and northern Atlantic Oceans in the vicinity of Iceland and Greenland. He found significant differences in the parasite densities of these two species between the Arctic and north Atlantic cod populations and in the densities of *P. decipiens* between the two north Atlantic populations (Iceland and Greenland). Platt (1976) suggested that the difference in *P. decipiens* could be used to determine the geographical origin of cod spawning off the southern and western coasts of Iceland.

On some occasions, the complete fauna of parasites must be examined in order to obtain suitable distributional information. Sindermann (1957), for example, studied the parasite distribution and abundance in herring, *Clupea harengus*, along the Atlantic coast of North America. The data suggested that there was little, if any, interchange between herring populations from the Gulf of Maine and the Gulf of Saint Lawrence. Moreover, he suggested that immature herring in the eastern and western parts of the Gulf of Maine could be separate populations. McGladdery and

Burt (1985) undertook a more comprehensive examination of the migration, feeding and spawning behaviour of herring. They found that six parasite species could be used as indicators of seasonal migration between the Bay of Fundy, the Nova Scotian Shelf and the Gulf of Saint Lawrence. The biological characteristics of these parasites indicated that the seasonal variations in prevalence could be better explained by changes in composition of herring stocks rather than by changes in the parasite component communities within the same stock. The presence of a protozoan, *Eimeria sardinae*, permitted the separation of 'races' of herring spawning at different times of the year, an important piece of information when attempting to manage commercial fisheries. MacKenzie (1985) also was able to follow the migrations of herring in the North Sea and the Scotland Shelf by tracking the presence of the metacercariae of two trematode species and a larval cestode. All three were long-lived parasites acquired by herring in their nursing grounds where prevalences varied significantly in the different locations.

8.5.2 Centres of origin and host dispersal as indicated by parasites

As previously mentioned, von Ihering (1891, 1902) was the first to use parasite data to establish relationships between hosts, centres of origin, dispersal and ancient hypothetical land connections. Finding the centre of origin for a taxon and then the dispersal routes of the derived taxa, has been one of the many applications for parasitological data on a global scale. It is assumed that hosts have a greater diversity of parasites in areas where they have occurred for a long time (their centre of origin) than in areas they have more recently colonized. Similarly, the dispersal of hosts into new areas may lead to the loss of their 'original' parasites, or the acquisition of new ones, either by host capture or speciation of the 'original' species. Thus, parasites present in the area of origin should be regarded as ancestral to those present in the newly colonized locations.

Parasites have been used to provide a more comprehensive insight, for the origin of dromus fishes such as *Oncorhynchus* spp. and *Anguilla* spp., as well as freshwater stingrays (Potamotrigonidae) in South American rivers. Margolis (1965) indicated that the specificity for Pacific salmonids (*Oncorhynchus* spp.) was greater among freshwater parasites than among marine species. Freshwater endoparasites included some with complex life cycles, suggesting a long association. There were only two marine parasites specific to salmon, a monogenean and a copepod, both with direct life cycles. These observations, in addition to other evidence, encouraged Margolis (1965) to conclude that Pacific salmon had a fresh-water origin and not a marine one as some ichthyologists had argued. In contrast, Manter (1955) examined the data on parasites of *Anguilla* spp.

from Europe, North America, Japan and New Zealand, and indicated that *Anguilla* probably had a marine and, more specifically, a Pacific, origin. He based his argument on the uniqueness of the Pacific parasites. Of nine species found in the Pacific, eight were specific for *Anguilla*; moreover, the only three parasite genera specific for this host were also in the Pacific. Atlantic eels that occur in Europe and North America possess a larger number of parasites than those in the Pacific, but these are mostly generalist trematode species common to other fishes. It was also suggested that the greater number of parasite species known from Atlantic eels was probably a result of their being more intensely studied and not because the Atlantic was their centre of origin.

Brooks, Thorson and Mayes (1981) devised a clever methodology, combining biogeographical and phylogenetic information, to determine the evolutionary origins of the potamotrigonid stingrays present in the rivers of South America. The basic question was directed at the evolutionary origin of the rays. It is believed that stingrays were secondary invaders of freshwater systems, being originally a marine group. Because all the drainage systems inhabited by potamotrigonids drain into the Atlantic side of South America, the common assumption was that these rays had an Atlantic origin. However, the closest relatives of their nematodes and cestodes (according to the phylogenetic analysis of these parasites) are present in marine urolophid rays along the Pacific coast of South America. This phylogenetic relationship, together with the geological history of the area, suggested that potamotrigonids migrated into freshwater rivers from the Pacific Ocean prior to the events that led to the formation of the Andes Mountains in the mid-Cretaceous. Accordingly, during the Andean orogeny, the river drainages were altered, a new Continental Divide was established, and the rays became isolated from the ancestral group along the Pacific coast of South America; they are presently relict taxa in the Atlantic drainage system.

There is a long-standing controversy regarding the geographical origin and dispersal routes of the species of *Merluccius* (hakes) (Fig. 8.4), close relatives of gadid fishes such as cod (*Gadus morhua*) and whiting (*Micromesistius merlangus*). Researchers have examined this problem from two different perspectives, one based on ichthyological data (Svetovidov, 1948; Inada, 1981) and another based on parasitological information (Szidat, 1955; Kabata and Ho, 1981; Fernández, 1985; Fernández and Durán, 1985). Svetovidov (1948) focussed his approach on the geographical origin of *Merluccius*, hypothesizing that it evolved in an area coincident with the present-day North Atlantic, from where it then migrated. He did not address, however, the question of dispersal routes. In contrast, Szidat (1955), based on the study of hake parasites, suggested that *Merluccius* originated in the North Pacific (Bering Sea) from where it spread, following two distinct routes. One was via the northern coast of North America into the Atlantic, subsequently moving south along the Atlantic coasts of both Europe and North America. The second route was south along the Pacific coasts of America until it reached the land bridge connecting Patagonia with the Antarctic. When this bridge disappeared in the Pleistocene, hakes moved again in two directions, one around the tip of South America along its Atlantic coast (Argentina), as well as across the Atlantic Ocean to the coast of South Africa. This hypothesis was based on Szidat's idea that the parasite fauna of the North Pacific hake was more primitive than that of the North Atlantic hake and that parasites of South Atlantic hake were more closely related to those of Pacific *Merluccius* than to those of North Atlantic.

More than 20 years later, Kabata and Ho (1981) and Inada (1981), reassessed the parasitological and ichthyological evidence available to that date and independently reached similar conclusions, but opposite to Szidat's. Thus, they argued that the genus *Merluccius* arose in the North Atlantic (*M. merluccius*) and spread south following two routes. One route followed the Atlantic coast of Europe and Africa, including a group that branched off into the Mediterranean (*M. merluccius, M. polli, M. senegalensis, M. capensis, M. paradoxus*); a second group migrated south along the American coast (*M. bilinearis*) and into the Pacific through the then submerged Isthmus of Panama. Once they reached the Pacific, they followed two different paths, one moving northward (*M. productus*) and the other southward (*M. gayi*). The only disagreement between the hypotheses of Inada (1981) and Kabata and Ho (1981) resides in the origin of the South Atlantic hake, *M. hubbsi*, and *M. australis* that occurs along the Pacific–Patagonian coast of Chile and New Zealand.

According to Kabata and Ho (1981), *M. hubbsi* also originated from the stock that split in the area of the Isthmus of Panama and migrated south down the Atlantic coast of South America (Fig. 8.12). Inada (1981), on the other hand, argued that *M. australis* (Patagonic population) and *M. hubbsi* originated in the Pacific and moved toward the Atlantic around the southern end of South America. Although Kabata and Ho (1981) did not know about the presence of *M. australis* on the Pacific coast of South America, they indicated that from a parasitological point of view, *M. australis* from New Zealand was more closely related to *M. hubbsi* from the South Atlantic coast than to *M. gayi* from the South Pacific (Fig. 8.12).

Fernández (1985) and Fernández and Durán (1985) studied the parasites of the Patagonic population of *M. australis* in an attempt to clarify some of these latter arguments. Based mainly on its specific copepods and the sanguinicolid trematodes (*Aporocotyle* spp.) in this and other hakes (see also Villalba and Fernández, 1986), they concluded that *M. australis* was more closely related to *M. hubbsi* on the Atlantic coast

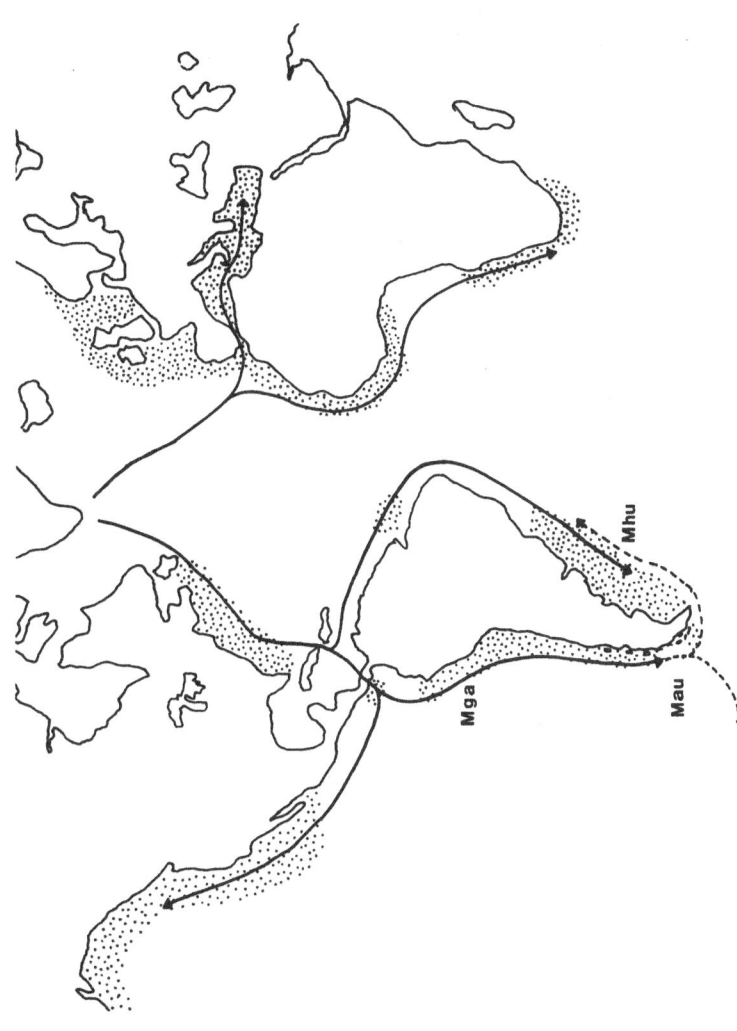

Figure 8.12 Origin and dispersal of the genus *Merluccius* according to the hypotheses of Kabata and Ho (1981) and Inada (1981). The major disagreement between these hypotheses relates to the dispersal of *M. hubbsi* (Mhu) and *M. australis* (Mau) from New Zealand. Continuous lines indicate dispersal of these latter species according to Kabata and Ho (1981) and broken lines according to Inada (1981).

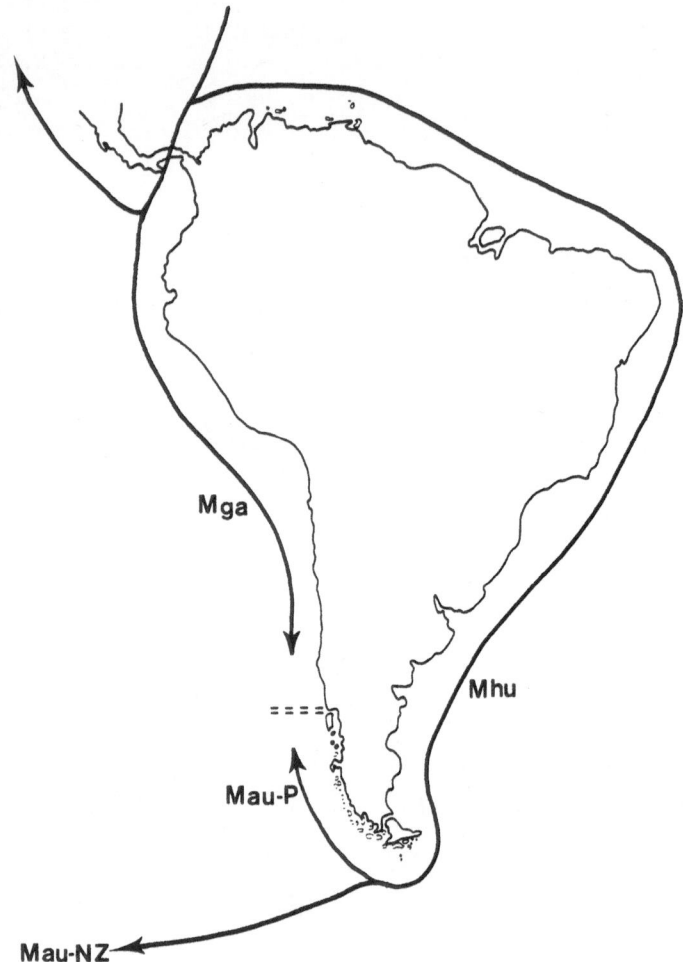

Figure 8.13 Possible dispersal of *Merluccius* around the South American continent based on the parasites of *M. australis* from the coast of Chile: Mau-P, *M. australis*, Patagonic population; Mau-NZ, *M. australis*, New Zealand population; Mhu, *M. hubbsi*; Mga, *M. gayi*.

and,from a parasitological perspective, unrelated to the sympatric- parapatric species, *M. gayi* (Fig. 8.13). This, then, supported Kabata and Ho's (1981) hypothesis regarding an Atlantic origin for *M. hubbsi* and *M. australis*.

In closing, it is worth noting that all of these observations and hypotheses are based on dispersal concepts and none considers the possibility of vicariant events. Moreover, none of these studies has employed phylogenetic methods for either hosts or parasites similar to the one devised by Brooks, Thorson and Mayes (1981). The final word regarding the origin and 'dispersal' of *Merluccius*, at least, must await these more refined analyses.

9 *Evolutionary aspects*

9.1 INTRODUCTION

Biological **evolution** is the process by which there is a change in gene frequency within a population from one generation to the next. The changes in gene frequency may be reflected in individual organisms through readily apparent morphological, behavioural, or physiological alterations, or sometimes via subtle factors that are not always easy to evaluate; in other cases, genetic change may not be apparent at all. It must be reiterated that unless these changes are transmitted from generation to generation, they cannot be considered to be biological evolution.

Evolution, as for many of the other phenomena already described in this book, is a scale-dependent process. Thus, **microevolution** is concerned with evolution as it occurs within a species while **macroevolution** addresses evolutionary questions above the species level. The temporal and spatial frames within which evolution can be studied at these two levels are very different; so are the methodologies employed. The individual species becomes the line of separation between micro- and macroevolutionary phenomena because genetic discontinuities between species are absolute. This means that genetic change in one species cannot be transferred to another because, for the most part, sexually reproducing organisms of different species cannot exchange genes due to reproductive barriers (the possibility of horizontal gene transfer between species will be considered at the end of this Chapter). In effect, most species are independent evolutionary units. There are exceptions and most of these are symbiotic in character; they would involve mutuals, commensals, parasites, **and** their partners.

9.2 MICROEVOLUTION

9.2.1 Introduction

In sexually reproducing organisms, a **species** can be defined as a group of potentially interbreeding individuals that are reproductively isolated from other groups of organisms and that produce reproductively viable offspring. Members of a species usually are not homogeneously distributed in space. Free-living organisms (hosts) are characteristically subdivided into smaller groups owing to environmental patchiness. Geographical areas having favourable habitats are intermixed with unfavourable ones; there is thus a tendency to reduce (or even eliminate) potential interbreeding among individuals occupying the different habitats. Behavioural variations (schooling, flocking, herding, colony formation or, in contrast, territoriality) may also enhance the effects of spatial heterogeneity on reproduction and gene flow. Locally interbreeding individuals of geographically isolated, or otherwise structured, groups are considered as populations or demes. The significance of the spatial distribution of populations rests with the idea that it is within these subunits where changes in gene frequencies are most likely to occur and, therefore, where evolution takes place. Because the various selection forces operating on these subunits may not be the same across the geographical range of a species, the population or deme must be considered as the basic unit of evolution. On the other hand, the precise geographical boundaries for these populations are not always fixed. For example, migration of individuals into and out of a habitat will considerably reduce the geographical and, therefore, the reproductive and genetic independence of populations.

The patterns of population structure associated with parasitic organisms are far more complex than those of their hosts. In part, this is related to several of the functional characteristics of parasitism itself. For example, parasites occur as infrapopulation subunits within which effective gene flow will take place for all, or part, of each sexually reproductive cycle. These infrapopulations may then disappear, only to reappear as completely new and genetically distinct infrapopulations in the next, sexually reproductive generation. Interestingly, there is no direct gene flow among parasites at the meta- or suprapopulation level. Intuitively, one might also predict that the high degree of overdispersion among parasite infrapopulations could uniquely contribute in some way to the population genetics of parasites, but this would be speculative. On the other hand, given that hosts are themselves overdispersed, perhaps the superimposition of parasite overdispersion on the overdispersion of the host could affect gene flow or, conversely, enhance drift via the constant renewal of potential founder effects with each new reproductive cycle. The notion of selection pressure imposed by the host's immune responsiveness is also an important aspect of the population genetics of a parasite species and will be considered in more detail later.

Price (1977, 1980) acknowledged the idea that populations were the basic units of evolution and emphasized that their population structure was a major factor in the evolutionary biology of parasitic organisms. On theoretical grounds, Price (1980) argued that because parasites are adapted to exploit small and discontinuous environments (hosts) in a matrix of inhospitable environments, they should have small and homogeneous populations with little gene flow between them. However, if host variability in time and space is considered, together with the specialized resource exploitation shown by the parasites, then three additional predictions are possible. Thus, if the environment (host) is stable in time and variable in space, then the local parasite population should be monomorphic and specialized, with several geographical races. But, if the environment is uniform in space and variable in time, the parasite population should be polymorphic, with several specialized types present on a cline within a geographical scale. Finally, when the host is variable in both space and time, Price (1980) predicted the formation of both geographical races and polymorphisms in the parasite population.

For evolution to occur, a population must have some degree of genetic variability; this will enhance the probability of changes in gene frequencies over time. Heritable variation in a population can arise through mutation and genetic recombination (but see section 9.4 for the possibility of horizontal gene transfer). Genetic drift, migration (gene flow) and natural selection are, on the other hand, the processes by which these variations are enhanced, reduced, or eliminated from a population. Although the mechanisms of reproduction will not affect allele frequencies, they have to be considered because they will influence genotype frequencies.

Genetic variation within a population frequently has been studied by the electrophoretic differentiation of **allozymes** (alternative enzyme forms encoded by different alleles at the same locus). Genetic variation is commonly quantified using **polymorphism** (P) and **heterozygosity** (H). The former is the proportion of polymorphic loci in a population, while the latter is the average frequency of heterozygous individuals per locus.

Theoretical research has provided numerous models for the study of population structure, gene flow, genetic drift, and differences in the size of the actual breeding population (the effective population size) using data from allele frequencies. Based on these simulations, it is possible to identify those factors that drive the evolution of various species. Fisher, as early as 1930, demonstrated mathematically that there was a direct correlation between the amount of genetic variation in a population and the rate of evolutionary change influenced by natural selection, as measured by individual **fitness** (the relative reproductive rate). Although numerous vertebrates and free-living invertebrates have been characterized at the population level (Avise and Aquadro, 1982; Nevo, Beiles and Ben-Shlomo, 1984), very few parasite species have been thus considered.

9.2.2 Genetic variability and metapopulation structure of parasites

Most parasite metapopulations have levels of genetic variation that are similar to those of free-living invertebrates (Tables 9.1 and 9.2). However, the data available regarding geographically separated infra- and meta-populations of a given species are not extensive. In a study of enzyme polymorphism and genetic variability in the nematode, *Ascaris suum*, Leslie *et al.* (1982) provided one of the first insights into the nature of variation in allele frequencies in relation to the extent of geographical separation. Levels of variation in *A. suum* were in the lower range for other parasites (Table 9.1); thus, $P = 0.207$, $H = 0.066$ in Iowa (USA) and $P = 0.167$, $H = 0.053$ in New Jersey (USA). Based on genotype frequency expectations from the Hardy-Weinberg law of equilibrium, worms within each locality appeared to be members of a single metapopulation, apparently sharing a common gene pool. Interestingly, comparisons between the Iowa and New Jersey metapopulations (a distance of nearly 1500 miles) did not reveal marked differences. With one exception, the same loci were polymorphic in Iowa and New Jersey populations and there were no significant differences in allele frequencies. Because these geographically distant metapopulations exhibited little genetic differentiation, it is likely that either gene flow between the two locations has been high historically, or cessation of the gene flow may have been a recent event. For many parasites of farm and other domesticated animals, the idea of genetically independent or distinct parasite metapopulations is probably unrealistic because gene flow may be enhanced easily through long-distance transport of farm stock, the spread of contaminated manure or by other husbandry practices.

Studies of genetic variability among different populations of ticks such as *Amblyomma americanus*, *Ornithodoros erraticus*, *O. senrai* in the USA or among different populations of six species of reptilian ticks in Australia (Wallis and Miller, 1983; Bull, Andrews and Adams, 1984; Hilburn and Sattler, 1986a, b), revealed a pattern similar to the one observed for *A. suum* (Leslie *et al.*, 1982). Thus, there was a low level of genetic variability within, as well as between, different geographical populations. In the American tick studies, both Nei's genetic distance and F-statistics were used to estimate the degree of inbreeding and divergence within and

Table 9.1 Genetic variability estimates for helminths. Criteria for defining polymorphism (P) and estimating average heterozygosity (H) vary by author. (Modified from Nadler, 1990).

Species	Locality	Number of loci surveyed	P	H	Reference[a]
Trematoda					
Paragonimus westermani	Mie (1), Japan	18	0.12	0.035	Agatsuma & Habe, 1985b
Paragonimus westermani	Mie (2), Japan	18	0.06	0.033	Agatsuma & Habe, 1985b
Paragonimus westermani	Ohita, Japan	18	0.0	0.0	Agatsuma & Habe, 1985b
Schistosoma mansoni	(mean of 22 strains) various localities	18	0.13	0.04	Fletcher et al., 1981
Schistosoma japonicum	Leyte, Phillippines	18	0.0	0.0	Woodruff et al., 1987
Schistosoma mekongi	Laso	15	0.0	0.0	Woodruff et al., 1987
Halipegus occidualis	North Carolina, U.S.A.	8	0.125	0.05	Goater et al., 1990
Fascioloides magna	Tennessee, U.S.A. [1]	22	0.643	0.09	Lydeard et al., 1989
Fascioloides magna	Tennessee, U.S.A. [2]	22	0.429	0.05	Lydeard et al., 1989
Fascioloides magna	South Carolina, U.S.A. [3]	22	0.429	0.13	Lydeard et al., 1989
Fascioloides magna	South Carolina, U.S.A. [4]	22	0.429	0.10	Lydeard et al., 1989
Cestoda					
Progamotaenia festiva (species complex)	Australia, various localities	16	n/a	0.03*	Baverstock et al., 1985
Echinococcus granulosus	Australia, mainland	20	0.15	0.02	Lymbery and Thompson, 1988
Echinococcus granulosus	Tasmania	20	0.15	0.06	Lymbery and Thompson, 1988
Nematoda					
Contracaecum sp. 'I'(larvae)	Mexico	11	0.54	0.141	Vrijenhoek, 1978
Contracaecum sp. 'II' (larvae)	Mexico	11	0.54	0.193	Vrijenhoek, 1978
Contracaecum osculatum 'B'	n/a**	21	0.62	0.10	Bullini et al., 1986
Contracaecum rudolphii 'A'	n/a	21	0.57	0.17	Bullini et al., 1986
Anisakis simplex (larvae)	North Atlantic Ocean	22	0.50	0.21	Nascetti et al., 1986
Anisakis pegreffii (larvae)	Mediterranean Sea	22	0.32	0.12	Nascetti et al., 1986

Table 9.1 (continued)

Species	Locality	Number of loci surveyed	P	H	Reference[a]
Anisakis physeteris (larvae)	n/a	22	0.50	0.11	Bullini et al., 1986
Phocascaris cystophorae	n/a	21	0.24	0.10	Bullini et al., 1986
Ascaris suum	Iowa, U.S.A.	38	0.21	0.066	Leslie et al., 1982
Ascaris suum	New Jersey, U.S.A.	38	0.17	0.053	Leslie et al., 1982
Ascaris suum	n/a	24	0.17	0.03	Bullini et al., 1986
Ascaris lumbricoides	n/a	24	0.25	0.02	Bullini et al., 1986
Parascaris equorum	Louisiana, U.S.A.	18	0.22	0.085	Nadler, 1986
Parascaris equorum	Central-eastern Europe	27	0.03	0.008	Bullini et al., 1978
Parascaris equorum	n/a	28	0.07	0.02	Bullini et al., 1986
Parascaris univalens	Central-eastern Europe	27	0.03	0.015	Bullini et al., 1978
Parascaris univalens	n/a	28	0.11	0.03	Bullini et al.,1986
Neoascaris vitulorum	n/a	18	0.11	0.04	Bullini et al., 1986
Toxocara canis	n/a	18	0.33	0.10	Bullini et al., 1986
Toxocara canis	Louisiana, U.S.A.	18	0.33	0.135	Nadler, 1986
Toxocara cati	Louisiana, U.S.A.	18	0.38	0.137	Nadler, 1986
Toxocara cati	n/a	18	0.17	0.05	Bullini et al., 1986
Toxascaris leonina	n/a	18	0.11	0.02	Bullini et al., 1986
Baylisascaris transfuga	n/a	18	0.17	0.05	Bullini et al., 1986

[a] References can be found in Nadler (1990)
* Mean for species studied
** n/a, not available
(1) Shelby Forest Wildlife Management Area; (2) Reelfoot National Wildlife Refuge; (3) Savannah River Plant; (4) Webb Wildlife Center

Table 9.2 Polymorphism (P) and heterozygosity (H) and their coefficient of correlation (r) for various organisms. (From Nevo, 1978.)

	Number of species	P Mean	P SD	H Mean	H SD	r(P,H)
Plants	15	0.259	0.166	0.0706	0.0706	0.206 -
Invertebrata (except insects)	27	0.399	0.275	0.1001	0.0740	0.788***
Insecta (except *Drosophila*)	23	0.329	0.203	0.0743	0.0810	0.680***
Drosophila	43	0.431	0.130	0.1402	0.0534	0.637***
Osteichthyes	51	0.152	0.098	0.0513	0.0338	0.883***
Amphibia	13	0.269	0.134	0.0788	0.0420	0.785***
Reptilia	17	0.219	0.129	0.0471	0.0228	0.605**
Aves	7	0.150	0.111	0.0473	0.0360	0.900**
Mammalia	46	0.147	0.098	0.0359	0.0245	0.838***
Total	242	0.263	0.153	0.0741	0.0510	0.678***
Reduced to three groups						
Plants (1)	15	0.259	0.166	0.0706	0.0706	0.206 -
Invertebrates (2–4)	93	0.397	0.201	0.1123	0.0720	0.710***
Vertebrates (5–9)	135	0.173	0.119	0.0494	0.0365	0.823***

Levels of significance: ** = P<0.01
*** = P<0.001
- = Non significant

between supposed populations. In the case of *A. americanus*, nine geographical populations were examined. Some of them exhibited quantitative and qualitative differences that argued in favour of a small degree of genetic structuring. However, the differences were not large enough to affect the measurements of genetic relatedness. Because the degree to which populations diverge genetically is primarily determined by the amount of gene flow between them, even infrequent exchange between populations will prevent extensive genetic divergence. Since ticks are relatively inactive when they are not on a host, the rate of gene exchange will depend on host vagility. The low host specificity and the number of hosts in the life cycle also will contribute to increasing, or decreasing, the amount of gene flow between geographical populations. For *A. americanus*, both of these factors are important. Each tick attaches and feeds on three hosts during its lifetime but, each time the ticks moult or lay eggs, they leave the host and remain in the ground. This behaviour enhances its chances of capturing a different individual or host species each time it re-attaches, thereby increasing its chances of dispersal and gene flow.

It thus appears that host vagility, range of distribution, the nature of the parasite's life cycle, and its host specificity, may all influence the degree of divergence between what otherwise appear to be geographically distinct metapopulations. Based on these observations and those of Leslie *et al.* (1982), it may be necessary to reconsider the original definition of a population from a genetic perspective. Therefore, in the cases of *A. suum* and the Australian ticks, what originally were thought to be distinct meta-populations based on geographical distributions, are apparently just groups of individuals of the same metapopulation from geographically distant areas. The questions regarding what constitutes a population, as expressed by Leslie *et al.* (1982), are worth considering, to wit, 'Does a population consist of all the worms in the pigs of a single farm?, In a county?, On a continent?'. It appears that the best way to resolve this dilemma would be to compare samples of parasites collected along a geographical transect as well as individually within each host (Nadler, 1990).

This latter approach was undertaken to some extent by Lydeard *et al.* (1989) in their study of the liver fluke, *Fascioloides magna*, in the white-tailed deer, *Odocoileus virginianus*, in the southeastern USA and by Lymbery and Thompson (1988, 1989) on the cestode, *Echinococcus granulosus*, in Australia. The former authors examined four, geographically separated metapopulations of *F. magna*, two from Tennessee and two from South Carolina. Levels of genetic variability (including both H and P) in the different locations were comparable to those reported for other species of parasites (Table 9.3). Genetic differences between localities were observed as the degree of differentiation (or genetic distance) increased with geographical separation. On partitioning the total genetic variation in *F. magna*, 83% of the variance was found among individuals within a sample population. Significant genetic heterogeneity among flukes was evident in the deficiency of heterozygous individuals within a population. In addition, there was a deficiency of heterozygous individuals within localities. The latter observation might suggest extensive inbreeding or, alternatively, spatial subdivision of the liver fluke population for each locality. Also, there could be a high degree of parasite relatedness in the deer as a consequence of the asexual reproduction (polyembryony) that takes place in the snail intermediate host. Accordingly, large numbers of genetically identical cercariae are released from an infected snail and encyst as metacercariae on vegetation within a narrowly defined geo-graphical area. As individual deer or small groups of deer browse, their chances of acquiring metacercariae derived from a single, genetically distinct, adult thus would be considerably enhanced. Significant spatial

Table 9.3 Allele frequencies, sample sizes (± 1 SE), alleles per locus, proportion of polymorphic loci (P), and heterozygosity (H, direct count) for four populations of *Fascioloides magna*. (From Lydeard et al., 1989.)[*]

Locus	SFWMA,TN	RNWR,TN	SRP,SC	WWC,SC
			Location[**]	
Allele				
$Acon^{100}$	–	–	0.34	0.38
$Acon^{60}$	0.92	1.00	0.32	0.30
$Acon^{38}$	0.08	–	0.24	0.19
$Acon^{null}$	–	–	0.10	0.13
$f\text{-}Est^{100}$	0.28	0.02	0.53	0.64
$f\text{-}Est^{89}$	0.72	0.98	0.47	0.36
Aat^{188}	0.13	0.05	0.14	0.11
Aat^{100}	0.87	0.95	0.86	0.89
Lap^{112}	0.01	–	–	–
Lap^{100}	0.99	1.00	1.00	1.00
$la\text{-}Pep^{120}$	0.10	–	–	–
$la\text{-}Pep^{100}$	0.90	1.00	1.00	1.00
Mdh^{200}	0.25	0.41	0.17	0.10
Mdh^{100}	0.75	0.59	0.83	0.90
Mnr^{184}	0.01	0.01	0.02	0.02
Mnr^{125}	–	–	0.01	0.01
Mnr^{108}	–	–	0.01	0.02
Mnr^{100}	0.99	0.99	0.96	0.95
Gpd^{123}	0.03	0.01	–	–
Gpd^{100}	0.97	0.99	1.00	1.00
$p\text{-}Est^{100}$	0.99	0.98	0.95	0.98
$p\text{-}Est^{96}$	0.01	0.02	0.05	0.02
Mean sample size per locus	278.9 ± 22.8	92.5 ± 3.1	1214.6 ± 81.8	55.0 ± 9.8
Mean number of alleles per locus	1.6	1.4	1.8	1.6
P	64.3	42.9	42.9	42.9
H	0.09 ± 0.04	0.05 ± 0.03	0.13 ± 0.06	0.10 ± 0.05

[*]*Acon* frequencies for the SRP and WWC samples were based on maximum likelihood estimations; the remaining are observed frequencies.
[**]SFWMA, Shelby Forest Wildlife Management Area; RNWR, Reelfoot National Wildlife Refuge; SRP, Savannah River Plant; WWC, Webb Wildlife Center.

variation in genetic structure was also evident between sites within a state (4%), as well as between states (13%). The low degree of divergence between localities within a state suggests some gene flow may occur among these populations.

The cestode, *Echinococcus granulosus*, is perhaps one of the most studied parasites with respect to genetic variability. Electrophoretic and DNA analyses have been combined to support the idea that *E. granulosus* consists of a complex of genetically distinct strains. Individual strains appear to be homozygous and monomorphic; certainly, the two strains examined intensively in the UK conform to this pattern. In the UK, one strain cycles between horses and hunting hounds and another completes its life cycle using sheep and sheep dogs (McManus and Smyth, 1979; MacPherson and McManus, 1982; McManus and Simpson, 1985).

The strains of *E. granulosus* identified on both the mainland of Australia and in Tasmania exhibit small levels of polymorphism and heterozygosity; both strains differ in allelic frequencies at two loci (Table 9.4), but diagnostic alleles for each one of these strains were not found, indicating that differentiation between the two strains was not well defined (Lymbery and Thompson, 1988). These patterns are thought to result from a combination of two processes. The first is associated with the method of reproduction by the worms. The parasite thus is presumed to be largely self-fertilizing (Smyth and Smyth, 1964, 1969) favouring the production of homozygous progeny. Moreover, since larvae (hydatid cysts) reproduce clonally, a large number of genetically identical protoscoleces will be produced as the result of infection of an intermediate host by a single egg (each protoscolex develops into an adult worm). Self-fertilization, in the absence of extensive mutation, will lead to a reduction in genetic diversity within strains, but an increase in genetic difference between strains (McManus and Smyth, 1979; Bryant and Flockhart, 1986). The second process associated with strain formation may be strongly influenced by the long-term and mostly restricted relationship of the parasite with specific kinds of domestic hosts. Practices in animal husbandry may have created artificial, but powerful, geographical and ecological barriers between parasite populations that are associated with different specific hosts. These

Table 9.4 Proportion of polymorphic loci (P) and total gene diversity (H) in samples of *Echinococcus granulosus* from the mainland of Australia and Tasmania, compared with mean values from 16 species of parasitic helminths and 27 species of free-living invertebrates (excluding insects). (From Lymbery and Thompson, 1989).

Sample	P	H
E. granulosus (mainland)	0.15	0.02
E. granulosus (Tasmania)	0.15	0.06
Parasitic helminths	0.32	0.08
Free-living invertebrates	0.39	0.10

various metapopulations easily could have diverged genetically to produce the strains in different domesticated host species (Rausch, 1985).

From this example, it is evident that an integration of factors across different scales is necessary to gain insight into the genetic structure of parasite metapopulations. However, despite the fact that *E. granulosus* is one of the most studied parasite species, there are still some scaling problems in delimiting their, and probably other parasite, meta-populations. Lymbery and Thompson (1989) warned about the inappro-priateness of some studies because genetic variation was analysed at a level where strain differences could not be recognized. These authors found isozyme differences between protoscoleces derived from different cysts in the same host, suggesting that a given host might possess hydatid cysts derived from genetically distinct adults, not unlike the pattern of variation observed for *F.magna* in deer (Lydeard *et al.*, 1989). Genetic variation on this scale must be addressed on a cyst-by-cyst basis rather than from pooled analysis of several cysts.

As can be seen, studies on the genetic structure and variability of parasite infra- and metapopulations are still in their infancy with respect to methodologies and interpretation. Although similar limitations were resolved at least a decade ago for free-living organisms, the extra spatial-temporal dimension added by the host(s) and the subtle character of many host–parasite interactions has made rapid advances difficult. On the other hand, examination of genetic structure and variability will be extremely helpful for a clear understanding of basic aspects of the population biology and epidemiology of parasites in general, and hel-minths in particular.

9.2.3 Correlates of genetic variability

It is evident from Tables 9.1 and 9.2, that different species and meta-populations have varying degrees of polymorphism and heterozygosity. Because of these variations, two important questions arise: what mechanisms account for the differences and how are these differences maintained within parasite metapopulations? Two hypotheses have been proposed to explain the differing degrees of polymorphism and hetero-zygosity. One is based on a neutralist perspective and the other on a selectionist approach. Although there has been a long-term debate between the two schools of thought, in reality they are not mutually exclusive. Indeed, Nadler (1990) has urged that the controversy should be more properly focussed on what proportion of the observed variations is maintained by one mechanism or the other.

The **neutralist hypothesis** is based on the observation that there is a relatively constant rate of molecular evolution in homologous molecules from different species lineages because most of the allelic changes that take place over time are selectively neutral. It is agreed by both sides that new alleles appear by mutation. However, the neutralists argue that if alternate alleles have identical fitness, then alterations in allelic frequencies will occur only by accidental change from generation to generation (drift). Genetic drift will be affected by modifying effective infrapopulation sizes via founder effects, bottlenecks, and departures from 1:1 sex ratios (Roughgarden, 1979; Nadler, 1987). The neutralist hypothesis recognizes, however, that deleterious mutations can be eliminated by natural selection.

The **selectionist hypothesis** assumes that the changes in gene frequencies are adaptive. As a consequence, adaptive variations will gradually increase in frequency over time at the expense of less adaptive ones. The selection hypothesis also explains the diversity of organisms because it promotes their adaptation to new and different life styles. According to the selectionist viewpoint, genetic variability among species and populations is often positively correlated with levels of ecological heterogeneity (Selander and Kaufman, 1973; Nevo, 1978). Species, or populations, from 'heterogeneous' environments should have increased genetic variability in response to diverse selection pressures. Similarly, in homogeneous environments or in organisms that experience the environment as 'fine grained', genetic variability should be lower because a specialized genotype is more likely to be selected. Data from free-living organisms seem to support this hypothesis (Powell, 1975; Selander and Kaufman, 1973; Nevo 1978). Thus, geographically widespread generalists occupying heterogeneous habitats have greater genetic variability than geographically restricted specialists with narrow niches.

Bullini *et al.* (1986) attempted to compare the genetic structure of several ascaridoid nematodes with single and multiple host life-cycles (Table 9.5). Their working hypothesis was based on the observed correlation between the genetic variation of free-living organisms and the degree of environmental variation they experienced (Powell, 1975; Selander and Kaufman, 1973; Nevo, 1978). The evidence indicated that species whose life cycles were completed in a single homeothermic host had lower genetic variability (both polymorphism and heterozygosity) than those species using both poikilothermic and homeothermic hosts in completing their life cycles (Table 9.5). They argued that in species with more than one host in their life cycle, a given allele may have greater fitness in one stage of the life cycle whereas another allele may function optimally in another stage. In a multiple-host life cycle then, heterozygosity would be favoured. Bullini *et al.* (1986) also argued that ascaridoid species with life cycles that were completed exclusively in homeothermic hosts were buffered against environmental variation. They

Table 9.5 Parameters of genetic variability in ascaridoid worms needing respectively one homeothermic host (upper group) or several hosts, both poikilothermic and homeothermic (lower group) to carry out their life cycles. The values in brackets were calculated on the 12 loci shared by all the species studied. t-tests were used to estimate the significance of the differences between the means. N_g = average number of genes sampled per locus; N_l = number of loci studied; H_c = expected mean heterozygosity per locus; P = proportion of polymorphic loci (1% criterion); A = mean number of alleles per locus (alleles with frequency <0.01 were not considered for computation). (From Bullini *et al.* 1986.)

	N_g	N_l	H_c		P		A	
Ascaris lumbricoides	92	24	0.02	[0.04]	0.25	[0.33]	1.38	[1.58]
Ascaris suum	445	24	0.03	[0.05]	0.17	[0.25]	1.29	[1.50]
Parascaris univalens	447	28	0.03	[0.08]	0.11	[0.25]	1.18	[1.50]
Parascaris equorum	131	28	0.02	[0.04]	0.07	[0.17]	1.11	[1.33]
Neoascaris vitulorum	253	18	0.04	[0.04]	0.11	[0.08]	1.11	[1.17]
Toxocara canis	157	18	0.10	[0.13]	0.33	[0.42]	1.50	[1.58]
Toxocara cati	44	18	0.05	[0.09]	0.17	[0.25]	1.22	[1.42]
Toxascaris leonina	46	18	0.02	[0.04]	0.11	[0.17]	1.11	[1.25]
Baylisascaris transfuga*	10	18	0.05	[0.09]	0.17	[0.25]	1.17	[1.33]
Average ± standard error			0.04 ± 0.01	[0.06 ± 0.01]	0.16 ± 0.03	[0.24 ± 0.04]	1.24 ± 0.05	[1.42 ± 0.05]
Anisakis simplex A	272	22	0.12	[0.16]	0.64	[0.67]	2.32	[2.83]
Anisakis simplex B	251	22	0.21	[0.19]	0.64	[0.58]	2.60	[2.67]
Anisakis physeteris	84	22	0.11	[0.14]	0.50	[0.67]	1.95	[2.50]
Phocascaris cystophorae*	12	21	0.10	[0.17]	0.24	[0.42]	1.33	[1.58]
Contracaecum osculatum A	78	21	0.12	[0.10]	0.48	[0.50]	2.14	[2.33]
Contracaecum osculatum B	169	21	0.10	[0.15]	0.62	[0.75]	2.71	[3.17]
Contracaecum rudolphii A	87	21	0.17	[0.18]	0.57	[0.67]	2.24	[2.58]
Contracaecum rudolphii B	65	21	0.21	[0.19]	0.62	[0.67]	2.33	[2.67]
Average ± standard error			0.15 ± 0.02	[0.16 ± 0.01]	0.58 ± 0.02	[0.64 ± 0.03]	2.33 ± 0.10	[2.68 ± 0.10]
Student's t			5.61	[5.57]	10.29	[8.31]	10.23	[11.44]
Probability			<0.001	[<0.001]	<0.001	[<0.001]	<0.001	[<0.001]

*Not considered in computing the averages, because of the small sample size.

did not require genetic flexibility in order to cope with environmental heterogeneity (it was implicit that the number and kind of hosts were considered to be the source of heterogeneity). On the other hand, those species using both poikilothermic and homeothermic hosts were subjected to greater environmental variability and thus benefited from increased genetic flexibility. They also suggested that nematodes with low host specificity as larvae (like *Anisakis simplex*, see also Chapters 5 and 8) were subjected to extreme ecological heterogeneity because of the many species of fishes acting as potential paratenic hosts and should also exhibit a greater degree of genetic variability.

Although these conclusions (Bullini *et al.*, 1986) are intuitively logical, they were not totally correct; it is suggested that their experimental design and analyses compromised their results. For example, to consider the number of hosts as a measure of habitat heterogeneity is an inaccurate assumption. Even parasites with a one-host life cycle may encounter a high degree of environmental heterogeneity if they are required to migrate extensively through different organs before reaching the final site of infection in a host. Habitat heterogeneity in these cases can be related to variations in temperature, pH, osmolarity, concentrations of dissolved oxygen, digestive enzymes, immunological responsiveness, and so on. An additional problem in their analysis was that all the species with a one-host life cycle were terrestrial, while the multiple-host species were all marine. Nevo (1978) showed that among free-living invertebrate species, those from marine environments had a higher degree of genetic variability than those from terrestrial environments. It is therefore possible that the differences observed by Bullini *et al.* (1986) could have been due to the quality of the habitat rather than the nature of the life cycle.

Finally, if enzyme polymorphisms and their correlation with environmental heterogeneity are used as arguments to explain their maintenance in the populations by natural selection, then comparative studies on the biochemical properties of different allozymes (under different temperatures, pH, etc.) are necessary to support such arguments. These studies should generate a better understanding of how selection is operating. On the other hand, the hypothesis of increased genetic heterozygosity and polymorphism can also be tested in terms of the neutralistic theory of molecular evolution. Thus, if heterozygosity is selected for, then allelic frequency distributions should depart from random expectations (the neutralist argument) to maximize **heterosis** (hybrid vigour). Nadler (1990) took this approach and compared the allelic frequency distributions of three nematode species (*Toxocara canis*, *T. cati* and *Parascaris equorum*) against the so-called **infinite allele–constant mutation model** (Chakraborty, Fuerst and Nei, 1980) that predicts expected

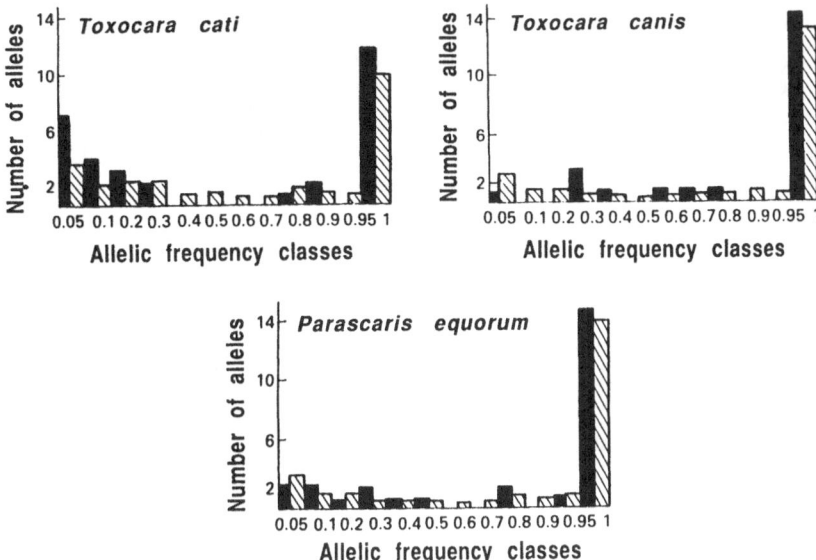

Figure 9.1 Observed (solid) and expected (hatched) distributions of number of alleles by frequency for three species of ascaridoid nematodes. Observed distributions based on results of protein-electrophoretic studies of individuals (Nadler, 1986). Expected distributions calculated using the infinite allele-constant mutation rate model (Chakraborty, Fuerst and Ne; 1980). (With permission, from Nadler, 1990.)

frequencies according to the neutralistic theory (Fig. 9.1). No differences were found between the observed and expected distributions of allele frequencies in these species, indicating that heterozygosity had not been selected for and that the levels observed may have been due to random genetic events.

9.3 EVOLUTION OF HOST–PARASITE INTERACTIONS

9.3.1 Genetics of host–parasite interactions

As defined previously in section 9.2.1, evolution is a change in gene frequency from one generation to the next. Three criteria must be met in order for evolution to occur. First, there must be genetic variation within the population. Second, the variation must be at least partially heritable.

And, third, the variation must affect the probability of leaving reproductively viable offspring to the next generation. In host–parasite systems, all three of these criteria must be met by both hosts and parasites for co-evolution to occur. Examples of the types of variation most commonly involved are resistance/susceptibility factors within the host population and virulence/avirulence factors in the parasite metapopulation.

Most evidence for the inheritance of resistance comes from observations that different strains of hosts vary in their degree of susceptibility to a specific parasite. In many cases, when artificial selection is applied for different levels of susceptibility, a satisfactory response is obtained. However, in most of these same cases, the basic genetics has not been determined (Taylor and Muller, 1976). One of the classic examples regarding changes in susceptibility/resistance patterns is the reciprocal genetic switch in the rabbit–myxoma virus interaction. Although it is generally accepted now that group selection accounted for the co-evolution of both the rabbit and the virus, the genetic ramifications underlying the process still are not fully understood (Barrett, 1984).

Inherited variation of pathogenicity in both plant and animal parasites also has been well documented. However, based on the different approaches to control infection in animals and plants, most of the information on the genetics of pathogenicity is restricted to plants. In man and other animals, the traditional approach for controlling parasites has been to improve hygiene and apply prophylactic measures to prevent infections. As a result, breakdowns in control are frequently associated with the evolution of evasive strategies by the parasite, e.g. the use of reservoir hosts, but the influence of the host on the evolution of the parasite is generally considered as minimal (Barrett, 1984). Any variation in the ability of the parasite to induce infection will be observed only during experimental manipulations. For example, the severity of infection induced in standard test animals by the protozoans, *Trypanosoma cruzi* and *Trichomonas vaginalis*, depends on the parasite strain used, as well as the strain of the test animal (Levine, 1973). Parasite–plant combinations, on the other hand, are genetically well known because the use of genetically mediated resistance by plants is a primary defence against pathogens. Accordingly, a breakdown in control due to any evolution in the ability to overcome this resistance can be directly attributed to a genetic component in the host–parasite interaction. Unlike animal systems, an understanding of the genetics of pathogenicity in plant parasites is a major objective in the control of plant diseases (Day, 1974).

The functional biology of host–parasite interaction is directly related to the quality of the interaction that occurs. For the most part, however, the genetics of these interactions have not been well studied. Most of the research in this area comes again from plant–parasite systems. Flor (1942, 1955, 1956), in a series of pioneering experiments, investigated the genetics of pathogenicity and resistance (simultaneously) between flax, *Linum usitatissimum*, and the flax-rust fungus, *Melampsora lini*. He found that resistance in the host was controlled by simple Mendelian genetics and that the ability of the fungus to attack different variants was controlled by genes that segregated in the F2 generation. He summarized his results in what has since become known as the **gene-for-gene hypothesis**. It states that for every gene controlling resistance in the host, there is a corresponding gene in the parasite affecting pathogenicity. These gene-for-gene systems have been shown to operate in several other plant–parasite combinations and are not restricted to plant–fungus interactions. However, the nature and extent of gene-for-gene interactions are at the present time under strong scrutiny because there is a suspicion they may be artefacts of man's agricultural activities (Day, 1974; Barrett, 1984).

One of the few animal systems in which host–parasite genetic variation has been studied (at least partially) involves the trematode, *Schistosoma mansoni*, and its intermediate and definitive hosts. Richards (1973, 1975a, b, 1976a, b, 1984) demonstrated that there were a number of genetic factors in the snail, *Biomphalaria glabrata*, governing susceptibility to infection with specific strains of *S. mansoni*. Although some aspects of host–parasite compatibility were under simple genetic control, genetic factors related to snail age altered the pattern of susceptibility/ unsusceptibility between juvenile and adult snails of a given strain. Studies concerning the infectivity of different schistosome strains to the snail host have been logistically more difficult to carry out, but the available evidence suggests that closely related parasites, at the species or strain level, can produce quite different reactions in the same snail strain (Rollinson and Southgate, 1985).

The genetic mechanisms of schistosome-definitive host interactions also are only partially known. There is evidence showing the importance of the host phenotype in susceptibility and pathological response to infection (Fanning *et al.*, 1981). Similarly, marked differences in the proportion of successful cercariae developing into adult worms have been found among different mice strains (Fanning and Kazura, 1984). Although the evidence was limited, the results suggested that susceptibility to primary infections by *S. mansoni* was a polygenic phenomenon. Moreover, studies in congenic mice indicated that the genes of the major histocompatibility complex did not play a significant role in the ability of *S. mansoni* cercariae to develop into adults.

Quantitative data relating to the frequencies of compatibility traits in natural host–parasite populations are virtually non-existent. Even in the well developed areas of plant–parasite interaction, changes in populations sizes of hosts and parasites as a result of their interaction have been largely ignored. Haldane (1949, 1954), while studying human blood groups, proposed that parasites could be a determining factor in the maintenance of certain blood group polymorphisms. He argued that if parasites were successful because they could not be distinguished antigenically from self by the host, then the success of a single parasite genotype would be reduced in a host population that was variable for blood protein antigens. Parasites would tend to evolve adaptations to attack the more common host genotypes and, as a result, the rare host genotypes would be at a selective advantage. This process would result in a continuous, frequency-dependent selection of both hosts and parasites as each responded to the influences of the other.

In addition to blood group polymorphisms, there is also circumstantial evidence that biochemical variation in the major tissue compatibility system in humans (HLA) may be involved in the genetics of host–parasite interactions. The HLA system has four loci and an increasing number of alleles have been identified at each locus (Bodmer and Thompson, 1977; Bodmer and Bodmer, 1978). In various human population studies, it has been shown that:

1. allele frequencies differ between racial groups;
2. linkage disequilibria for some haplotype combinations exist in some populations;
3. a cline in haplotype frequency occurs in Europe; and
4. some HLA genotypes are implicated in autoimmune diseases as well as with susceptibility to some infectious diseases. These observations regarding the HLA are consistent with patterns derived from systems subjected to density-dependent selection such as the one predicted by Haldane (1949, 1954).

Along the same lines, Clark (1979) speculated that successful parasite infections require a certain level of 'fine tuning' by the parasite to the physiology and biochemistry of the host. If this is so, then a high level of protein polymorphism in a host population would reduce the possibilities of a single parasite genotype being completely compatible with the majority of the host population. This type of system would, again, produce frequency-dependent selection on each of the protein polymorphisms.

It is apparent from the scarcity of data that an analysis of host–parasite genetic interactions is mostly confined to theoretical arguments and examination of a few, partially documented cases. As Barrett (1984) has emphasized, 'what is lacking are data from field studies specifically designed to examine both ecological and genetic components of host–parasite systems'.

9.3.2 Host–parasite co-evolution

One of the most challenging areas in the evolutionary biology of parasites is how the interspecific interaction of hosts and parasites has influenced the rates of evolution and the patterns of adaptive radiation of the organisms involved (Futuyma, 1986). The term **co-evolution** was coined by Ehrlich and Raven (1964) for their discussion of the evolutionary influence that plants and herbivorous insects have had on each other. Since its introduction, the term has been widely applied, but presently there is no agreement on its exact definition. For example, Roughgarden (1976) defined it as the type of evolution in which the fitness of each genotype depends on the population densities and genetic composition of the species itself and the species with which it interacts. A more specific definition was provided by Janzen (1980); it requires that each interacting species change its genetic structure in response to a genetic change in its partner. This definition implies both specificity and reciprocity among the evolving traits. A still more restrictive definition of co-evolution would include the requirement that both partners evolve at the same time.

9.3.2.1 Phylogenetic aspects of co-evolution

As discussed earlier, a fundamental postulate of co-evolution is that interacting species influence each other's evolution. Because present day associations are, in part, products of historical developments, it is necessary to understand the history of the evolutionary events between hosts and parasite. The extent to which host and parasite lineages have been associated through time and have experienced similar patterns of speciation, can be estimated directly by using phylogenetic systematic techniques (Brooks, 1989). The purpose of these procedures is to assess the degree to which the genealogy of a parasite 'fits' the phylogeny of its host. In general, if host and parasite phylogenies parallel each other, they are **congruent** and there is an indication of historical association. In contrast, **incongruent** areas in their phylogenies represent episodes during which a parasite species evolved in one host species and then colonized another, distantly related, but ecologically similar host (**host switching**).

In the case of congruent phylogenies, hosts and parasites have been similarly affected by the same speciation events, that is, parasites speciated every time that the host did. Alternatively, some parasites might have speciated at a faster or slower rate than the host. If the former occurred, then one host taxon will harbour more than one parasite taxon, and each parasite species present will be each other's closest relative. If parasite speciation was slower than that of the host, then closely related host taxa will harbour a particular taxon. In these situations, even though host and parasite phylogenies are not totally congruent, they are still consistent. An estimate of consistency (Kluge and Farris, 1969), even after host switching or colonization events, can be obtained by using a **consistency index (CI)**. For helminth–vertebrate systems, the degree of consistency between the two phylogenies ranges between 50 and 100% (Brooks, 1985, 1988).

Critical protocols have been developed for the uniform analysis of phylogenetic relationships and many of the potential problems have been identified and resolved. For example, the use of a considerable number of characters allows the identification of instances of convergent evolution because it is unlikely that all traits will show the same pattern of co-variation. Host switching can be recognized by analysing several parasite groups in the same host. Again, it is unlikely that all parasite taxa will show the same pattern of colonization at the same point in time or space. Finally, indirect information regarding the ages of the species involved as might be obtained from the geological history of an area, may permit detection of spurious congruence, but an overlap of host, parasite and geographical phylogenies is required (Brooks, 1989). By incorporating molecular traits into the phylogenetic analysis, it is also possible to obtain time estimates based on the principle of molecular clocks. Hafner and Nadler (1988, 1990) developed a protocol in which estimates of genetic distances can be used to explore relative rates of genetic change (evolution) in hosts and parasites as well as the relative timing of the co-speciation events in the host–parasite assemblage.

Several authors have investigated co-evolutionary histories of hosts and parasites by means of phylogenetic techniques. Brooks and Glen (1982), for example, examined the co-evolution of primates and their oxyurid nematodes (*Enterobius* spp.). The analysis of 13 species of *Enterobius*, based on 31 morphological characters, supported the notion that pin-worms and primates have co-evolved (Fig. 9.2). Only the relationship between *E. vermicularis* and humans was in conflict, suggesting that its presence in humans was a result of host switching or that man has been misclassified among the great apes. Additional studies (Glen and Brooks, 1986), including a comprehensive phylogenetic analysis of hookworms (*Oesophagostomum* spp.) and presence/absence check lists of primate helminths, again showed the same pattern of congruence among all groups except in the great apes.

The phylogenetic relationship between the ancyrocephalid mono-geneans, *Ligictaluridus* spp., and their hosts, ictalurid fishes, was investigated using a traditional cladistic approach (Klassen and Beverley-Burton, 1987). Host and parasite cladograms were only partially congruent (66%). They concluded that two-thirds of these host–parasite associations were co-evolutionary. Host switches among ecologically, but

Enterobius species   Primate hosts

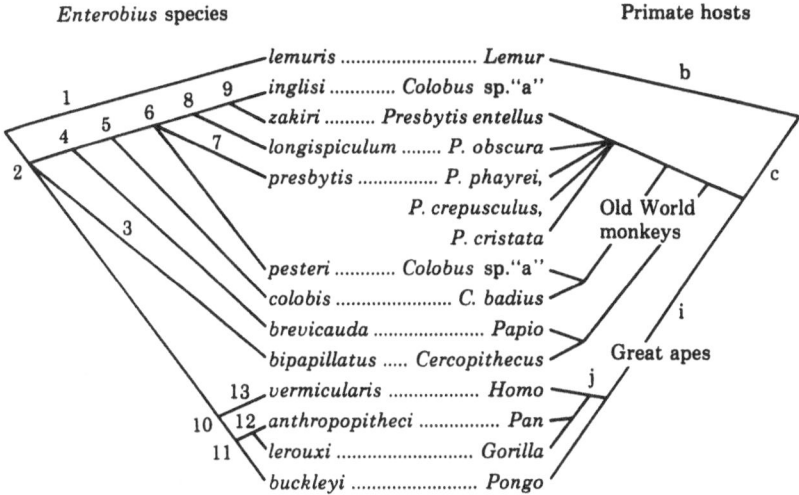

Figure 9.2 Phylogenies of the pinworm genus *Enterobius* (Nematoda) and its primate hosts according to Brooks and Glenn (1982). (With permission, from Mitter and Brooks, 1983.)

not phylogenetically, related hosts would explain the remaining host–parasite associations. Unfortunately, to reconstruct the parasites' phylogeny they only considered six characters, all related to the male copulatory apparatus, which reduces the reliability of the proposed associations. Deets (1987) undertook a phylogenetic analysis and revision of the kroyernd copepods parasitic on chondrichthyan fishes (elephant-fishes, skates, rays and sharks). Although he did not pursue a phylogenetic analysis of the hosts, the relationships obtained from mapping the hosts into the parasite's phylogenetic tree were consistent with recent hypotheses of chondrichthyan phylogeny (Fig. 9.3). Unfortunately, the latter two studies offer only limited criteria for the optimum analysis of host–parasite co-evolution because of their rather narrow approach. Deets (1987) recognized the limitations and stated that 'any single parasite taxon ... like any single host character transformation series ... may not resolve the phylogenetic relationship of the host group'. This was particularly true in his investigation because of the restricted distribution of the parasite in relation to the host taxa and provided just a partial view of the association.

Unlike previous workers, Hafner and Nadler (1988) approached the analysis of phylogenetic relationships between pocket gophers and their chewing lice using only biochemical data of both host and parasites. Using standard electrophoretic procedures, 31 enzyme loci in eight species of

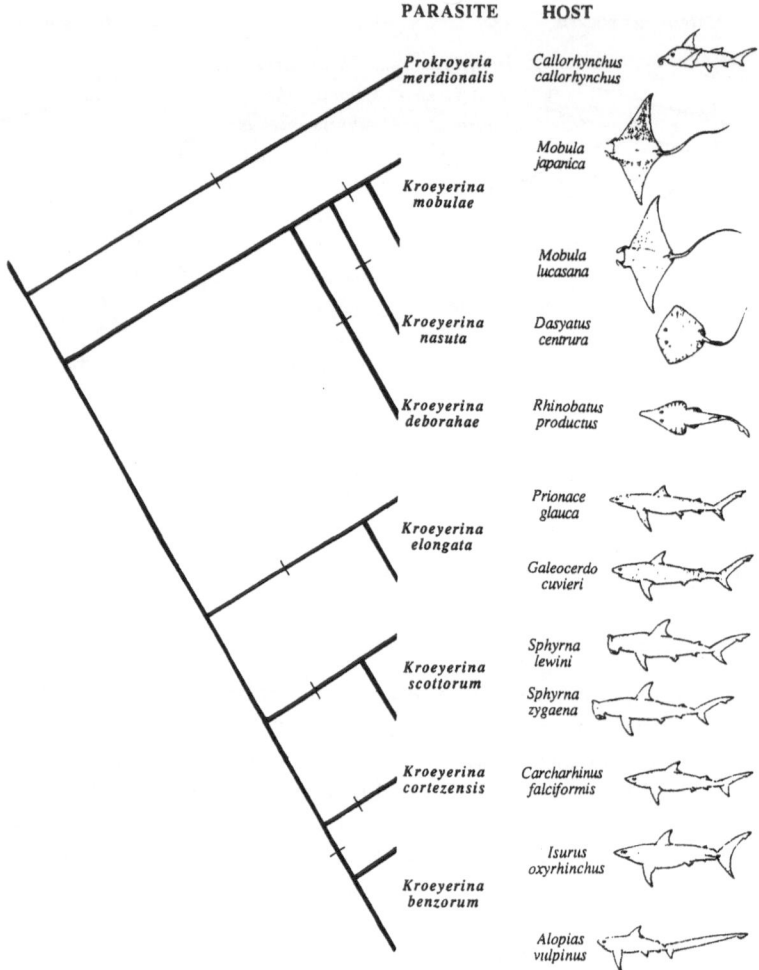

Figure 9.3 Cladograms of the phylogenetic relationships among kroyeriid copepods parasitic on chondrichthyan fishes, with the hosts mapped on the tree. The results place the Holocephala (*Callorhynchus callorhynchus*) as the plesiomorphic sister group to the elasmobranchs and resolve separate lineages for skates, rays, and sharks. (With permission, modified from Deets, 1987.)

pocket gophers (Rodentia: Geomyidae) and 14 loci in 10 species of chewing lice (Mallophaga: Trichodectidae) were examined. Host and parasite phylogenetic trees were topologically identical and highly congruent (Fig. 9.4). Sister taxa and branching sequences above the species level were also congruent. Host switches apparently occurred three times (see asterisks in Fig. 9.4) and, in all cases, the new hosts were sympatric.

Moreover, estimates of genetic distances showed that the length of the branches in most cases were similar, indicating that speciation of the hosts and their ectoparasites was contemporaneous. In only one case (node F in Fig. 9.4), speciation in lice occurred after host speciation. A continuous gene flow between parasites after host speciation may explain the lag time in this particular case.

9.3.2.2 Ecological and evolutionary considerations

Although co-evolutionary relationships occur during the course of a variety of biological interactions (predator–prey, competitors, plant–pollinators, algal–fungal mutualisms, etc.), the interaction between hosts and parasites is the most appealing because of its asymmetrical and

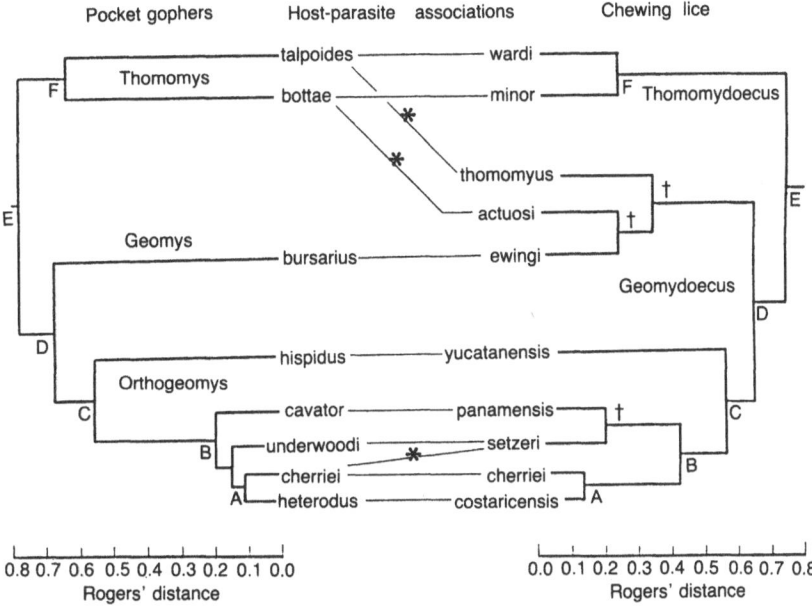

Figure 9.4 Phylogenetic relationships between pocket gophers and their chewing lice based on standard starch gel electrophoretic procedures, using Rogers' genetic distance. Data were clustered using unweighted pair-groups mean average (UPGMA). Host–parasite associations are indicated by lines connecting extant species. Asterisks indicate instances in which host switching probably took place. The louse, *G. setzeri*, is found on two host species; certain hosts (*T. talpoides*, *T. bottae* and *O. cherriei*) harbour two species of lice. In the latter cases, the louse species are usually sympatric on an individual pocket gopher. Daggers identify nodes on the louse tree that are absent on the pocket gopher tree. (With permission, from Hafner and Nadler, 1988.)

obligatory nature. Considerable discussion exists regarding the ways in which hosts and parasites have co-evolved; the ecological and evolutionary scenarios are practically unlimited. However, most of the current viewpoints on host–parasite co-evolution can be summarized within the context of three models: **mutual aggression, prudent parasite**, and **incipient mutualism**. The models include both parasite (*vis-a-vis* infectivity, exploitation of the host, and benefits to the host) as well as host characteristics (*vis-a-vis* susceptibility and defence against parasites) in terms of the selective pressures placed upon them. As in many other characterizations of biological relationships, a continuum is likely to exist within the framework of the three models.

In the mutual aggression model, hosts and parasites are involved in an evolutionary arms race, with each partner constantly evolving in an aggressive manner toward the other. Selection by the parasite is always for greater exploitation of the host; selection by the host is always for a more efficient exclusion of the parasite (Dawkins and Krebs, 1979). Successful hosts evolve effective genetic, behavioural, physiological, and immunological mechanisms to combat the invading parasite. Conversely, the parasite must evolve ways of avoiding such defences. This form of co-evolution is assumed to occur within the context of the gene-for-gene hypothesis (as previously discussed). It is also central to van Valen's (1973) **Red Queen hypothesis** that predicts continued evolutionary change under conditions of constancy in the physical environment or, as quoted from *Alice Through the Looking Glass*, 'Now here, you see it takes all the running you can do to keep in the same place'. This model is the one that best agrees with the strictest definition of co-evolution.

In the prudent parasite model, evolution of the parasite is in the direction of selecting characteristics that limit harm to the host, thereby increasing the life span of both the host and the parasite. Negative pressures by the parasite on the host will be reduced over time and there will be a reduction in pathogenicity, or virulence, or both. This model represents the classic view of an aggressive system evolving toward a more benign or neutral relationship.

The third model, incipient mutualism, considers co-evolution as an actively cooperative process, with the host and the parasite evolving mutualistic attributes. The outcome will be for each partner to promote the continued presence of the other (Lincicome, 1971; Davies *et al.*, 1980; Wilson, 1980). Such a model has been used by some (Margulis, 1981) to explain the relationship between mitochondria and chloroplasts and their host cells and is a central feature in a widely accepted model for the evolution of life as it is known on earth today. Indeed, Margulis (as quoted by Mann, 1991) states, 'the major source of evolutionary novelty is the acquisition of symbionts – the whole thing is then edited by natural selection. It is never just the accumulation of mutations'.

Whether host–parasite co-evolution lies somewhere on a continuum that ranges from antagonism to mutualism or not is, in part, highly dependent on the mode of parasite transmission and reproduction (May and Anderson, 1983). If successful reproduction or transmission, or both, by the parasite is somehow related to host survival (and vice versa) then selection should favour mutualistic adaptations. If parasite reproduction or transmission, or both, are not dependent on host survival (and vice versa), then selection may favour antagonistic adaptations.

One of the strongest indications of the importance of reciprocal adaptation between parasites and their hosts is the way in which parasites have developed mechanisms to evade the immune system of their hosts. Bloom (1979) and Mitchell (1979) considered evasive actions within the context of three different strategies. The first involved a reduction in parasite antigenicity but also encompassed the ideas of molecular mimicry, antigen masking, etc. The second included a modification of the extracellular environment such as the inhibition of vacuole fusion, escape from phagolysosomes, inhibition of digestive enzymes, etc. The third was identified as a modulation of the host immune response through immunosuppression, degradation of host antibodies, etc. Some of these mechanisms, with examples of those postulated for helminths, are identified in Table 9.6. Each of these mechanisms could be interpreted in terms of reciprocal co-evolution because each is a response by the parasite to a specific capacity associated with the host, presumably one that has co-evolved in part as a response to the parasite. It should be emphasized, however, that resistance to parasite infection is not entirely an immunological phenomenon. Physiological, behavioural and environmental factors are also involved (Holmes, 1983). In the context of immune reactions, the high diversity of antigens within the **major histocompatibility system** (that portion of the genome that regulates cellular interactions such as the cell-mediated immune response and self recognition) has been interpreted as being the result of diversifying selection in an attempt by the host to counteract the parasite's evasion strategies (Damian, 1979; Davies *et al.*, 1980).

The conventional wisdom regarding host–parasite interactions is that parasites evolve to be harmless to their hosts. The basic assumption is that, all else being equal, a parasite that inflicts little damage would represent an advantage both to the individual host and to the parasite. Most of the evidence supporting this notion relies on the lower pathogenicity or virulence of parasites in the 'traditional host' and the higher, sometimes extreme, virulence in 'newly' acquired hosts. For example, in Africa

Table 9.6 Mechanisms of evasion of host responses by parasites with examples of those known or postulated for helminths. (From Holmes, 1983).

Mechanisms	Parasite[a]	Host	References[b]
Reduced antigenicity			
Anatomical inaccessibility	Trichinella spiralis*	Mammals	Mitchell (1979b)
Antigenic variation	Fasciola hepatica	Sheep	Hanna (1980)
Modulation of antigens	Nippostrongylus brasiliensis	Rat	Ogilvie (1974)
Molecular mimicry	Schistosoma mansoni	Mouse	Damian (1979)
Blocking antibodies	Taenia taeniaeformis*	Mouse	Richard (1974)
Non-Ig masking of antigen	Schistosoma mansoni	Mouse	Smith and Kusel (1974)
Loss of MHC antigens on parasitized cell	Schistosoma mansoni	Mouse	Smith and Kusel (1974)
Modification of intramacrophage environment			
Modulation of host immune response			
Immunosuppression			
Lymphoid tissue disruption	Mesocestoides corti*	Mouse	Mitchell and Handman (1978)
Clonal deletion	Echinococcus granulosus*	Mouse	Ali-Khan (1978)
Mitogens, antigenic competition and increased antibody turnover	Nippostrongylus brasiliensis	Rat	Jarrett and Bazin (1974)
Nonspecific suppressor cells	Heligmosomoides polygyrus	Mouse	Brown et al. (1976)
Specific active suppression	Trichinella spiralis	Mouse	Barriga (1980)
Cytotoxic parasite molecules	Schistosoma mansoni	Human	Ottesen and Poindexter (1980)
Effector cell blockade	Echinococcus granulosus*	Cattle	Annen et al. (1981)
Degradation of antibodies	Schistosoma mansoni	Rat	Mazingue et al. (1980)
Anticomplementary effects	Schistosoma mansoni	Rat	Auriault et al. (1980)
Antinflammatory effects	Taenia taeniaeformis*	Rat	Hammerberg et al. (1980)
	Trichinella spiralis	Rat	Castro et al. (1980)

[a]Asterisks indicate larval stages
[b]References can be found in Holmes (1983).

indigenous ruminants suffer mild infections with insignificant morbidity due to the haemoflagellate, *Trypanosoma brucei*, whereas domestic ruminants introduced into these endemic areas suffer infections that are typically fatal. Wild rats in cities that have had recent plague epidemics with *Yersinia pestis* show higher survival after challenge infections with the plague bacillus than rats from cities without a recent history of plague (Levin *et al.*, 1982). Dobson (1983 *sensu* May and Anderson, 1983), in a survey of parasite virulence in some 300 host–parasite associations (mainly invertebrates), found a general tendency for parasites than are 'older' in evolutionary time to be less virulent than the more recent forms. Similarly, Holmes (1983) observed that parasite infections appear to exert a more efficient regulatory role among newly introduced plants or animals, or when the parasites are introduced into new geographical localities.

Although this type of evidence at first glance seems to be fairly supportive of the prudent parasite model, it could also be used as an argument for the antagonistic or mutual aggression model. Genetic evidence from hosts and parasites (see previous section) indicates that local strains or populations are mutually adapted and in equilibrium in relation to patterns of resistance/susceptibility and virulence/avirulence. If these factors are genetically controlled in most species, then it would be feasible to assume that the lower degree of pathogenicity is due to a balanced equilibrium of these allelic systems in the scaled interactions of an arms race. Thus, increased pathogenicity in unfamiliar hosts would be determined by the different frequency of these alleles in both populations. With time, the reciprocal interaction between the alleles would tend to move toward a state of equilibrium.

The adaptations exhibited by hosts and parasites resulting in mutually adapted and equilibrial local populations can also be influenced by trade-offs among patterns of susceptibility/resistance or virulence/avirulence and other important traits in the context of their life histories. For example, although variability in susceptibility is known to occur in *Biomphalaria glabrata* against *Schistosoma mansoni*, resistant snails are not predominant in nature and apparently have not been at a selective advantage. Wright (1971) suggested that this resistance may be associated with disadvantageous or physiological defects. Minchella and Lo Verde (1983), using stocks of *B. glabrata* of known susceptibility, were able to demonstrate that resistant snails had a lower reproductive output than susceptible snails when the former were exposed to parasite pressure or the presence of susceptible snails. A similar phenomenon involving a trade-off between resistance and susceptibility and some fitness components was observed by Lensky and Levin (1985) in a bacteria-phage system. In this case, a trade-off between resistance and competitive ability of the bacteria permitted the phage to persist in an equilibrial situation even though phages were unable to infect the resistant hosts.

One of the main arguments favouring selection toward a mutualistic relationship is based on the premise that reduced reproduction (or virulence) of the parasite (in forms that reproduce and remain within the host) may actually increase the survival and long-term propagation of parasite infrapopulations within a host population. For any specific level of multiplication then, parasites should generally benefit from genes that reduce pathogenicity. Ewald (1983), and subsequently supported by Thompson (1986b), argued that the different levels of parasite reproduction and pathogenicity will depend on how the costs and benefits of extensive reproduction vary among different parasite species. For example, because hosts are temporary islands for parasites, effective transmission from one host to another is essential for the survival of parasite genes. Extensive reproduction within a host will increase the number of parasites available to reach other hosts. If this occurs, reproduction by the host may be adversely affected; immobilization or death of the host would reduce the chances of the parasites to infect other hosts. On the other hand, if parasites are transmitted by biting arthropods, then this scenario is not necessarily the most appropriate. Immobilization of the host should increase the chances of being bitten and extensive reproduction may be more beneficial because it increases the possibility that a biting vector will obtain an infective dose. Ewald (1983) supported these ideas with data from human diseases. Using mortality rates as an estimate of pathogenicity, it was shown that diseases transmitted by insect vectors were significantly more lethal than those transmitted by other means. Similarly, by using immobilization of the host as an indicator of pathogenicity, vector-transmitted diseases were also significantly more lethal. He acknowledged, however, that because his results were correlational and comparative rather than experimental, he could not exclude the possibility that other correlates of vector transmission and of specialization on human hosts were contributing to the observed associations.

Selection on parasites toward mutualism or reduced levels of pathogenicity is probably effective only under a restrictive set of ecological and genetic conditions (Anderson and May, 1982; Levin, 1983; Thompson, 1986b). The behavioural, ecological, immunological, and physiological adaptations of parasites and their hosts do not represent the culmination of the evolutionary process; instead, 'they reflect the current status of an ongoing series of manipulations and retaliations' (Smith-Trail, 1980). Other factors that are usually ignored when examining interactions among organisms include elements of the physical and biotic environment where the interactions occur. Thompson (1986a, 1986b, 1987) argued that interactions across different environments can have different outcomes even when the genotypes are held constant. The range of variations was termed the **interaction norm** (Fig. 9.5) to parallel the **reaction norm** of genotypes in population genetics. In the latter case, the same genotypes could generate different phenotypes in different environments (Lewontin, 1974). The widespread contention that parasites and their hosts tend to evolve towards a mutualistic relationship is an overly simplistic view of host–parasite co-evolution.

The models and evidence discussed to this point incorporate certain epidemiological and genetic components, although in most cases they are biased one way or another. Despite these biases, or limitations, a common finding is that the outcome or trajectory of a particular host–parasite association will depend on the way virulence and the production of transmission stages of the parasite are interlinked on the one hand and the cost to the host of evolving resistance on the other.

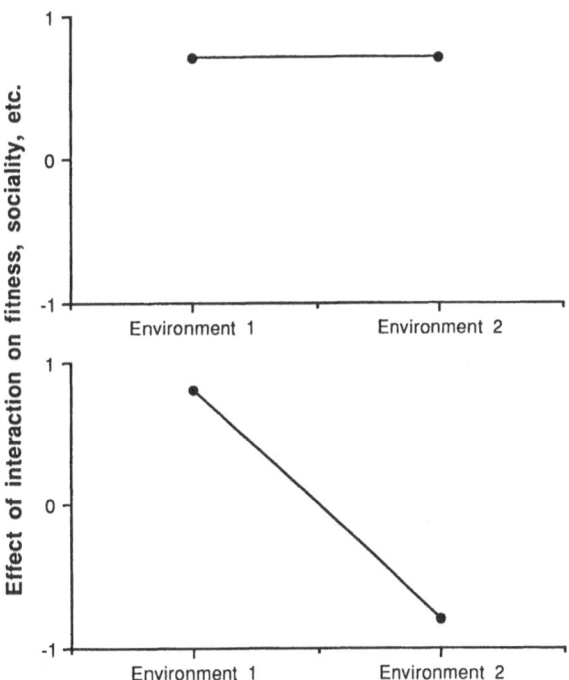

Figure 9.5 Interaction norms, showing the variation in outcome of interactions across different physical or biotic environments as genotypes are held constant. (With permission, from Thompson, 1986b.)

9.4 PARASITE INFLUENCE ON THE EVOLUTIONARY BIOLOGY OF THE HOST

The evolution of host–parasite interactions is not restricted to co-evolutionary considerations. Outside the boundaries of historical association or strict phylogeny, there is a wide array of host–parasite interactions that deserve attention from an evolutionary point of view. Some host–parasite interactions have been shown to be strictly related to evolutionary change and selection pressures, e.g., parasites and host sexual selection. Others have direct ecological implications that may affect the evolution of ecological systems and their components, e.g. parasite-mediated competition.

Parasites may influence the population biology of an organism in many ways. These would include changes in growth rates and population size, changes in spatial–temporal dynamics, and changes in patterns of co-existence and competition. A number of field and laboratory observations have shown that parasites may act as mediators of various ecological interactions among different host species or populations. The outcomes could influence community structure, interspecific interactions, and distributional patterns.

In a now classical experiment, Park (1948) demonstrated the influence of *Adelina tribolii* (a sporozoan) on the population size and competitive interactions of two species of flour beetle, *Tribolium confusum* and *T. castaneum*. When uninfected beetles were grown in monoculture, the populations of both species included about 20% more adults than when infected beetles were grown in monoculture. Moreover, in single species cultures, the equilibrium population size of uninfected *T. castaneum* was almost three times higher than in infected cultures, while the population size of *T. confusum* was not influenced by the parasite. When uninfected cultures of the two species were mixed, *T. castaneum* usually became dominant, with *T. confusum* becoming extinct. However, when *Adelina* was present, *T. castaneum* became extinct and *T. confusum* survived. The parasite thus affected the age structure, the equilibrium numbers (carrying capacity), and the outcome of competitive encounters between these two species.

A similar form of interaction occurs between the meningeal nematode, *Pneumostrongylus tenuis* (*Parelaphostrongylus tenuis*), and its cervid hosts. The normal host for this parasite is the white-tailed deer, *Odocoileus virginianus*, and is tolerant of the infection. Moose, pronghorn antelope, mule deer and woodland caribou are also potential hosts, but all will suffer severe neurological disease even with very small numbers of nematodes in the brain (Anderson, 1972). When the northern forests of North America were opened by man, the white-tailed deer came in contact with moose and other major cervids, replacing them in these areas because of the lethal effects of the nematode in the new hosts. Price (1980), however, has suggested that a new dynamic state presently exists between moose in parts of its southern range, and white-tailed deer. Thus, moose occupy refugia on high ground and other forested sites; in effect, they are physically isolated from deer, at least during spring, summer and fall when parasite transmission occurs. In the winter, the moose are forced to descend to the lowland areas occupied by white-tailed deer but, at this time, transmission of the parasite is virtually nil, facilitating the co-existence of both cervid species. The extent of negative interaction created by the parasite between white-tailed deer and other great cervids is such that attempts to introduce these latter species into their original habitats have proved unsuccessful (Anderson, 1976, Anderson and Prestwood, 1979).

The types of host interactions mediated by parasites in ecological communities were recognized by Barbehenn (1969) who suggested that parasites may be considered as effective 'weapons of competition'. This role of parasites in host communities or populations can be highly significant to the development of theories in zoogeography and speciation, and in the structure and function of ecosystems (Barbehenn, 1969). Most of the evidence regarding this kind of interaction comes from host species that have changed their normal patterns of distribution because of human activities. These activities would include the introduction of new species to new geographical localities (voluntarily or by accident), the opening of cross-country canals or navigation routes, the clear-cutting of forests, etc. Additional cases would include, for example, the introduction of the mosquito, *Culex pipiens fatigans*, into the Hawaiian archipelago and the resulting spread of malaria to native birds, and the exclusion of the roof rat, *Rattus rattus*, by the Norway rat, *Rattus norvegicus* (Warner, 1968; Crosby, 1972). Under these circumstances, human intervention may be providing an undesired source of evidence for the otherwise obscure significance of parasitism in the evolution of various host communities. The extent to which some of the present distributional or interactive patterns among different host species are the result of parasite influence, is unknown. On the other hand, it is likely that some of the negative host–parasite interactions presently observed as a consequence of human activities will moderate over time. The outcome will be a more equilibrial situation as a result of selection for more resistant hosts, or less virulent parasites, or both.

Most of the previous research in population and community ecology has focussed on the ecological interactions between potential hosts without considering the potential occurrence of other levels of interaction such as parasites. For example, there are several cases in which current ecological theory does not provide appropriate explanations for the patterns observed. Barbehenn (1969) studied the patterns of diversity and abundance of *Rattus* spp. on mainland Malaya and nearby islands. He found that on the mainland, diversity was greater than in the islands while total rat density was lower, conflicting with the predictions from current theories on co-existence, species packing, and species diversity. He proposed that the most parsimonious explanation for his observations was that in areas where diversity was high, cross-infection of parasites among the host species would be also high, keeping the realized niches much smaller than the fundamental niches. This would amplify their spatial segregation and cause distributional gaps between species (Barbehenn, 1969; Price, 1980). In another case, distributional gaps that seemed suitable for colonization by birds, were observed by Cornell (1974). Price (1980) suggested that the gaps between species could be maintained by vectors travelling between populations; the vector would be carrying a highly pathogenic parasite for one species but innocuous for another. The result would be to create an area where vectors with highly pathogenic parasites for both species would be present, thereby generating distributional gaps among potential hosts. The occurrence of *Trypanosoma brucei* in certain areas of Africa and the exclusion of domestic cattle from such areas may be a similar phenomenon, although unidirectional in this case.

Parasites may not only be mediators for interactions, they also have been implicated in interdemic selection for populations of the red-spotted newt, *Notophthalmus viridescens*. Thus, Gill (1978) observed that trypanosomes were strongly involved with the failure of reproduction by several newt populations and argued that they might play a significant role in their evolution.

So far, most of the host interactions described have been mediated by parasites that have had a negative effect on one or more of the interacting species. However, there are cases where parasites favour or facilitate inter-specific interactions without having a detrimental effect on either host. Price *et al.*, (1986) described, for example, the acquisition of new bio-chemical capabilities by the olive fly after associating with certain symbionts. As a result, the fly was able to exploit a new type of resource with the potential for adaptive radiation (Hagen, 1966). The bruchiid beetle, *Caryedes brasiliensis*, relies on the presence of symbiotic bacteria in their gut to tolerate L-canavanine, a potent plant allelochemical present in the seeds upon which the beetles feed (Rosenthal, 1983). Finally, parasites have also been involved with the evolution of relationships such as the cleaning symbioses among tropical fishes.

In all these examples, parasites are not necessarily the primary agent in selection and host evolution. However, many times they have played a major role in affecting some of the host's biological and ecological characteristics that, in turn, may ultimately influence the evolutionary biology of the host.

Several years ago, Hamilton (1980) introduced the rather novel idea that because parasites are short-lived and evolve rapidly, at least compared with their hosts, they could represent an evolutionary pressure sufficiently widespread to account for the occurrence of sex. He developed two genetic models. The first was a one-locus, diploid selection model and the second was a two-locus, haploid selection model. He was able to show that with an appropriate intensity of frequency-dependent selection, the models generated cycles; at certain points in these cycles, sexual species obtained higher long-term fitness than any monotypic asexual species.

Subsequently, Hamilton and Zuk (1982) incorporated Hamilton's (1980) ideas into an hypothesis regarding the evolution of secondary sexual characters. They proposed that the bright colours and elaborate displays of many animal species allowed females to assess a potential reproductive partner's ability to resist parasites. Then, by mating with such males, females would assure the passage of resistance genes to their offspring. The hypothesis was based on three assumptions. First, the display of secondary sexual traits of individual animals is related to their general health and vigour. Second, hosts co-evolve with their parasites and, as a result, they constantly generate heritable resistance to parasites. Finally, parasites have a negative effect on host viability.

Based on these assumptions, there are several intra- and interspecific predictions for the Hamilton and Zuk hypothesis. Møller (1990) summarized them as follows:

1. females will obtain mates with few or no parasites by choosing males with fully expressed, sexual ornamentation;
2. males with inferior ornamentation will be less desirable partners and, therefore, take longer to be noticed;
3. assortative mating occurs with respect to parasite densities;
4. females that mate with males having poorly developed, secondary characters will copulate outside their pair bond with males showing more fully expressed ornamentation; and
5. in species that are particularly vulnerable to parasitism, sexual selection should favour greater development of 'health-certifying' traits, allowing females to more accurately assess the parasite load of a male.

Møller (1990) provided a comprehensive review of the experimental evidence for parasite-driven host sexual selection. Comparative studies testing the interspecific prediction (prediction number 5) are equivocal, mainly because methodological problems compromise the results. Thus, parasite prevalence may be affected by sampling bias, reliability of host displays is not always assessed, host species are normally used as independent observations despite the fact that they are phylogenetically related and, finally, developmental constraints and adaptation to common environments may affect the extent of the display.

Hamilton and Zuk (1982) supported their hypothesis with observations that a positive association existed between the prevalence of blood infections (*Leucocytozoon, Haemoproteus, Plasmodium, Trypanosoma, Toxoplasma,* and nematode microfilariae) and three types of host displays (male brightness, female brightness, and song complexity) in passerine bird species. A re-analysis of the data (Read, 1988), controlling for the variables previously mentioned, confirmed some of the Hamilton and Zuk (1982) findings and ruled out the possibility that the correlation could have been produced by taxonomic association of the hosts. A similar study was conducted across families of British and Irish freshwater fishes (Ward, 1988, 1989). In this case, the degree of sexual dimorphism was positively associated with parasite loads after controlling for various confounding variables. However, recent criticisms of Ward's conclusions indicate that the correlation in his data was to a large extent due to uncontrolled ecological variables. Thus, Chandler and Cabana (1991) found that dichromatism was not correlated with the mean number of parasite species per individual host, but it was correlated with the total number of parasite genera per host species, which can become an extremely biased measurement due to unequal study efforts. Indeed, Chandler and Cabana (1991) showed that the correlation between dichromatism and parasite genera per host species disappears when study effort is statistically controlled.

The relationships observed between the degree of sexual selection and parasite load in these two different host taxa are significant from an evolutionary point of view because they suggest that the associations result from phenomena common to both groups. While the lack of comparative studies in more host species does not permit definitive conclusions regarding the Hamilton and Zuk (1982) hypothesis, some recent investigations (Møller, 1990) dealing with the intraspecific predictions have been supportive. Thus, the assumption that an expression of sexual ornamentation will reflect a certain level of parasitism was supported by eight of 10 intraspecific tests involving fishes, frogs, lizards and birds (Table 9.7). Similarly, the assumptions regarding heritability of host resistance to parasites, the prediction that females choose males with fewer parasites, and the negative effect of parasites on host fitness were supported by all the studies in which these questions were investigated (Table 9.7).

General evolutionary theory considers evolution to be the product of variation generated by random mutation and recombination, with natural

Table 9.7 Results of intraspecific tests of the assumptions and predictions of the Hamilton and Zuk hypothesis. (From Møller, 1990.) *

	Field crickets	Guppies	Three-spined sticklebacks	Gray tree frogs	Fence lizards	Pheasants	Junglefowl	Sage grouse	Rock doves	Barn swallows	Stain bowerbirds
Assumptions:											
Expression of ornaments depends on parasite load	o	O	E	O	E	E	E	E	?	E	o
Heritable resistance	-	O?	-	-	-	E	E	-	-	E	O
Negative effect of parasites on host fitness	O	O	E	-	O	E	E	-	?	E	O
Predictions:											
Females choose males with fewer parasites	e	O	E	O	-	E	E	E	O/E?	O	O
References (Please note that these references are in Møller (1990).)	Zuk (1987a, b, c, d, 1988)	Kennedy et al. (1987)	Milinski and Bakker (1990)	Hausfater et al. (1990)	Ressel and Schall (1989)	Hillgarth (1990)	Zuk et al. (1990)	Johnson and Boyce (1990) Spurrier et al. (1990)	Clayton (1990)	Møller (1990 a, b)	Borgia (1986), Borgia and Collis (1989)

* Evidence is listed as either observational (O or o), when based on correlational evidence, or as experimental (E or e), when based on field or lab experiments. Capitals are used when the evidence confirms the assumptions or the predictions and small case letter when the evidence is contradictory.

selection acting on the variation produced. In recent years, however, the idea of horizontal **gene transfer** also has received attention. This phenomenon is known to occur mainly in prokaryotic organisms in which it seems to be fairly common and, as such, has some strong evolutionary implications. Although less common (or perhaps less studied?), gene transfer is also known from some of the lower eukaryotes, probably mediated by retroviruses (Gruskin, Smith and Goodman, 1987; Benviste and Todaro, 1974; Lang, 1984; Waring *et al.*, 1984). Certain plants and their symbiotic bacteria are also known to exchange genes horizontally (Drummond, 1979; Carlson and Chelim, 1986). Transfer of the gene for the enzyme copper/zinc superoxide dismutase from the ponyfish to the symbiotic bacterium, *Photobacterium leiognathi*, has likewise been reported (Lewin, 1985).

The idea of gene transfer in host–parasite systems introduced by Howell (1985) was based primarily on the ability of parasites to produce host-like molecules (antigens) and thus to avoid recognition by the host. Such phenomena would certainly be possible if parasites have genomes containing DNA sequences that are identical or closely related to those of their hosts (DNA homology). From a neo-Darwinian point of view, this kind of homology could be the product of either long-term conservation, or a more recent adaptive change through mutation, recombination and selection. However, Howell (1985) argued that these DNA homologies also may be the product of direct incorporation of host genetic material into the parasite genome. This process of interspecies gene transfer could be a two-way flow from host to parasite and vice versa. Incorporation of parasite genetic material by the host may lead to tolerance towards the parasite if the information is expressed and recognized by the host as self. On the other hand, incorporation of host DNA into the parasite genome may lead to the expression of host macromolecules by the parasite, facilitating the avoidance of immune responses. Possible mechanisms for such two-way flow require the presence of retroviruses. These are RNA viruses that, upon infection of a host cell, can copy their viral RNA into DNA by means of the enzyme, reverse transcriptase. This DNA then enters the cell nucleus and integrates into the host DNA where it is transcribed and its product translated into proteins. In these terms, their potential to act as a vehicle for exchange of genetic material between hosts and parasites is based on their ability to transduce DNA, cross species boundaries, and enter the germ lines of their hosts.

At the present time, the evidence for this phenomenon between host and parasites is scarce. Although virus or virus-like particles similar to C-type (RNA) viruses have been observed in several parasite groups, from protozoans to nematodes (Terzakis, 1972; Gibbs, 1973; Harrison, 1973; Mueller and Strano, 1974a, b; Byram *et al.*, 1975; Dougherty *et al.*,

1975), there is no direct evidence of gene exchange. Circumstantial evidence, for example, includes the similarity between human and *Ascaris* collagens (Michaeli *et al.*, 1972), the ability of *Fasciola hepatica* to synthesize human blood group substances (Ben-Ismail *et al.*, 1982), and the numerous instances of antigen-sharing by host and parasite (Capron *et al.*, 1968; Damian, 1979). In the latter case, however, the question is still open to discussion. It is not clear yet if the antigen is acquired passively by the parasite or if it is synthesized from messenger RNA captured from the host instead of being synthesized *de novo* by the parasite (Smyth, 1973).

In summary, horizontal gene transfer between hosts and parasites is a potential source of variation upon which natural selection can operate, in addition to mutation and recombination. Even if the frequency of this phenomenon is low, Rohde (1990) suggested that it may have a profound impact on the phylogeny of many groups. In this context, then, phylogenetic systematics as a method to determine phylogenetic relationships between organisms would not necessarily be reliable. In his view, **mosaic evolution** (with gene transfer functioning as a source of isolated evolutionary units) might explain, for example, the conflicting evidence regarding the relationships of many parasitic taxa.

10 Summary

10.1 INTRODUCTION

By this point, it should be clear that the extent of parasitism in nature is staggering, whether the focus is on plant or animal hosts, including man. In the latter case, the prevalence of disease-producing parasites, especially in Third World countries, continues to rise. This is despite the fact that many of these diseases are not only preventable, most are treatable. It appears that an unfortunate combination of public and private ignorance and negligence, and deplorable socioeconomic conditions has promoted the spread of the enteric, as well as the vector-borne, parasites of mankind. In this regard, it is almost paradoxical that many of man's domesticated animals are probably more free of parasitic disease than man himself.

While parasitism is difficult to define, it is reasonably easy to describe. Many of the more traditional definitions have correctly included the ideas of physiological and genetic interdependence between the two partners of the symbiotic relationship. On the other hand, a problem with some of the older approaches is that they inevitably allude to the induction of harm by the parasite. Since harm is a qualitative assessment, these definitions are largely subjective in character and thereby introduce an untenable appraisal of the relationship between a host and its parasites.

With the publication of Crofton's (1971a, b) approach to parasitism, there was at least an effort to quantify the relationship. His scenario also had its pitfalls, however, because he included a deterministic component regarding mortality in his model for parasite population dynamics. The problem was compounded when he indicated that 'heavily infected' hosts were killed; after all, what is meant by heavily infected? Crofton's ideas are almost automatically linked to another concept, the one regarding regulatory interaction between host and parasite populations. This principle later was emphatically championed by Anderson and May (1979) and May and Anderson (1979). Indeed, these authors declared that parasites are as significant as predators and insect parasitoids in regulating host populations. The problem with this view is that there is not enough evidence to support it. In fact, a literature review would suggest it is doubtful that many protozoan and helminth parasites have the capacity to kill directly at all. They are much more likely to either produce morbidity or affect the host's behaviour in such a way as to increase the likelihood of successful transmission to the next host in the cycle. In the former case, the outcome serves to emphasize the insidious character of parasitism. The latter, on the other hand, emphasizes the successful consummation of co-evolutionary interactions between hosts and parasites, *vis-à-vis* the ability of parasites to capitalize on essential predator–prey interactions that have evolved simultaneously with host–parasite relationships.

The concept of the host–parasite relationship is an umbrella construct that applies to, among other things, the myriad of adaptations employed by different parasites to ensure success in this highly specialized way of life. Many species have evolved complex life cycles which ensure that residual infective stages remain in a habitat after the adult parasites have moved on to another habitat. Others, such as the cestodes, have lost their digestive system, with the gut being replaced by an external 'brush border' that is capable of acquiring nutrients in a highly efficient manner. Metabolically, many parasites have evolved the capacity to fix carbon dioxide; in the process, additional energy is captured that might otherwise be lost. The behavioural modifications exhibited by parasites, or induced in their hosts, are truly spectacular. Consider that the behaviour of a few metacercariae of *Dicrocoelium dendriticum* increases the probability of the ant intermediate host being eaten by sheep and thus of transmission of the parasite to the definitive host. Or, recall the effect of *Schistocephalus solidus* plerocercoids in altering the growth rate of its definitive host so that the stunted and swollen fish host is caused to swim upside down, thereby losing its protective colouration and becoming vulnerable to predation.

The behavioural adaptations used for inducing predation and parasite transmission are no more remarkable than the range of sensory adaptations employed by a parasite in migrating within a host in order to reach an appropriate site of infection. The examples used earlier (*Fasciola hepatica* and *D. dendriticum*) illustrate both how complex and how simple the routes used by parasites can be in reaching a site of infection. These species are representative of an enormously wide range of elegant sensory adaptations that have successfully evolved over time. This sort of behaviour is fundamental to the concept of site specificity or site fidelity exhibited by parasites.

The immune systems of vertebrate animals include a most complicated series of cellular and humoral capacities. On the other hand, a variety of mechanisms have been evolved by parasites to successfully impugn the immune system, or to otherwise alter it in some manner. Some protozoans, for example, can avoid the immune response by becoming sequestered within host cells, frequently the very cells that would otherwise be involved in killing the parasites. Other parasites modulate the host's immune system by shutting it down or by suppressing it. Some parasites hide from the host by becoming covered with host antigens so that the host is unable to recognize them as foreign. These, and other, mechanisms are probably more extensive in nature than is presently known because so much of the research effort to date has focussed on but a few host–parasite models, or upon those parasites that have a great impact on mankind from a public health perspective. It also should be noted that invertebrates are capable of mounting immune responses and, while not as sophisticated as those associated with vertebrate animals, they are no less effective in their own way. On the other hand, it is just as likely that some parasites have been effective in developing avoidance responses to the immune systems of invertebrates.

In summary, the scope and success of parasitism in nature is incredible; nonetheless, a comprehensive understanding of the phenomenon is still lacking. On the other hand, significant research strides have been made in areas ranging from molecular biology and immunology to ecology and epidemiology and more are on the horizon, promising exciting advances in the future.

10.2 POPULATION CONCEPTS

The population biology of animal parasites had a distinctly natural history flavour until the publication of Crofton's papers (1971a, b). This is not to infer that the work preceding that of Crofton was less than acceptable scientifically. The significance of Dogiel's efforts, for example, and that of many other early workers is certainly recognized and appreciated at the present time. In fact, it includes a core of useful information and a methodology that continues to be employed by many working in the area.

In 1975, Esch, Gibbons and Bourque identified a gap in the emerging area of parasite population biology. They recognized the heuristic value of a conceptual organization in which parasite populations are structured from a hierarchical standpoint. They thus developed the infra- and suprapopulation concepts from which the metapopulation idea emerged later (Riggs *et al.*, 1987). Those working in the area of parasite community ecology subsequently applied the hierarchical approach to their systems (Bush and Holmes, 1986a; Holmes and Price, 1986) and, in the process, defined the structure of infra-, component and compound communities. The subsequent application of these concepts has encouraged a more orderly approach to the study of parasite population and community biology.

10.3 FACTORS INFLUENCING PARASITE POPULATIONS

Factors that affect parasite infra-, meta-, and suprapopulations occur both internally and externally to hosts. Externally, such forces could include any one, or a combination, of the physiochemical components in a given environment. Some of these factors, for example, may influence the host-finding capabilities of parasites having free-living life-cycle stages. More recent studies have drawn attention to the idea of scaling as it applies to temporal and spatial distributions of parasites and their hosts. The emphasis on scaling potentially offers significant opportunities for extending and refining the epizootiological or epidemiological insight for many, already well-studied, parasite species. Further, it has been suggested that the application of additional epidemiological findings could be used in developing appropriate strategies for the effective treatment of certain parasitic diseases of man.

Primary among the internal factors affecting parasite infrapopulation dynamics are inter- and intraspecific competition, diet, immunity, and the genetics of the host. It is difficult to assess the relative importance of each but, historically, diet has been the main focus of attention by many investigators. It now seems likely that generating an understanding of the population genetics of both hosts and parasites will be a fundamental objective in future research.

The actual dynamics of parasite infrapopulation growth still are not well understood. In part, this is because most studies on the growth dynamics of parasite infra- and metapopulations require the 'kill and count' method for enumerating parasite numbers. Unless sampling protocols are carefully designed, undesirable side-effects may be generated with such procedures.

The concept of mutual, regulatory interaction between host and parasite populations has been promulgated mainly by Anderson and May (1979) and May and Anderson (1979), but there is not much direct evidence to support it in the literature, certainly as it applies to the more 'traditional' protozoan and helminth parasites. The approach taken by most investigators has been to examine the effects of density-dependent or density-independent forces on the biology of parasite populations at all three hierarchical levels, or to study the effects of high parasite densities on host population biology.

A number of reasonable generalizations can be made about the density-independent control of parasite populations (in the absence of density-dependent effects on parasite populations most would argue that there is no regulation, a position that has been followed here). For these species, transmission processes and parasite reproduction are almost always influenced by some external factor such as temperature or photoperiod. Because of short, or even long-term, vagaries in climate within a given geographical locality, parasites under the influence of density-independent constraints may be subject to radical shifts in density or prevalence. Consequently, many of these same species will be more vulnerable to local extinction. Synchronization of the life cycles of many parasites and hosts that are affected by density-independent factors must be matched closely; such synchronization also is important in considering the co-evolution of various host–parasite systems. This is not to say that 'regulated' host–parasite systems do not have synchronized life-cycle patterns because they do, but many times the patterns of the former have the appearance of being in closer lock-step.

The case histories of organisms described in Chapter 3 are but a few of the examples of the many that could have been selected to illustrate the influence of density-independent factors on parasite infra- and meta-populations. *Caryophyllaeus laticeps* was chosen because it was studied in a number of different host species and because, in each, it behaved differently; it also represents an excellent example of the way in which temperature may influence a parasite's population biology. *Bothriocepalus acheilognathi*, on the other hand, is not only affected directly by annual temperature cycles, there is evidence that if the parasite is recruited by definitive hosts of small size, intraspecific competitive interactions will occur with a concomitant reduction in infrapopulation sizes and apparently even host mortality. This thus represents a highly complicated host–parasite relationship since it suggests the influence of density-independent factors as well as density-dependent regulation. The *Eubothrium salvelini* study can almost be considered as a classic because it ran for 15 years, making it one of the longest involving parasites. In this case, as for *Diphyllobothrium sebago*, there is a clear indication that both hosts and parasites are in tight synchrony with regard to their annual reproductive cycles. Each of these studies serves well in illustrating the nature of density-independent factors that affect parasite population biology. While they may appear as 'natural history-ish', they are no less important than the more sophisticated quantitative approaches taken by many other investigators, including the mathematical modellers. Indeed, these kinds of studies are the grist for modellers and must be encouraged in the future.

Density-dependent regulation in the field is difficult to demonstrate, but a few studies have provided some limited evidence for its occurrence. Perhaps one of the best examples was that described by Wassom, Guss and Grundmann (1973), Wassom, DeWitt and Grundmann (1974) and Wassom *et al.* (1986). In these investigations, Wassom and his co-workers clearly showed that the genetically-based expulsion of *Hymenolepis citelli* from deer mice in Utah (USA) was dose-dependent, being directly related to the number of cysticercoids recruited. In contrast, however, while working with a similar host–parasite system in Wisconsin (USA), Munger, Karasov and Chang (1989) could not confirm the results of the Utah study. While they reported the same genetically-based, expulsion phenomenon, little else in the Wisconsin system was the same as it was in Utah.

Interspecific competition also is an important component of density-dependent regulation and has been reported at both the infrapopulation and infracommunity levels of organization. While there is clear evidence that it can be demonstrated under controlled laboratory conditions, its extent in nature is uncertain. The classic studies by Holmes (1961, 1962a, b) on *Moniliformis dubius* and *Hymenolepis diminuta* under laboratory conditions revealed that both intra- and interspecific competition can occur. Field evidence reported by Chappell (1969) suggested competitive interaction between a tapeworm and an acanthocephalan in a natural population of sticklebacks in England. The investigation by Holmes (1961, 1962a, b) proposed that competition was related to limited nutrient resources, probably carbohydrate, as had been suggested earlier by Read (1959).

Developmental arrest is an apparently unique, but relatively widespread, form of density-dependent regulation associated with several species of mammalian nematodes. A form of diapause, developmental arrest has been linked with several environmental factors that, in this case, also make it a density-independent transmission strategy. There is, however, ample evidence from Russell, Baker and Raizes (1966) and Michel, Lancaster and Hong (1975a, b) that some species of nematodes will respond to dose-dependent stimuli and that the resulting density plateaus should be considered in terms of 'biomass or immunogenic thresholds'.

At the suprapopulation level, there are some clear problems for any investigator because of the tremendous biological complexity associated with most host–parasite systems. However, the studies by Hairston (1965), Holmes, Hobbs and Leong (1977), Dronen (1978), Jarroll (1979), Aho and Kennedy (1987), Riggs and Esch (1987), and Riggs, Lemly and Esch (1987) combine to provide sufficient information to formulate a couple of important generalizations. First, it appears that the transmission efficiencies of helminth parasites are in a delicate balance with both the host's and the parasite's reproductive capacities. If the parasite gets 'too good' at reaching the next host in the cycle, then it could kill the host. If there are 'too few' infective stages produced, then the parasite's chances of reaching the next host in the cycle could be jeopardized; the chances for transmission success by an individual life cycle stage are certainly not high under any circumstance. A second collective observation regarding studies on parasite suprapopulation dynamics relates to the significance of reservoir hosts in the life cycles of many parasite species. Perhaps it would be better stated to say that reservoir hosts have always been recognized as important but, with some of the studies just cited, their significance in maintaining a given parasite within a habitat acquired even greater meaning.

Over the years, there has been a great deal of research on the ways in which abiotic factors affect the metapopulation biology of parasites in aquatic systems, primarily those in northern temperate areas of the world where seasonal temperature changes are typically of some magnitude. It is contended, however, that a major short-coming in this area is the almost total lack of studies in the tropics or sub-tropics. It would seem as though field studies in the tropics and sub-tropics would be of real value in characterizing and further defining the basic role of abiotic factors in influencing parasite population dynamics.

10.4 INFLUENCE OF PARASITES ON HOST POPULATIONS

A central component in Crofton's paradigm was in combining the concept of overdispersion with that of parasite-induced host mortality among heavily infected hosts. The negative binomial model then provided an appropriate framework for comparing theoretical and observed frequency distributions. As was subsequently pointed out by May (1977), however, a difficulty with Crofton's approach was his application of the so-called 'Lethal Level' concept. This approach, as May (1977) noted, introduced a deterministic flavour to the notion of parasite-induced host mortality, one that is unrealistic in the real world. There is simply too much biological variability to allow for the application of a 'Lethal Level' concept to host–parasite interactions.

The phenomenon of overdispersion is certainly not the sole domain of parasites since most organisms are distributed contagiously in space. On the other hand, overdispersion becomes exceedingly important in terms of its potential for the parasite's induction of host mortality, especially among individuals that are heavily infected. It is certainly not difficult to understand why overdispersion is important from the standpoint of population regulation. The question then becomes, how extensive is the regulation of host populations by parasites? As alluded to earlier, the answer to this question based on an examination of the literature is, not very. Simply stated, parasites typically do not kill their hosts. Obviously, there are exceptions, with malaria and Chagas' disease among humans as good examples. The morbidity, on the other hand, produced by both helminth and protozoan parasites of man and other hosts is extensive. For example, according to Quinnell, Medley and Keymer (1990) at least one billion people are infected with *Ascaris lumbricoides* alone. Because diseases such as ascariasis are as common as they are and because they appear to be on the increase, especially in Third World countries, they have begun to receive substantially more attention in the past decade from the standpoint of treatment and control.

The first application of the Crofton (1971a) paradigm to the epidemiology of human parasitic diseases can be attributed to Croll and Ghadarian (1981). Based on these and subsequent studies, some epidemiologists advocate the traditional approach of massive or widespread treatment of human enteric parasites. Others have argued that this is unnecessary, if not ineffective or even wasteful, from the standpoint of the limited resources that are available. The latter group holds to two propositions that make them believe a limited approach would be more effective. First, because of overdispersion, a relatively small fraction of people in a population or a given village can be considered as 'wormy'; that is, a few individuals invariably carry most, by far, of the parasites. Thus, some would advocate that only the 'wormy' people should be treated. Certainly in terms of cost effectiveness, this approach would seem reasonable. The second proposition is related to why these particular individuals are 'wormy' and not others? Herein lies the basis for a serious debate among some epidemiologists with an interest in understanding the nature of parasite transmission processes. There are those who hold that transmission of many of the human enteric helminths is largely stochastic in character. Others, on the other hand, believe that some people are predisposed to infection, but there are also differences in the way predisposition is viewed within this group. Some assert that predisposition is genetic, or that some individuals are inherently more susceptible to infection. Another group feels that predisposition is related to socioeconomic status at the individual level or to variations in non-genetic cultural behaviour. If predisposition is genetic, then it makes sense to treat only the 'wormy' individuals. Moreover, there are some modellers who contend that if 'wormy' individuals are de-wormed, the population dynamics of the parasite will become unstable and that transmission of the parasite would decline perceptibly. Again, however, the problem is that not enough data of the right sort exist; this makes the idea of genetic predisposition impossible to assess at the present time. On the other hand, if genetic predisposition can be demonstrated, then the notion of treating only the 'wormy' individuals in a population has real potential for success and should be given serious attention.

Early use of mathematical modelling in the development of treatment protocols for human parasite diseases can be attributed to Hairston (1965); the quote from Hairston (section 4.6) regarding the application of models to the control of disease is particularly appropriate in this regard. The mathematical model in both human and non-human host–parasite systems has been useful in many ways. The most widely employed model for describing parasite frequency distributions is the negative binomial. Its usefulness is clear but, far too often, when a fit is generated nothing more is done with the data or the study. It must be emphasized that curve fitting, as an end in itself, is a useless exercise. On the other hand, there now appears to be some question about the usefulness of certain parameters within the negative binomial model itself. Thus, Scott (1987b) has argued that the $S^2/\overline{X}$ ratio is more useful in interpreting the nature of frequency distributions than the value of k in the negative binomial model. In this sense, it almost appears that we have come full circle because Crofton (1971b) argued it was overdispersion by itself, and not the type of theoretical model, that was the most important component in his quantitative approach to parasitism.

As previously indicated, modelling is useful because it encourages the development of quantitative approaches to the study of host–parasite relationships. For this reason, if for no other, additional modelling should be encouraged. Moreover, it should be extended to the level of the parasite community, although Dobson (1990) has made a start in that direction. The mathematical model is in general a sophisticated mathematical hypothesis, but most models have not been subjected to testing under field conditions and this constitutes another serious short-coming that needs to be corrected in the future.

10.5 LIFE-HISTORY STRATEGIES

Reproduction by parasitic helminths includes both gametic and agametic methods. Gametic reproduction is common among the adults of many parasitic taxa. These organisms may be either monoecious or dioecious. Protandry has been shown to occur in some species. Agametic reproduction is common among the intramolluscan stages of digenetic trematodes where cyclical parthenogenesis is believed to represent the main reproductive process. Among the trematodes, then, there is mictic reproduction associated with adults and parthenogenesis with the snail stages. Strobilization (modular iteration according to Hughes (1989)) among adult tapeworms is a process that involves the continuous production of new proglottids, each of which is packaged as a complete reproductive unit with respect to male and female genitalia. Some cestodes, primarily several species in three families of the order Cyclophyllidea, employ an asexual method of reproduction as larvae in their intermediate hosts.

While many parasitic organisms have evolved highly complex life cycles, high fecundity is also an important feature in their reproductive repertoire. Although it is clear that high fecundity has a genetic basis, it is not entirely clear whether it has been selection forces operating in an evolutionary time scale that favour a parasite's high reproductive capacity, or if it is more simply a function of the extensive resource base that is available to most parasites. Competition, host genetic factors, and immunity also are known to affect parasite reproductive capacities.

The colonization strategies evolved by parasites must be considered within the framework of spatial and temporal scales. In effect, both the parasite and the host must be in the right place at the right time in order for successful transmission to occur. In general, colonization or trans-mission strategies can be separated into those that employ active processes and those that are passive; many species use one or another of these processes at different stages in their life-cycles. Active transmission implies that the parasites possess locomotory structures or abilities. Those that are passive will frequently depend on predator–prey interaction to assure transmission and, in some cases, may even actively alter a host's behaviour to increase the probability of predation and transfer of an infective agent to the next host in the life cycle. Parasites that employ passive mechanisms for transmission may depend upon unique spatial relationships between hosts and parasites to ensure successful completion of their life cycles.

Life-history strategies have been the target of a large number of 'theoretical' papers over the last 40 years and a number of efforts have been made to apply them to host–parasite systems. The r- and K-correlates of Pianka (1970), for example, viewed life-history strategies as a continuum, ranging from the r- end of the spectrum in which there are no density effects and no competition, to the K- extreme in which the environment is saturated and density effects are the greatest. Esch, Hazen and Aho (1977) argued that parasites exhibited characteristics of both r- and K- life-history strategies but, as a group, they tended more toward the r-end of the continuum.

Sibly and Calow (1986) predicted that survivorship probabilities were likely to be important in the evolution of parasite life cycles. Southwood (1988) argued that defence, migratory abilities, size of the organism, longevity, and the capacity to cope with environmental harshness, were all involved with the evolution of life-history strategies in response to habitats of increasing adversity and habitat disturbance. The current application of life-history 'theory' to the parasitic life style is not completely satisfactory, however. There are several attendant problems. These are variously related to the fact that parasites live in or on another host and, over evolutionary time, this has meant that parasites have had to adjust to the immunological adversity imposed by one or more hosts in their life cycle. Another problem is that some parasites clearly have co-evolved with their hosts. In other words, as the life-history strategy of the host changed within an evolutionary time frame, that of the parasite has also adjusted, otherwise it would have been necessary for the parasite to find a new host or become extinct. It is submitted that these kinds of problems are unique to the parasite life style and that they are not adequately handled by any of the current life-history paradigms.

In terms of the life-history strategies and colonization problems that still require resolution, perhaps the most difficult to approach will be that of understanding the sensory physiology involved with the internal migration of parasitic organisms once they have gained access to their host. At the present time, knowledge of the operational mechanisms for internal migration can be likened to the proverbial 'black box'. It is known, for example, that *Fasciola hepatica* ends up in the bile ducts and gall bladder of sheep definitive hosts, and that the internal migration route has been determined. But, what is the exact nature of the stimulus, or stimuli, and how does the parasite identify these stimuli? Moreover, what are the sensory mechanisms by which the parasite responds to these stimuli? The nervous systems in many worms, including *F. hepatica*, have been described, but the precise stimulus–response mechanisms remain unknown. Since virtually all protozoan and helminth parasites exhibit strong site fidelity, the questions asked for *F. hepatica* are no different from those which could be asked of any enteric species which remains in a particular location within the gut after hatching or excysting. The enteric parasites are probably responding to chemical gradients of some sort, e.g. pH, bile salt concentrations, etc., but the exact operational mechanisms are not known with any precision. The concept of site specificity thus raises many important questions. The obvious complexity of the phenomenon, however, suggests that answers to many of these, and related, questions will remain elusive for some time to come.

10.6 INFRACOMMUNITY DYNAMICS

An analysis of the evolution of parasite communities presents problems that are somewhat similar to those associated with life-history strategies.

There are presently four models that address the issue. At the two extremes are the co-speciation model in which communities are non-interactive and competition is presently not a factor in structuring the community and the non-equilibrium model which contends that communities are unsaturated, with many empty niches. The non-asymptotic model also argues that parasite communities are non-interactive and that they increase in size linearly through time. At the other extreme is the asymptotic model (Theory of Island Biogeography). This model holds that the evolution of communities sequentially involves both non-interactive and interactive components, followed by assortative and evolutionary phases. The structuring forces operating at this level include both stochastic and biotic events such as competition; there may have been an interactive component in its evolution or competition is presently a determinant of community structure.

In general, it can be said that a wide range of factors will influence the organization and structure of parasite infracommunities in definitive hosts. As much as possible, these have been summarized in Table 7.1 (page 179 and 180). Biotic interactions such as competition and the host's immune response probably play a central role in structuring some infra-communities, but certainly not all. Other factors include the thermal physiology of the host, complexity of the gut, composition of diet, breadth of diet, host genetics and ontogeny, an aquatic versus a terrestrial life style and host vagility. Individually and collectively, these factors will influence the number of parasites in a given host, the number of species present, the mix of core and satellite species, the proportion of specialist and generalist species and site specificity. The outcome will be the establishment of either an interactive, or an isolationist, infracommunity.

A comparison of interactive and non-interactive enteric infracom-munities in fishes and birds was made by Kennedy, Bush and Aho (1986). They predicted that freshwater fish infracommunities should be isola-tionist and non-interactive; some birds, they asserted, should have species rich and interactive infracommunities. If their predictions are restricted to freshwater fishes, several species of passerines and most species of shore-birds, wading birds, and waterfowl, then they would be largely correct. They attributed the differences in birds and freshwater fishes to the fact that most of the latter are not particularly vagile, they are ectothermic, their intestines are rather simple in terms of niche diversity, and they have a relatively restricted dietary range. According to Aho (1990), most amphibians and reptiles also possess depauperate and non-interactive infracommunities. In contrast, while many mammalian infracommunities cannot be described as depauperate, most of those examined do not appear to be interactive either. Montgomery and Montgomery (1990) believe that their studies on *Apodemus sylvaticus* infracommunities support the notion of non-equilibrial conditions for parasite communities as espoused by Brooks (1980), Price (1980), and Rohde (1982).

Infracommunities in intermediate hosts are influenced by most of the same factors that affect infracommunities in definitive hosts. There are a couple of exceptions. The first is related to the imposition of predation by some larval trematodes on other trematodes in molluscan intermediate hosts. In certain species of molluscs, predation by trematode rediae is so extensive that it has even encouraged the creation of an elaborate scheme of dominance hierarchies (Kuris, 1990). Second, and while not entirely restricted to trematode–molluscan systems, there is evidence that some species of flukes partition molluscan resources both spatially and temporally (Fernández and Esch, 1991a, b). By doing so, they avoid antagonistic interactions and the occurrence of interactive communities.

Manipulation of most parasite infracommunities is virtually impossible. The intractable nature of this problem has meant that many of the conclusions drawn from community studies are inferential at best. The species-rich infracommunities of birds, for example, present tremendous practical difficulties not only from the standpoint of rearing potential hosts, but in obtaining infective stages of various parasites. Until these sorts of problems can be resolved, it is doubtful that significant new information can be generated with respect to the structuring mechanisms of parasite infracommunities in most vertebrate animals.

10.7 COMPONENT AND COMPOUND COMMUNITIES

Component and compound parasite communities have also received attention, although not to the extent infracommunities have been studied. Enteric component communities in most freshwater and marine fishes appear to differ substantially. As already noted, the former are largely depauperate, while many of the latter are relatively species rich. The richness of component communities is due, in part at least, to the relatively high degree of vagility associated with marine fishes, or perhaps to the vagility of the parasite's intermediate hosts. Another factor contributing to component community richness in some marine fishes is the relatively low level of host specificity exhibited by many of these parasites, making them generalists and capable of developing to maturity in a greater range of host species. Ectoparasitic communities, primarily monogenetic trematodes on the gills of marine fishes, are reportedly (Rohde, 1982) non-interactive and possess many empty niches. Rohde (1991) provides a strong case that ectoparasitic communities on the gills of fishes are not structured by interspecific competition and that narrow niche sizes for many of these species enhance opportunities to mate.

Component communities in amphibians and reptiles are similar to those in freshwater fishes in that they are mostly depauperate. Many birds, in contrast, possess highly structured, and species rich, component communities. It has been argued that these communities are influenced by a complex set of screens and filters. The communities in mammals appear not to be as complex as they are in many birds. To a very great extent, they are influenced by the habitat variability of the definitive hosts, as well as other determinants such as vagility. An important idea here is that the range of potential intermediate hosts in the diet will certainly affect the composition of the community, just as it frequently does in waterfowl and shore birds.

Ecological succession is a phenomenon that affects many ecosystems. It has been described as the process by which an ecosystem matures; it is directly influenced by the organisms in the environment and is non-cyclical in character. In freshwater systems, the process of succession may be affected by man and is known as cultural eutrophication. Interest in eutrophication and how it impacts on host–parasite relationships go back to the work of Wisniewski (1955, 1958). It is not surprising that eutrophication should have an effect on the community structure of the parasites within an ecosystem since it has such a profound effect on the community dynamics of the free-living organisms.

A question that has been posed frequently is whether there is a relationship between the nature of a parasite compound community and the trophic status of the lake in which it is found? The answers to this question and its corollary, what actually occurs during succession, are still largely equivocal. Some believe that the faunistic transformations which occur during succession, or the differences observed between fauna in lakes with different trophic conditions, are due to changes, or differences, in the dominant predators. Others assert that the observed patterns are more directly linked with differences that occur in predator–prey relationships within the lakes of changing trophic status. This is, in turn, tied to the allogenic–autogenic species concept and to the nature of colonization strategies employed by different parasite species. Allogenic species are characteristic of many wading and shore birds and are alleged to be more often associated with eutrophic systems with extensive littoral zones suited for the foraging of such species. Autogenic species, on the other hand, are suggested to occur with greater frequency in systems that are 'closed' in the sense that foraging by wading and shore birds in the littoral zone is more restricted.

Research on component and compound parasite communities has been fairly extensive especially as it applies to freshwater fishes and aquatic birds. Less attention has been directed at amphibians, reptiles, and mammals. Moreover, in almost all host taxa that have been studied to date, the data have been summed before analysis. In other words, individual hosts have not been treated as replicate habitats. Future studies should consider this oversight and provide appropriate analyses so that more accurate comparisons can be made.

Another short-coming in this area rests with the insufficient attention which has been given to component and compound communities in marine fishes. The work which has been accomplished, i.e., that of Campbell, Haedrich and Munroe (1980), Campbell (1983, 1990), Rohde (1982, 1984) and Holmes (1990), has been both of high standard and revealing, but it too must be extended if we are to have a better under-standing of how these parasite assemblages fit within the overall frame-work of parasite community biology.

10.8 BIOGEOGRAPHICAL ASPECTS

Biogeography has a long history as a sub-discipline. Humbolt and von Ihering in the 19th century were certainly pioneers in the distributional analysis of free-living groups while Metcalf and Manter in the present century led the way for parasites. It is interesting to note, however, that von Ihering actually first recognized the value of using parasites in studying the zoogeography of host species over 100 years ago. The emergence of the 'von Ihering method' for the identification of phylo-genetic relationships has been an important element for many of the subsequent, distributional analyses.

On a global scale, natural geographical barriers, the fragmentation of land masses, the evolutionary time frame and the vagility of hosts, are all important to the geographical distribution of parasites. As the scale size is reduced, however, a wider range of biotic and abiotic forces create a highly complex matrix of variables that influence the distribution of species in space and time. Moreover, these matrices have changed and continue to change under the influence of man, as well as local and global climatic vagaries.

In freshwater systems, as already noted, both cultural and natural eutrophication are known to profoundly affect the abundance and diversity of hosts and parasites. Drainage patterns and water flow are im-portant factors in influencing distributional patterns. Salinity and basic water chemistry have been shown time and again to affect the zoo-geography of various host–parasite systems. Many of these same factors also will affect so-called 'terrestrial' parasite species since many of them are tied to freshwater at some point in their life cycles. In marine systems, temperature–salinity (T–S) profiles appear to be among the most significant determinants of host–parasite distributions. The distinctive heterogeneity in marine systems also is linked to T–S profiles, geographical location, depth, and land barriers. All of these physical and chemical factors are important to the dispersal and distribution of parasites and their hosts. Several studies also have identified the presence of clearly grouped assemblages of parasites in association with specific taxa of marine fishes such as the striped mullet, hake, and herring.

A number of investigators, but primarily Brooks, Thorson and Mayes (1981) have combined biogeographical and phylogenetic approaches in examining host–parasite relationships from an historical perspective. While these inputs have been important from the standpoint of understanding the lineages of several host and parasite groups, the results also have provided a strong suggestion of various co-evolutionary events, as well as evidence for the proposed timing and patterns in the fragmentation of the supercontinents. In some cases, dispersal of the definitive host and phylogenetic relationships among parasites seem more important for parasite distribution than vicariant events.

For some taxonomic groups of marine organisms, there appears to be a latitudinal gradient for species richness. This certainly seems to be the case for monogenetic and digenetic trematodes in fishes, with numbers increasing closer to the equator. The same case cannot be made for parasites of marine mammals as species richness is greatest in colder, more polar waters. This is probably because the diversity of marine mammals is also greater in polar latitudes.

Based on available evidence, it appears as though the monogenean parasite fauna of the Pacific Ocean is greater than that of the Atlantic. Rohde (1982, 1986) suggested that this was because the Pacific has had a much longer evolutionary history than the Atlantic and that in the last glaciation episode, vast areas in the Atlantic were covered by ice. He indicated that this could have led to the extinction of many monogeneans, thereby reducing species richness.

It also seems that parasite diversity is affected by depth in marine environments. This is true for both benthic and pelagic fishes. Diversity is greater in the former than among the latter species. Changes in parasite diversity with depth appear to be related to the size and degree of interaction of the ecosystem components and the permanence of the host's association with particular invertebrate communities. In general, host density constrains the establishment of parasites with direct life cycles, while overall faunal diversity constrains the presence of parasites with complex life cycles. On the other hand, the lower parasite diversity in pelagic fishes is apparently related to differences in food supply and life span of the host. Even though mesopelagic fishes maximize their energy sources by means of vertical migration, this is not enough to override the parasite diversity of benthic fishes at similar depths. Similarly, the shorter life span of mesopelagic fishes (up to 2 years) means briefer periods of exposure to parasites when compared to the longer life span of benthic fishes.

Another question concerns that of species–area relationships. Is species diversity a function of the geographical area occupied by a host, a host's range, a host's size, or perhaps host diversity within a habitat? Rohde (1989) concluded that host range and host size were not important, at least for marine monogeneans, but that phylogenetic diversity of the host group was significant. The data presented for *Mugil cephalus* also support this conclusion. In some terrestrial and freshwater systems, host geographical range and parasite diversity seems to be correlated, but the evidence is equivocal since, in some cases, phylogenetic and historical processes have not been considered.

The economic value of some marine fisheries has led to the use of parasites as 'biological markers' (Margolis, 1956, 1990). This technique employs parasites to identify the geographical origin and the feeding and reproductive areas of anadromous salmonid fishes, as well as some strictly marine species. In order to be useful, the parasite markers must have restricted geographical distributions at some point in their life cycle, they must not be capable of interhost transfer while at sea, they must be able to survive marine migration, and they should have long life spans. Generally, most ectoparasites and many endoparasites do not meet these requirements. Larval helminths that are sequestered in host tissues are the best markers because they are not subjected to the hazards of the external environment or to potential physiological changes in the host's intestine that might be induced by variations in diet during migration.

Interestingly, although biogeographical investigations of host–parasite systems began over a hundred years ago, and a great deal of attention has been given to parasites of marine fishes presumably because of the economic importance of these hosts, not many comprehensive studies exist. Much less is known about parasites in terrestrial hosts and freshwater fishes. Moreover, much of what is known about the latter two groups is based on studies conducted in northern temperate areas of the world and relatively little focus has been given to the tropical and subtropical localities. Clearly, these are fertile areas for future work.

10.9 EVOLUTIONARY ASPECTS

Understanding the evolution of any species is a difficult proposition, but it becomes even more so when dealing with soft-bodied organisms such as parasites. This difficulty is compounded by the fact that in considering the evolution of a parasite, the issue is not just that of an individual parasite species. Parasitism, as a life style, also must be scrutinized. Moreover, it is the evolution of a life style that always involves two organisms in widely separated taxa. Further exacerbating the problem, most parasites have complicated life cycles involving two or more hosts. Not surprisingly, there are a limited number of studies in this area. On the other hand, the use of electrophoresis and of the other genetic and biochemical probes have permitted the development of some new and exciting insights regarding the evolution of certain parasite species and species groups.

The genetic variability among several species of parasitic helminths has been considered from a number of perspectives. Current evidence suggests that the extent of genetic variability in parasites is similar to the average level of variability among free-living invertebrates. Surprisingly, however, the amount of genetic divergence among some species is not affected by geographical separation as much as it is by host isolation. For example, *Ascaris suum* was examined from hog populations separated by a distance of 1500 miles, yet the parasites were virtually indistinguishable genetically. These similarities were attributed to increased gene flow that may, it was suggested, be related to the husbandry practices involved with modern farming operations. In contrast, *Echinococcus granulosus* appears to be a complex of genetically distinct strains. This species is reported to be mostly self-fertilizing which, in combination with the fact that asexual reproduction in hydatid cysts is clonal, appears to reduce genetic variation. Superimposed on this tendency has been the development of ecological barriers that have led to isolation of specific strains of the parasite in unrelated, but behaviourally similar, host species.

The neutralist and selectionist hypotheses have been advanced to explain polymorphisms and heterozygosity levels within various populations. The former concept holds that most allelic changes are selectively neutral and that there is a relatively constant rate of molecular evolution of homologous molecules in the lineages of different species. The selectionist hypothesis, on the other hand, argues that changes in gene frequencies are adaptive and that, accordingly, adaptive variations will gradually increase at the expense of less adaptive ones over time. The application of these two approaches to parasitic organisms has two drawbacks. The first is the lack of adequate data. The second is related to the question of what is being compared? Bullini *et al.* (1986), for example, examined several ascaridoid nematodes from the perspective of single- versus multiple-host life cycles. They concluded that increased genetic flexibility in parasites could be related to life-cycle complexity, with genetic variability increasing as the life-cycle complexity increased. The difficulty here is that the number of hosts in a life cycle is not the only measure of life-cycle complexity. Other factors such as internal migration routes and external environmental vagaries can combine to create extreme selection pressures and possibly influence genetic variability.

As previously stressed, when dealing with parasites in terms of evolution, the host must also be considered. Therefore, one must be concerned with co-evolutionary processes. While recognizing that co-evolution of some hosts and parasites has occurred, there is no agreement yet as to the exact meaning of the term. It is generally agreed, however, that a basic formulation in the concept is that interacting species will influence each other's evolution. The extent to which some host and parasite lineages have been intermingled over time has been estimated using phylogenetic systematic techniques (Brooks, 1989). Brooks (1985, 1988) estimates that the level of consistency between the phylogenies of helminth and vertebrate systems ranges from 50 to 100%, a reasonably high degree of co-evolutionary overlap.

The actual mechanism for co-evolution between hosts and parasites is summarized in three models, namely: mutual aggression, prudent parasitism, and incipient mutualism. The first model most closely coincides with the strictest definition of co-evolution; it views the relationship as an 'arms race', with each partner constantly evolving in an aggressive manner toward the other. In the second model, selection pressures exerted by the parasite on its host are reduced over time and selection on individual parasites will be toward reduction in virulence. The third model has less support and is based on the idea that both hosts and parasites evolve mutualistic characteristics that actually promote the presence of each other.

Several studies have shown that parasites can act as mediators of a number of ecological interactions that ultimately affect community structure, interspecific interactions and host distributions. The *Tribolium* spp. –*Adelina* and *Pneumostrongylus*–cervid systems are good cases in point. Based on these and several other examples, parasites have been considered by some as 'weapons of competition' from the standpoint of influencing geographical distributions and host speciation.

Hamilton (1980) proposed that parasites have even been instrumental in the evolution of sex. Hamilton and Zuk (1982) subsequently hypothesized that parasites could have had a role in the evolution of secondary sexual characteristics. They suggested that bright colours and elaborate displays permit females to assess the potential of males to resist parasites. The hypothesis was supported by examination of blood parasites in a number of passerine bird species. Other researchers extended this type of investigation to British and Irish fishes, but their results appear equivocal.

Finally, the idea of horizontal gene transfer through the intervention of retroviruses has been proposed as a way of increasing genetic diversity without mutation. The transfer of genes from host to parasite would certainly explain the molecular similarities known to exist for several host–parasite combinations. It also would explain, in the view of Rohde (1990), conflicting evidence regarding the taxonomic relationships of many parasite species. He considered that gene transfer may function as a source of isolated evolutionary units upon which natural selection could act in addition to mutations, calling it 'mosaic evolution'.

10.10 WHAT NEXT FOR THE EVOLUTION AND ECOLOGY OF PARASITISM?

A number of obvious gaps exist in our understanding of ecological parasitology. For example, while the development of mathematical models for parasite population ecology is impressive, the lack of application of these models to host–parasite systems under field conditions is equally impressive. Thus, while these models are sophisticated and potentially useful mathematical hypotheses, few of them have been tested. There are certainly promising opportunities, especially as related to the epidemiology of many of the human helminthiases. This point was strongly made in the papers by Anderson and May (1979) and May and Anderson (1979) and since has been reiterated by others.

Another problem area is the almost total absence of long-term parasite population studies. There are a few (Smith, 1973; Esch *et al.*, 1986; Marcogliese, Goater and Esch 1990), but these are quite limited in their scope. While they have some heuristic value from the standpoint of population biology, they are not comprehensive enough to permit solid predictions regarding the long-term patterns of parasitism. Long-term studies of parasite communities are totally non-existent. It would, therefore, seem appropriate that some of the studies on parasite communities conducted 10–20 years ago be repeated on the same systems at the present time.

Studies on the population genetics of parasitic helminths and their hosts are in their infancy. Some species have been examined fairly extensively, e.g., *Echinococcus granulosus*, *Schistosoma mansoni* and *Biomphalaria glabrata*, among others. In many of these cases, however, the efforts have focussed on interactions occurring in the laboratory. More attention needs to be directed toward conditions in the field. Additional studies also should be directed at host–parasite relationships in marine ecosystems.

A number of observations strongly suggest a wide divergence in the rates of speciation among many helminths and protozoans, including those affecting man. Is there a way of accounting for these differences? Other questions arising from these kinds of studies would include, what constitutes a population from a genetic perspective? Thus, how does one account for the clear differences in the population genetics of *Ascaris suum*, *Echinococcus granulosus*, or *Fascioloides magna*? Is genetic variability among parasites a function of life-cycle complexity or have selection forces associated with the host's immune responsiveness, for example, played an overriding role in affecting genetic variability? How extensive are parasites as 'weapons of competition' or, stated in another way, how extensive has parasitism been in affecting host speciation and geographical distribution? Is the Hamilton and Zuk (1982) hypothesis regarding the role of parasitism in the evolution of sex valid, either partially or wholly? Answers to questions such as these would constitute a significant core of basic information regarding the ecology and evolution of parasitism as a life style. Hopefully, these answers will be forthcoming in the next several years.

Rohde (1982) has stated that our understanding of the zoogeographical aspects of parasites is exceedingly limited. He was writing primarily about marine systems, but the same can be said for parasites in terrestrial and freshwater systems. Certainly, additional comprehensive studies will go far in understanding the underlying patterns of host and parasite distributions and in the processes which created these patterns.

There is much known about the internal migratory behaviour of parasitic organisms and about the way in which parasites can modify host behaviour, but not much is known about how these behaviours actually function or how they are triggered. More effort should be made to link behaviour with the molecular and cellular interactions involved with host location and site specificity. An intriguing question along the same line is related to the phenomenon of self-recognition among predatory intramolluscan stages of digenetic trematodes. The relationship must be fairly sophisticated because cannibalism has never been reported among these species.

A question that persists in the literature concerns the extent to which parasites are capable of regulating host populations? This question requires serious attention, but it needs to be extended away from the vertebrate definitive host level to that of the intermediate host. The likelihood of regulation is much greater for the latter group of hosts than it is for definitive hosts because the pathology induced in these organisms is frequently, though not always, much more severe.

Another area of study in which research has not been conducted very often involves parasite suprapopulations. When they have, however, they have usually produced insights that were most revealing about the way in which transmission dynamics were actually operating and, in a couple of cases, in ways that could not have been predicted a priori. It would seem that additional investigations at this level would be most productive. Similarly, more work on compound parasite communities would be an appropriate undertaking as well.

The concept of spatial and temporal scaling holds real promise for the further elucidation of transmission dynamics for certain parasite species.

Such studies may become quite important in understanding resource partitioning by parasites in both time and space. It was suggested by Croll and Ghadarian (1981) that the epidemiology of certain helminths of man may be much more subtle than is currently believed. It is proposed that knowledge of these subtleties, if they indeed exist, may provide a key to the manner in which these diseases are treated on a global and a long-term basis, as well as locally and in the near-term. Although not a new idea, the concept of landscape ecology or epidemiology on a micro-scale may have broader application than is currently envisaged by many investigators.

Another gap in ecological parasitology relates to the non-human, parasitic protozoans. While a few studies have been undertaken, not much is really known. This group of organisms represents a real opportunity for future work.

In general, knowledge of ecological parasitology has expanded significantly since the work of Dogiel nearly 50 years ago. Much of the research has been focussed on field studies and most of these are strictly 'observational' in character. It is suggested that whenever feasible, future investigations should employ combinations of field *and* laboratory research and, moreover, that systems should be experimentally manipulated when possible. The use of more sophisticated experimental approaches and technologies in ecological parasitology will enhance greatly our understanding of host–parasite relationships and ultimately create a more solid foundation for the discipline.

References

Adams, D.B. (1986) Developmental arrest of *Haemonchus contortus* in sheep treated with corticosteroid. *International Journal for Parasitology*, **16**, 659–664.

Aho, J.M. (1990) Helminth communities of amphibians and reptiles: comparative approaches to understanding patterns and processes. In *Parasite Communities: Patterns and Processes* (eds G.W. Esch, A.O. Bush and J.M. Aho), Chapman and Hall, New York, pp. 157–195.

Aho, J.M. and Kennedy, C.R. (1987) Circulation pattern and transmission dynamics of the suprapopulation of the nematode *Cystidicoloides tenuissima* (Zeder) in the River Swincombe, England. *Journal of Fish Biology*, **31**, 123–141.

Allison, F.R. and Coakley, A. (1973) Two species of *Gyrocotyle* in the Elephant fish, *Callorhynchus milii* (Bory). *Journal of the Royal Society of New Zealand*, **3**, 381–392.

Anderson, R.C. (1972) The ecological relationships of meningeal worm and native cervids in North America. *Journal of Wildlife Diseases*, **8**, 304–310.

Anderson, R.C. (1976) Helminths. In *Wildlife Diseases*, (ed. L.A. Page), New York, Plenum, pp. 35–43.

Anderson, R.C. and Prestwood, A.K. (1979) Lungworms. In *Diseases of the White-tailed Deer* (ed. F. Hayes), U.S. Department of the Interior, US Wildlife Service, Washington, DC.

Anderson, R.M. (1974) Population dynamics of the cestode *Caryophyllaeus laticeps* (Pallas, 1781) in the bream (*Abramis brama* L). *Journal of Animal Ecology*, **43**, 305–321.

Anderson, R.M. (1976) Seasonal variation in the population dynamics of *Caryophyllaeus laticeps*. *Parasitology*, **72**, 281–305.

Anderson, R.M. (1978) The regulation of host population growth by parasitic species. *Parasitology*, **76**, 119–157.

Anderson, R.M. (1989) Transmission dynamics of *Ascaris lumbricoides* and the impact of chemotherapy. In *Ascariasis and Its Prevention and Control* (eds D.W.T. Crompton, M.C. Nesheim and Z.S. Pawlowski), Taylor and Francis, London, pp. 253–273.

Anderson, R.M. and Gordon, D.M. (1982) Processes influencing the distribution of parasite numbers within host populations with special emphasis on parasite-induced host mortalities. *Parasitology*, **85**, 373–398.

Anderson, R.M. and May, R.M. (1978) Regulation and stability of host–parasite interactions. I. Regulatory processes. *Journal of Animal Ecology*, **47**, 219–247.

Anderson, R.M. and May, R.M. (1979) Population biology of infectious diseases: Part I. *Nature*, London, **280**, 361–367.

Anderson, R.M. and May, R.M. (1982) Coevolution of parasites and hosts. *Parasitology*, **85**, 411–426.

Anderson, R.M. and Schad, G.A. (1985) Hookworm burdens and faecal egg counts: an analysis of the biological basis of variation. *Transactions of the Royal Society of Tropical Medicine and Hygiene*, **79**, 812–825.

Andrews, C., Chubb, J.C., Coles, T. and Dearsley, A. (1981) The occurrence of *Bothriocephalus acheilognathi* Yamaguti, 1934 (*B. gowkongensis*) (Cestoda: Pseudophyllidea) in the British Isles. *Journal of Fish Diseases*, **4**, 89–93.

Arai, H.P. and Mudry, D.R. (1983) Protozoan and metazoan parasites of fishes from the headwaters of the Parsnip and McGregor rivers, British Columbia: a study of possible parasite transfaunation. *Canadian Journal of Fisheries and Aquatic Sciences*, **40**, 1676–1684.

Arme, C. and Owen, R.W. (1967) Infections of the three-spined stickleback, *Gasterosteus aculeatus* L., with the plerocercoid larvae of *Schistocephalus solidus* (Muller, 1776), with special reference to pathological effects. *Parasitology*, **57**, 301–314.

Avise, J.C. and Aquadro, F. (1982) A comparative summary of genetic distances in the vertebrates. Patterns and correlations. In *Evolutionary Biology*, Vol. 15 (eds M.K. Hecht, B. Wallace and G.T. Prance), Plenum Press, New York, pp. 151–185.

Bailey, R.E. and Margolis, L. (1987) Comparison of parasite fauna of juvenile sockeye salmon (*Oncorhynchus nerka*) from southern British Columbian and Washington State lakes. *Canadian Journal of Zoology*, **64**, 420–431.

Bakke, T.A. (1972a) Studies on the helminth fauna of Norway XXIII: The common gull, *Larus canus* L., as a final host for Digenea (Platyhelminthes). II. The relationship between infection and sex, age and weight of the common gull. *Norway Journal of Zoology*, **20**, 189–204.

Bakke, T.A. (1972b) Studies on the helminth fauna of Norway XXII: The common gull, *Larus canus* L., as a final host of Digenea (Platyhelminthes). II. The ecology of the common gull and infection in relation to season and the gull's habitat, together with the distribution of the parasites in their intestines. *Norway Journal of Zoology*, **20**, 165–188.

Barbehenn, K.R. (1969) Host–parasite relationships and species diversity in mammals: an hypothesis. *Biotropica*, **1**, 29–35.

Barnard, C.J. and Behnke, J.M. (eds) (1990) *Parasitism and Host Behaviour*, Taylor and Francis, New York.

Barrett, J. (1981) *Biochemistry of Parasitic Helminths*, Macmillan, London.

Barrett, J.A. (1984) The genetics of host–parasite interaction. In *Evolutionary Ecology* (ed. B. Shorrocks), Blackwell, Oxford, pp. 275–294.

Bartlett, C.M. and Anderson, R.C. (1989a) Taxonomic descriptions and comments on the life history of new species of *Eulimdana* (Nematoda: filarioidea) with skin-inhabiting microfilariae in the Charadriiformes (Aves). *Canadian Journal of Zoology*, **67**, 612–629.

Bartlett, C.M. and Anderson, R.C. (1989b) Some observations on *Pseudomenopon pilosum* (Amblycera, Menoponidae), the louse vector of *Pelecitus fulicaeatrae* (Nematoda: Filarioidea) of coots, *Fulica americana* (Aves: Gruiformes). *Canadian Journal of Zoology*, **67**, 1328–1331.

Beaver, P.C. (1989) *Mesocestoides corti*: Mouse type host, uncharacteristic or questionable? *Journal of Parasitology*, **75**, 815.

Befus, A.D. (1975) Secondary infections of *Hymenolepis diminuta* in mice: effect of varying worm burdens in primary and secondary infections. *Parasitology*, **71**, 61–75.

Begon, M., Harper, J.L. and Townsend, C.R. (1990) *Ecology: Individuals, Populations and Communities*, 2nd ed., Blackwell Scientific Publications, Boston.

Ben-Ismail, R., Mulet-Clamagirand, C., Carme, B. and Gentillini, M. (1982) Biosynthesis of A, H and Lewis blood group determinants in *Fasciola hepatica*. *Journal of Parasitology*, **68**, 402–407.

Benviste, R.E. and Todaro, G.J. (1974) Evolution of C-type viral genes: inheritance of exogenously acquired viral genes. *Nature* (London), **252**, 456–459.

Beverly-Burton, M. and Early, G. (1982) *Deretrema philippinensis* n. sp. (Digenea: Zoogonidae) from *Anomalops katoptron* (Bericiformes: Anomalopidae) from the Philippines. *Canadian Journal of Zoology*, **60**, 2403–2408.

Bloom, B.R. (1979) Games parasites play: how parasites evade immune surveillance. *Nature* (London), **279**, 21–26.

Boddington, J.F. and Mettrick, D.R. (1976) Seasonal changes in the biochemical composition and nutritional state of the immigrant triclad *Dugesia polychroa* (Platyhelminthes: Turbellaria) in Toronto Harbor, Canada. *Canadian Journal of Zoology*, **53**, 1723–1734.

Boddington, M.J. and Mettrick, D.F. (1981) Production and reproduction in *Hymenolepis diminuta* (Platyhelminthes: Cestoda). *Canadian Journal of Zoology*, **59**, 1962–1972.

Bodmer, W.F. and Bodmer, J.G. (1978) Evolution and function of the HLA system. *British Medical Bulletin*, **34**, 309–316.

Bodmer, W.F. and Thomson, G. (1977) Population genetics and evolution of the HLA system. In *HLA and Disease* (eds J. Dausset and A. Svejgaard), Munksgaard, Copenhagen, pp. 280–295.

Boray, J.C. (1969) Experimental fascioliasis in Australia. *Advances in Parasitology*, **7**, 96–210.

Bradley, D.J. (1972) Regulation of parasite populations. A general theory of the epidemiology and control of parasitic infections. *Transactions of the Royal Society of Tropical Medicine and Hygiene*, **66**, 697–708.

Brass, W. (1958) Simplified methods of fitting the truncated negative binomial distribution. *Biometrika*, **45**, 59–68.

Brener, Z. and Krettli, A. U. (1990) Immunology of Chagas' disease. In *Modern Parasite Biology: Cellular, Immunological and Molecular Aspects* (ed. D.J. Wyler), W.H. Freeman and Company, New York, pp. 247–261.

Brooks, D.R. (1977) Evolutionary history of some plagiorchioid trematodes of anurans. *Systematic Zoology*, **26**, 277–289.

Brooks, D.R. (1979) Testing hypothesis of evolutionary relationships among parasites: the digeneans of crocodilians. *American Zoologist*, **19**, 1225–1238.

Brooks, D.R. (1980) Allopatric speciation and non-interactive parasite community structure. *Systematic Zoology*, **30**, 192–203.

Brooks, D.R. (1985) Phylogenetics and the future of helminth systematics. *Journal of Parasitology*, **71**, 719–727.

Brooks, D.R. (1988) Macroevolutionary comparisons of host and parasite phylogenies. *Annual Review of Ecology and Systematics*, **19**, 235–259.

Brooks, D.R. (1989) Coevolution of helminths and vertebrates. In *Current Concepts in Parasitology* (ed. R.C. Ho), Hong Kong University Press, pp. 255–267.

Brooks, D.R. and Glen, D.R. (1982) Pinworms and primates: a case study in coevolution. *Proceedings of the Helminthological Society of Washington*, **49**, 76–85.

Brooks, D.R., Thorson, T.B. and Mayes, M.A. (1981) Fresh-water stingrays (Potamotrygonidae) and their helminth parasites: testing hypothesis of evolution and coevolution. In *Advances in Cladistics*, Vol. 1 (ed. V.A. Funk and D.R. Brooks), New York, New York Botanical Garden, pp 147–175.

Brooks, J.L. and Dodson, S.I. (1965) Predation, body size and composition of plankton. *Science* (NY), **150**, 28–35.

Brown, J.H. (1988) Species diversity. In *Analytical Biogeography* (eds A.A. Myers and P.S. Giller), Chapman and Hall, New York, pp. 57–89.

Brown, J.H. and Gibson, A.C. (1983) *Biogeography*, C.V. Mosby, St. Louis.

Bryant, C. and Behm, C. (1989) *Biochemical Adaptation in Parasites*. Chapman and Hall, London.

Bryant, C. and Flockhart, H.A. (1986) Biochemical strain variation in helminths. *Advances in Parasitology*, **25**, 275–319.

Bull, C.M., Andrews, R.H. and Adams, M. (1984) Patterns of genetic variation in a group of parasites, the Australian reptile ticks. *Heredity*, **53**, 509–525.

Bull, P.C. (1964) Ecology of helminth parasites of the wild rabbit *Oryctolagus cuniculus* (L.) in New Zealand. Bulletin 158, New Zealand Department of Scientific and Industrial Research, 147 pp.

Bullini, L., Nascetti, G., Paggi, L., Orecchia, P., Mattiucci, S. and Berland, B. (1986) Genetic variation of ascaridoid worms with different life cycles. *Evolution*, **40**, 437–440.

Bundy, D.A.P. (1988) Population ecology of intestinal helminth infections in human communities. *Philosophical Transactions of the Royal Society of London*, B, **321**, 405–420.

Bundy, D.A.P. Cooper. E.S., Thompson, D.E., Didier, J.M., Anderson, R.M. and Simmons, I. (1987) Predisposition to *Trichuris trichiura*. *Epidemiology and Infection*, **98**, 65–71.

Burn, P.R. (1980) Density dependent regulation of a fish trematode population. *Journal of Parasitology*, **66**, 173–174.

Busher, H.N. (1965) Dynamics of the intestinal helminth fauna in three species of ducks. *Journal of Wildlife Management*, **29**, 772–781.

Buscher, H.N. (1966) Intestinal helminths of the blue-winged teal, *Anas discors* L., at Delta, Manitoba. *Canadian Journal of Zoology*, **44**, 113–116.

Bush, A.O. (1990) Helminth communities in avian hosts: determinants of pattern. In *Parasite Communities: Patterns and Processes* (eds G.W. Esch, A.O. Bush and J.M. Aho), Chapman and Hall, New York, pp. 197–232.

Bush, A.O., and Holmes, J.C. (1986a) Intestinal parasites of lesser scaup ducks: patterns of association. *Canadian Journal of Zoology*, **64**, 132–141.

Bush, A.O. and Holmes, J.C. (1986b) Intestinal parasites of lesser scaup ducks: an interactive community. *Canadian Journal of Zoology*, **64**, 142–152.

Butterworth, (1990) Immunology of schistosomiasis. In *Modern Parasite Biology: Cellular, Immunological and Molecular Aspects* (ed. D.J. Wyler), W.H. Freeman and Company, New York, pp. 262–288.

Byram, J.E., Ernst, S.C., Lumsden, R.D. and Sogendares-Bernal, F. (1975) Virus like inclusions in the cecal epithelial cells of *Paragonimus kellicotti* (Digenea, Troglotrematidae). *Journal of Parasitology*, **61**, 253–264.

Cable, R.C. (1972) Behaviour of digenetic trematodes. In *Behavioural Aspects of Parasite Transmission* (eds E.U. Canning and C.A. Wright), *Zoological Jornal of the Linnean Society*, Supplement 1, Academic Press, London, pp. 1–18.

Callinan, A.P.L. (1979) The ecology of free-living stages of *Trichostrongilus vitrinus*. *International Journal for Parasitology*, **9**, 133–136.

Calow, P. (1983a) Patterns and paradox in parasite reproduction. *Parasitology*, **86**, 197–207.

Calow, P. (1983b) Energetics of reproduction and its evolutionary implications. *Biological Journal of the Linnean Society*, **20**, 153–165.

Campbell, R.A. (1983) Parasitism in the deep sea. In *The Sea*, Vol. 8 (ed. G.T. Rowe), John Wiley and Sons, Inc., pp. 473–552.

Campbell, R.A. (1990) Deep water parasites. *Annals de Parasitologie Humaine et Comparée*, **65**, Supplement 1, 65–68.

Campbell, R.A., Haedrich, R.L. and Munroe, T.A. (1980) Parasitism and ecological relationships among deep-sea benthic fishes. *Marine Biology*, **57**, 301–313.

Capron, A., Biguet, J. Vernes, A. and Afchain, D. (1968) Structure antigénique des helminthes. Aspects immunologiques des relations hôte–parasite. *Pathologie et Biologie*, Paris, **16**, 121–138.

Carlson, T.A. and Chelim, B.K. (1986) Apparent eukaryotic origin of glutamine synthetase II from the bacterium *Bradyrhizobium japonicum*. *Nature*, (London), **322**, 568–570.

Caswell, H. (1978) Predator mediated coexistence: a nonequilibrium model. *American Naturalist*, **112**, 127–154.

Caswell, H. (1985) The evolutionary demography of clonal reproduction. In *Population Biology and Reproduction of Clonal Organisms* (eds J.B.C. Jackson, L.W. Buss and R.E. Cook), Yale University Press, New Haven pp. 187–224.

Chabaud, A.G. (1981) Host range and evolution of nematode parasites of vertebrates. *Parasitology*, **82**, 169–170.

Chakraborty, R., Fuerst, P.A. and Nei, M. (1980) Statistical studies on protein polymorphism in natural populations. III. Distribution of allele frequencies and the number of alleles per locus. *Genetics*, **94**, 1039–1063.

Chandler, A.C. and Read, C.P. (1961) *Introduction to Parasitology*, 10th ed., John Wiley and Sons, Inc., USA.

Chandler, M. and Cabana, G. (1991) Sexual dichromatism in North American freshwater fish: do parasites play a role? *Oikos*, **60**, 322–328.

Chang, K.P. (1990) Cell biology of *Leishmania*. In *Modern Parasite Biology: Cellular, Immunological and Molecular Aspects* (ed. D.J. Wyler), W.H. Freeman and Company, New York, pp. 79–90.

Chappell, L.H. (1969) Competitive exclusion between two intestinal parasites of the three-spined stickleback *Gasterosteus aculeatus* L. *Journal of Parasitology*, **55**, 775–778.

Cheng, T. (1986) General Parasitology, 2nd ed., Academic Press, Inc., Orlando, Florida.

Christensen, N.O., Nansen, P. and Fransen, F. (1978) The influence of some physico-chemical factors on the host finding capacity of *Fasciola hepatica* miracidia. *Journal of Helminthology*, **52**, 61–67.

Chubb, J.C. (1963) On the characterization of the parasite fauna of the fish of Llyn Tegid. *Proceedings of the Zoological Society of London*, **141**, 609–621.

Chubb, J.C. (1970) The parasite fauna of British freshwater fish. *Symposia of the British Society for Parasitology*, **8**, 119–144. A.E.R. Taylor and R. Muller, Blackwell Scientific Publications, Oxford.

Clapham, P.A. (1942) On the identification of *Multiceps* spp. by measurement of the large hook. *Journal of Helminthology*, **20**, 31–40.

Clark, B. (1979) The evolution of genetic diversity. *Proceedings of the Royal Society of London*, B, **205**, 453–474.

Cohen, L.M. and Eveland, L.K. (1988) *Schistosoma mansoni*: characterization of clones maintained by the microsurgical transplantation of sporocysts. *Journal of Parasitology*, **74**, 963–969.

Cohen, J.E. (1977) Mathematical models of Schistosomiasis. *Annual Review of Ecology and Systematics*, **8**, 209–233.

Cole, L.C. (1954) The population consequences of life history phenomena. *Quarterly Review of Biology*, **29**, 103–137.

Collard, S.B. (1970) Some aspects of host–parasite relationships in mesopelagic fishes. In *A Symposium of Diseases of Fishes and Shellfishes* (ed. S.F. Snieszko), American Fisheries Society, Washington DC, USA, pp. 41–46.

Conn, D.B. (1990) The rarity of asexual reproduction among *Mesocestoides* tetrathyridia (Cestoda). *Journal of Parasitology*, **76**, 453–455.

Conn, D.B. and Etges, F.J. (1983) Maternal transmission of asexually proliferative *Mesocestoides corti* tetrathyridea (Cestoda) in mice. *Journal of Parasitology*, **69**, 922–925.

Connell, J.H. (1980) Diversity and the coevolution of competitors, or the ghost of competition past. *Oikos*, **35**, 131–138.

Connor, E.F., and McCoy, E.D. (1979) The statistics and biology of the species–area relationship. *American Naturalist*, 113, 791–833.

Cornell, H.V. (1974) Parasitism and distributional gaps between species. *American Naturalist*, 108, 880–883.

Cornwell, G.W. and Cowan, A.B. (1963) Helminth populations of the canvasback (*Aythya valisineria*) and host–parasite–environmental interrelations. *Transactions of the North American Wildlife and Natural Resources Conference*, 28, 173–198.

Cort, W.W., McMullen, D.B. and Brackett, S. (1937) Ecological studies on the cercariae of *Stagnicola emarginata angulata* (Soweby) in the Douglas Lake region, Michigan. *Journal of Parasitology*, 23, 504–532.

Cram, E.B. (1943) Studies on oxyuriasis. XXVIII. Summary and conclusions. *American Journal of Diseases of Children*, 65, 46–59.

Crews, A. and Esch, G.W. (1986) Studies on the population biology of *Halipegus occidualis* (Hemiuridae) in the snail host *Helisoma anceps* (Pulmonata). *Journal of Parasitology*, 72, 646–651.

Crofton, H.D. (1971a) A quantitative approach to parasitism. *Parasitology*, 62, 179–193.

Crofton, H.D. (1971b) A model for host–parasite relationships. *Parasitology*, 63. 343–364.

Croll, N.A. and Ghadarian, E. (1981) Wormy persons: Contributions to the nature and patterns of overdispersion with *Ascaris lumbricoides*, *Ancylostoma duodenale*, *Necator americanus* and *Trichuris trichiura*. *Tropical and Geographical Medicine*, 33, 241–248.

Croll, N.A., Anderson, R.M., Gyorkos, T.W. and Ghadarian, E. (1982) The population biology and control of *Ascaris lumbricoides* in a rural community in Iran. *Transactions of the Royal Society of Tropical Medicine and Hygiene*, 76, 187–197.

Crompton, D.W.T. (1987) Host diet as a determinant of parasite growth, reproduction and survival. *Mammalogical Review*, 17, 117–126.

Crompton, D.W.T. (1989) Prevalence of ascariasis. In *Ascariasis and its Prevention and Control* (eds D.W.T. Crompton, M.C. Nesheim and Z.S. Pawlowski), Taylor and Francis, London, pp. 45–69.

Crompton, D.W.T. and Nesheim, M.C. (1976) Host–parasite relationships in the alimentary tract of domestic birds. *Advances in Parasitology*, 14, 95–194.

Crosby, A.W. (1972) *The Columbian Exchange: Biological and Cultural Consequences of 1492.* Greenwood, Westport, Connecticut.

Cumbie, P.M. and Van Horn, S.L. (1978) Selenium accumulation associated with fish mortality and reproductive failure. *Proceedings of the Annual Conference of the Southeastern Fish and Wildlife Agencies*, 32, 612–624.

Custer, J.W. and Pence, D.G. (1981) Ecological analyses of the helminth populations of wild canids from the Gulf coastal prairies of Texas and Louisiana. *Journal of Parasitology*, 67, 289–307.

Damian, R.T. (1964) Molecular mimicry: antigen sharing by parasite and host and its consequences. *American Naturalist*, 98, 129–149.

Damian, R.T. (1979) Molecular mimicry in biology adaptation. In *Host–Parasite Interfaces* (ed. B.B.Nickol), Academic Press, New York, pp. 103–126.

Damian, R.T. (1987) Presidential address. The exploitation of host immune responses by parasites. *Journal of Parasitology*, **73**, 1–13.

Davies, A.J.S., Hall, J.G., Targett, G.A.T. and Murray, M. (1980) The biological significance of the immune response with special reference to parasites and cancer. *Journal of Parasitology*, **66**, 705–721.

Davis, D.J. (1936) Pathological studies on the penetration of the cercariae of the strigeid trematode *Diplostomum flexicaudum*. *Journal of Parasitology*, **22**, 329–338.

Dawkins, R. and Krebs, J.R. (1979) Arms races between and within species. *Proceedings of the Royal Society of London, B*, **205**, 489–511.

Day, P.R. (1974) *Genetics of Host–Parasite Interaction*, W.H. Freeman, San Francisco.

Deets, G.B. (1987) Phylogenetic analysis and revision of *Kroeyerina* Wilson, 1932 (Siphonostomatoida: Kroyeriidae), copepods parasitic on chondrichthyans, with descriptions of four new species and the erection of a new genus, Prokroyeria. *Canadian Journal of Zoology*, **65**, 2121–2148.

De Jong, N. (1976) Helminths from the mallard (*Anas platyrhynchos*) in the Netherlands. *Netherlands Journal of Zoology*, **26**, 306–318.

Dienske, H. (1968) A survey of the metazoan parasites of the rabbit-fish, *Chimaera monstrosa L.* (Holocephali). *Netherlands Journal of Sea Research*, **4**, 32–58.

Dietz, R.S. and Holden, J.C. (1970) The break-up of Pangaea. *Scientific American*, **223(4)**, 30–41.

Dineen, J.K., Donald, A.D. and Wagland, B.M. (1965) The dynamics of the host–parasite relationship. III. The response of sheep to primary infection with *Haemonchus contortus*. *Parasitology*, **55**, 515–525.

Dobson, A.P. (1985) The population dynamics of competition between parasites. *Parasitology*, **91**, 317–347.

Dobson, A.P. (1990) Models for multi-species parasite–host communities. In *Parasite Communities: Patterns and Processes* (eds G.W. Esch, A.O. Bush and J.M. Aho), Chapman and Hall, New York, pp. 261–288.

Dobzhansky, T. (1950) Evolution in the tropics. *American Scientist*, **38**, 208–221.

Dogiel, V.A. (1964) *General Parasitology*, Oliver and Boyd, Edinburgh.

Dogiel, V.A., Petrushevski, G.K. and Polianski, Yu. I. (eds) (1961) *Parasitology of Fishes*, Oliver and Boyd, Endinburgh and London.

Donges, J. (1964) Der Lebenszyklus von *Posthodiplostomum cuticola* (v. Nordmann 1832) Dubois 1936 (Trematoda, Diplostomatidae). *Zeitschrift für Parasitenkunde*, **24**, 169–248.

Dougherty, R.M., Distefano, H., Feller, U. and Mueller, J.F. (1975) On the nature of particles lining the excretory ducts of pseudophyllidean cestodes. *Journal of Parasitology*, **61**, 1006–1015.

Dronen, N.O. (1978) Host-parasite population dynamics of *Haematoloechus coloradensis* Cort, 1915 (Digenea: Plagiorchidae). *American Midland Naturalist*, **99**, 330–349.

Drummond, M. (1979) Crown gall disease. *Nature*, London, **281**, 343–347.

Dunkley, L.C. and Mettrick, D.F. (1969) *Hymenolepis diminuta*: effect of quality of host dietary carbohydrate on growth. *Experimental Parasitology*, **25**, 146–161.

Dunkley, L.C. and Mettrick, D.F. (1977) *Hymenolepis diminuta*: migration and the host intestinal and blood plasma glucose levels following carbohydrate intake. *Experimental Parasitology*, **41**, 213–228.

Eckert. J., Vonbrandt, T. and Voge, M. (1969) Asexual multiplication of *Mesocestoides corti* (Cestoda) in the intestine of dogs and skunks. *Journal of Parasitology*, **55**, 241–249.

Edwards, D.D. and Bush, A.O. (1989) Helminth communities in avocets: importance of the compound community. *Journal of Parasitology*, **75**, 225–238.

Ehrlich, P.R. and Raven, P.H. (1964) Butterflies and plants: a study in coevolution. *Evolution*, **18**, 586–608.

Ekman, S. (1953) *Zoogeography of the Sea*, Sidgwick and Jackson, London.

Elkins, D.B., and Haswell-Elkins, M. (1989) The weight/length profiles of *Ascaris lumbricoides* within a human community before mass treatment and following reinfection. *Parasitology*, **99**, 293–299.

Elliot, J.M. (1983) *Some Methods for the Statistical Analysis of Samples of Benthic Invertebrates*, Freshwater Biological Association, Cumbria, UK, Scientific Publication No. 25.

Enriques, F.J., Zidian, J.L. and Cypess, R.H. (1988) *Nematospiroides dubius*: genetic control of immunity to infections of mice. *Experimental Parasitology*, **67**, 12–19.

Erasmus, D. A. (1960) The migration of cercaria X Baylis (Strigeida) within the fish intermediate host. *Parasitoloty*, **49**, 173–190.

Esch, G.W. (1971) Impact of ecological succession on the parasite fauna in centrarchids from oligotrophic and eutrophic ecosystems. *American Midland Naturalist*, **86**, 160–168.

Esch, G.W. (1982) Abiotic factors: an overview. In *Parasites - Their World and Ours* (eds D.F. Mettrick and S.S. Desser), Proceedings of the Fifth International Congress of Parasitology (ICOPA V), Toronto, Canada, Elsevier Biomedical Press, Amsterdam, pp. 279–288.

Esch, G.W. (1983) The population and community ecology of cestodes. In *Biology of the Eucestoda* (eds P. Pappas and C. Arme), Academic Press, New York, pp. 81–137.

Esch, G.W., Bush, A.O. and Aho, J.M. (eds.) (1990) *Parasite Communities: Patterns and Processes*, Chapman and Hall, New York.

Esch, G.W. and Gibbons, J.W. (1967) Seasonal incidence of parasitism in the painted turtle, *Chrysemys picta marginata* Agassiz. *Journal of Parasitology*, **53**, 818–821.

Esch, G.W., Gibbons, J.W. and Bourque, J.E. (1975) An analysis of the relationship between stress and parasitism. *American Midland Naturalist*, **93**, 339–353.

Esch, G.W., Gibbons, J.W. and Bourque, J.E. (1979a) Species diversity of helminth parasites in *Chrysemys s. scripta* from a variety of habitats in South Carolina. *Journal of Parsitology*, **65**, 633–638.

Esch, G.W., Gibbons, J.W. and Bourque, J.E. (1979b) The distribution and abundance of enteric helminths in *Chrysemys s. scripta* from various habitats on the Savannah River plant in south Carolina. *Journal of Parasitology*, **65**, 624–632.

Esch, G.W., Hazen, T.C. and Aho, J.M. (1977) Parasitism and r- and K-selection. In *Regulation of Parasite Populations* (ed. G.W. Esch), Academic Press, New York, pp. 9–62.

Esch, G.W., Hazen, T.C., Marcogliese, D., Goater, T.M. and Crews, A.E. (1986) Long-term studies on the population biology of *Crepidostomum cooperi* (Allocreadidae) in the burrowing mayfly, *Hexagenia limbata* (Ephemeroptera). *American Midland Naturalist*, **116**, 304–314.

Esch, G.W., Johnson, W.C. and Coggins, J.R. (1975) Studies on the population biology of *Proteocephalus ambloplitis* (Cestoda) in smallmouth bass. *Proceedings of the Oklahoma Academy of Sciences*, **55**, 122–127.

Esch, G.W., Kennedy, C.R., Bush, A.O. and Aho, J.M. (1988) Patterns in helminth communities in freshwater fish in Great Britain: alternative strategies for colonization. *Parasitology*, **96**, 519–532.

Esch, G.W. and Self, J.T. (1965) A critical study of the taxonomy of *Taenia taeniaeformis* Bloch, 1780; *Multiceps multiceps* (Leske, 1780); and *Taenia pisiformis* Batsch, 1786. *Journal of Parasitology*, **51**, 932–937.

Esch, G.W., Shostak, A.W., Marcogliese, D.J. and Goater, T.M. (1990) Patterns and processes in helminth parasite communities: an overview. In *Parasite Communities: Patterns and Processes* (eds G.W. Esch, A.O. Bush and J.M. Aho), Chapman and Hall, New York, pp. 1–19.

Eure, H. (1976) Seasonal abundance of *Proteocephalus ambloplitis* (Cestoidea: Proteocephalidae) from largemouth bass living in a heated reservoir. *Parasitology*, **73**, 205–212.

Evans, N.A. (1983) The population biology of *Hymenolepis tenerrima* (Linstow, 1882) Fuhrmann, 1906 (Cestoda, Hymenolepididae) in its intermediate host *Herpetocypris reptans* (Ostracoda). *Zeitschrift für Parasitenkunde*, **69**, 105–111.

Ewald, P.W. (1983) Host–parasite relations, vectors, and the evolution of disease severity. *Annual Review of Ecology and Systematics*, **14**, 465–485.

Eysker, M. (1981) Experiments on inhibited development of *Haemonchus contortus* and *Ostertagia circumcincta* in the Netherlands. *Research in Veterinary Science*, **30**, 62–65.

Fanning, M.M. and Kazura, J.W. (1984) Genetic-linked variation in susceptibility of mice to *Schistosoma mansoni*. *Parasite Immunology*, **6**, 95–103.

Fanning, M.M., Peters, P.S., Davis, R.S., Kazura, J.N. and Mahmoud, A.A.F. (1981) Immunopathology of murine infection with *Schistosoma mansoni*: relationship of genetic background to hepatosplenic disease and modulation. *Journal of Infectious Diseases*, **144**, 148–153.

Fernández, J. (1982) *Estudio Parasitólogico de* Merluccius australis (Hutton, 1872): *Aspectos Estadísticos, Sistemáticos y Zoogeográficos*. Thesis de Licenciatura, Universidad de Concepción, Concepción, Chile.

Fernández, J. (1985) Estudio parasitológico de Merluccius australis (Hutton, 1872) (Pisces: Merluccidae): aspectos sistemáticos, estadísticos y zoogeográficos. *Boletín de la Sociedad de Biología de Concepción*, **56**, 31–41.

Fernández, J. (1986) *Los Parásitos de la lisa*, Mugil cephalus L., *en Chile: Sistemática, Estructura Poblacional y Afinidades Zoogeográficas*. MSc Thesis, Universidad de Concepción, Concepción, Chile.

Fernández, J. (1987) Los parásitos de la lisa, *Mugil cephalus* L., en Chile: sistemática y aspectos poblacionales (Perciformes: Mugilidae). *Gayana, Zoologiá*, **51**, 3–58.

Fernández, J. and Durán, L. (1985) *Aporocotyle australis* n. sp. (Digenea: Sanguinicolidae), parásito de *Merluccius australis* (Hutton 1872) en Chile y su relación con la filogenia de *Aporocotyle Odhner*, 1900 en *Merluccius* spp. *Revista Chilena de Historia Natural*, **58**, 121–126.

Fernández, J. and Esch, G.W. (1991a) Guild structure of larval trematodes in the snail *Helisoma anceps*: patterns and processes at the individual host level. *Journal of Parasitology*, **70**, 528–539.

Fernández, J. and Esch, G.W. (1991b) The component community structure of larval trematodes in the pulmonate snail *Helisoma anceps*. *Journal of Parasitology*, **70**, 540–550.

Fernández, J., Villalba, C.S. and Albina, A. (1983) Parásitos del pejegallo, *Callorhynchus callorhynchus* (L.), en Chile: aspectos biológicos y sistemáticos. *Biología Pesquera*, **15**, 63–73.

Fernando, M.A., Stockdale, P.H.G. and Ashton, G.C.(1971) Factors contributing to the retardation of development of *Obeliscoides cuniculi* in rabbits. *Parasitology*, **63**, 21–29.

Fischer, H. and Freeman, R.S. (1969) Penetration of parenteral plerocercoid of *Proteocephalus ambloplitis* (Leidy) into the gut of smallmouth bass. *Journal of Parasitology*, **55**, 766–774.

Fisher, A. (1991) A new synthesis. II. How different are humans? *Mosaic*, **22**, 11–17.

Fisher, R.A. (1930) *The Genetical Theory of Natural Selection*, Clarendon, Oxford.

Fleming, M.W. (1988) Size of inoculum dose regulates in part worm burdens, fecundity, and lengths in ovine *Haemonchus contortus* infections. *Journal of Parasitology*, **74**, 975–978.

Flor, H.H. (1942) Inheritance of pathogenicity in *Melampsora lini*. *Phytopathology*, **32**, 653–669.

Flor, H.H. (1955) Host–parasite interaction in flax rust – its genetics and other implications. *Phytopathology*, **45**, 680–685.

Flor, H.H. (1956) The complementary genic systems in flax and flax rust. *Advances in Genetics*, **8**, 29–54.

Forrester, D.J., Conti, J.A., Bush, A.O., Campbell, L.D. and Frolich, R.K. (1984) Ecology of helminth parasitism of bobwhites in Florida. *Proceedings of the Helminthological Society of Washington*, **51**, 255–260.

Frank, G.R., Herd, R.P., Marbury, K.S., Williams J.C. and Willis, E.R. (1988) Additional investigations on hypobiosis of *Ostertagia ostertagi* after transfer between northern and southern U.S.A. *International Journal for Parasitology*, **18**, 171–177.

Fried, B. and Roberts, T.M. (1972) Pairing of *Leucochloridiomorpha constantiae* (Mueller, 1935) (Trematoda) in vitro, in the chick and on the allantois. *Journal of Parasitology*, **58**, 88–91.

Fulton, R.S. III. (1984) Predation, production and the organization of an estuarine copepod community. *Journal of Plankton Research*, **6**, 399–415.

Futuyma, D.J. (1986) *Evolutionary Biology*, 2nd ed, Sinauer Associates, Inc., Sunderland, Massachusetts.

Gartner, J.V. and Zwerner, D.E. (1989) The parasite faunas of meso- and pathypelagic fishes of Norfolk Submarine Canyon, western North Atlantic. *Journal of Fish Biology*, **34**, 79–95.

Gause, G.F. (1934) *The Struggle for Existence*. Hafner, New York. (Reprinted in 1964 by Williams and Wilkins, Baltimore, MD).

Gibbs, A.J. (1973) Other invertebrates. In *Frontiers of Biology 31, Viruses and Invertebrates*, (ed. A.J. Gibbs), North-Holland, Amsterdam, pp. 526–530.

Gill, D.E. (1978) The metapopulation ecology of the red-spotted newt, *Notophthalmus viridescens* (Rafinesque). *Ecological Monographs*, **48**, 145–166.

Gleason, L.N. (1984) Population composition and dispersal pattern of *Pomphorhynchus bulbocolli* in *Hypentelium nigricans* from the West Fork of Drake's Creek, Kentucky. *American Midland Naturalist*, **112**, 273–279.

Gleason, L.N. (1987) Population dynamics of *Pomphorhynchus bulbocolli* in *Gammarus pseudolimnaeus*. *Journal of Parasitology*, **73**, 1099–1101.

Glen, D.R. and Brooks, D.R. (1986) Parasitological evidence pertaining to the phylogeny of the hominoid primates. *Biological Journal of the Linnean Soiciety*, **27**, 331–354.

Goater, C.P. and Bush, A.O. (1988) Intestinal helminth communities in the long-billed curlews: the importance of congeneric host-specialists. *Holarctic Ecology*, **11**, 140–145.

Goater, T.M. (1989) *The Morphology, Life History, Ecology and Genetics of* Halipegus occidualis *(Trematoda: Hemiuridae) in Molluscan and Amphibian Hosts*. PhD dissertation, Wake Forest University, Winston-Salem, North Carolina, USA.

Goater, T.M. (1990) Helminth parasites indicate predator–prey relationships in desmognathine salamanders. *Herpetological Review*, **21**, 32–33.

Goater, T.M., Browne, C.R. and Esch, G.W. (1990) The structure and function of the cystophorous cercariae of *Halipegus occidualis* (Trematoda: Hemiuridae). *International Journal for Parasitology*, **20**, 923–934.

Goater, T.M., Esch, G.W. and Bush, A.O. (1987) Helminth parasites of sympatric salamanders: ecological concepts at infracommunity, component and compound community levels. *American Midland Naturalist*, **118**, 289–300.

Goodchild, C.G. and Moore, T.L. (1963) Development of *Hymenolepis diminuta* in mice made obese by aurothioglucose. *Journal of Parasitology*, **49**, 398–402.

Gordon, D.M. and Rau, M.E. (1982) Possible evidence for mortality introduced by the parasite *Apatemon gracilis* in a population of brook sticklebacks (*Culaea inconstans*). *Parasitology*, **84**, 41–47.

Grabda, J. (1974) The dynamics of the nematode larvae, *Anisakis simplex* (Rud.) invasion in the south-eastern Baltic herring (*Clupea harengus* L.). *Acta Ichthyologica et Piscatoria*, **4**, 3–21.

Granath, W.O. and Esch, G.W. (1983a) Temperature and other factors that regulate the composition and infrapopulation densities of *Bothriocephalus acheilognathi* (Cestoda) in *Gambusia affinis* (Pisces). *Journal of Parsitology*, **69**, 1116–1124.

Granath, W.O. and Esch, G.W. (1983b) Survivorship and parasite-induced host mortality among mosquitofish in a predator-free, North Carolina cooling reservoir. *American Midland Naturalist*, **110**, 314–323.

Granath, W.O. and Esch, G.W. (1983c) Seasonal dynamics of *Bothriocephalus acheilognathi* in ambient and thermally altered areas of a North Carolina cooling reservoir. *Proceedings of the Helminthological Society of Washington*, **50**, 205–218.

Granath, W.O., Lewis, J.C. and Esch, G.W. (1983) An untrastructural examination of the scolex and tegument of *Bothriocephalus acheilognathi* (Cestoda: Pseudophyllidea). *Transactions of the American Microscopical Society*, **102**, 240–250.

Greenslade, P.J.M. (1972) Evolution in the staphylinid genus *Priochirus* (Coleoptera). *Evolution*, **26**, 203–220.

Gregory, R.D. (1990) Parasites and host geographic range as illustrated by waterfowl. *Functional Ecology*, **4**, 645–654.

Grime, J.P. (1986) Global trends in species-richness in terrestrial vegetation: a view from the Northern Hemisphere. In *Organization of Communities Past and Present* (eds J.H.R. Gee and P.S. Giller), 27th Symposium of the British Ecological Society, Aberystwyth, Blackwell, Oxford, pp. 99–118.

Gruskin, K.D., Smith, T.F. and Goodman, M. (1987) Possible origin of a calmodulin gene that lacks intervening sequences. *Proceedings of the National Academy of Sciences, USA*, **84**, 1605–1608.

Guyatt, H.L. and Bundy, D.A.P. (1990) Are wormy people parasite prone or just unlucky? *Parasitology Today*, **6**, 282–283.

Hafner, M.S. and Nadler, S.A. (1988) Phylogenetic trees support the coevolution of parasites and their hosts. *Nature*, London, **332**, 258–259.

Hafner, M.S. and Nadler, S.A. (1990) Cospeciation in host–parasite assemblages: comparative analysis of rates of evolution and timing of cospeciation events. *Systematic Zoology*, **39**. 192–204.

Hagen, K.S. (1966) Dependence of the olive fly, *Dacus oleae*, larvae on symbiosis with *Pseudomonas savastanoi* for the utilization of olive. *Nature*, London, **209**, 423–424.

Hair, J.D. (1975) *The Structure of the Intestinal Helminth Communities of Lesser Scaup* (Aythya affinis). PhD Thesis, University of Alberta, Edmonton.

Hairston, N.G. (1965) On the mathematical analysis of schistosome populations. *Bulletin of the World Health Organization*, **33**, 45–62.

Hairston, N.G., Tinkle, D.W. and Wilbur, H.M. (1970) Natural selection and the parameters of population growth. *Journal of Wildlife Management*, **34**, 681–698.

Haldane, J.B.S. (1949) Disease and evolution, *La Ricerca Scientifica* (Suppl.), **19**, 68–76.

Haldane, J.B.S. (1954) The statics of evolution. In *Evolution as a Process* (eds J. Huxley, A.C. Hardy and E.B. Ford), Allen and Unwin, London, pp. 109–121.

Halvorsen, O. (1971) Studies on the helminth fauna of Norway. XVIII. On the composition of the parasite fauna of coarse fish in the River Gloma, southeastern Norway. *Norwegian Journal of Zoology*, **19**, 181–192.

Halvorsen, O. and Williams, H.H. (1967/68) Studies on the helminth fauna of Norway, IX. *Gyrocotyle* (Platyhelminthes) in *Chimaera monstrosa* from Oslo Fjord, with emphasis on its mode of attachment and regulation in the degree of infection. *Nytt Magasin for Zoologi*, **15**, 130–142.

Hamilton, W.D. (1980) Sex versus non-sex versus parasite. *Oikos*, **35**, 282–290.

Hamilton, W.D. and Zuk, M. (1982) Heritable true fitness and bright birds: a role for parasites? *Science*, **218**, 384–387.

Hanski, I. (1982) Dynamics of regional distribution: the core and satellite species concept. *Oikos*, **38**, 210–221.

Hare, G.M. and Burt, M.D.B. (1976) Parasites as potential biological tags of Atlantic salmon (*Salmo salar*) smolts in the Miramichi River System, New Brunswick. *Journal of the Fisheries Research Board of Canada*, **33**, 1139–1143.

Harrison, B.D. (1973) Viruses and nematodes. In *Frontiers of Biology 31, Viruses and Invertebrates* (ed. A.J. Gibbs), North-Holland, Amsterdam pp. 513–525.

Hassell, M.P. (1976) Arthropod predator–prey systems. In *Theoretical Ecology: Principles and Applications* (ed. R.M. May), W.B. Saunders, Philadelphia, pp. 71–93.

Hassell, M.P. (1978) The *Dynamics of Arthropod Predator–Prey Systems*, Princeton University Press, Princeton, New Jersey.

Haswell-Elkins, M.R., Elkins, D.B. and Anderson, R.M. (1987) Evidence for predisposition in humans to infection with *Ascaris*, hookworm, *Enterobius* and *Trichuris* in a South Indian fishing community. *Parasitology*, **95**, 323–337.

Haswell-Elkins, M.R., Elkins, D.B., Manjula, K., Michael, E. and Anderson, R.M. (1988) An investigation of hookworm infection and reinfection following mass anthelmintic treatment in the South Indian fishing community of Vairavankuppam. *Parasitology*, **96**, 565–577.

Haukisalmi, V., Henttonen, H. and Tenora, F. (1987) Parasitism by helminths in the grey-sided vole (*Clethrionomys rufocanus*) in Northern Finland: influence of density, habitat and sex of host. *Journal of Wildlife Diseases*, **23**, 233–241.

Hewitt, G.C. and Hine, P.M. (1972) Checklist of parasites of New Zealand fishes and of their hosts. *New Zealand Journal of Marine and Freshwater Research*, **6**, 69–114.

Heyneman, D. (1962) Studies on helminth immunity. II Influence of *Hymenolepis nana* (Cestoda: Hymenolepididae) in dual infections with *H. diminuta* in white mice and rats. *Experimental Parsitology*, **12**, 7–18.

Hilburn, L.R. and Sattler, P.W. (1986a) Electrophoretically detectable protein variation in natural populations of the lone star tick, *Amblyomma americanum* (Acari: Ixodidae). *Heredity*, **56**, 67–74.

Hilburn, L.R. and Sattler, P.W. (1986b) Are tick populations really less variable and should they be? *Heredity*, **57**, 113–117.

Hildrew, A.G. and Townsend, C.R. (1987) Organization in freshwater benthic communities. In *Organization of Communities Past and Present* (eds J.H.R. Gee and P.S. Giller), 27th Symposium of the British Ecological Society, Aberystwyth, 1986, Blackwell, Oxford, pp. 347–472.

Hine, P.M. and Kennedy, C.R. (1974) Observations on the distribution, specificity and pathogenicity of the acanthocephalan *Pomphorhynchus laevis* (Müller). *Journal of Fish Biology*, **6**, 521–535.

Hirsch, R.P. and Gier, H.T. (1974) Multi-species infections of intestinal helminths in Kansas coyotes. *Journal of Parasitology*, **60**, 650–653.

Hobbs, R.P. (1980) Interspecific interactions among gastrointestinal helminths in pikas of North America. *American Midland Naturalist*, **103**, 15–25.

Hoberg, E.P. (1986) Aspects of ecology and biogeography of Acanthocephala in Antarctic seabirds. *Annales de Parasitologie Humaine et Comparée*, **61**, 199–214.

Hoeppli, R.J.C. (1925) *Mesocestoides corti*, a new species of cestode from the mouse. *Journal of Parasitology*, 12, 91–97.

Hoffman, G.L. (1967) *Parasites of North American Freshwater Fishes*, University of California Press, Berkeley, CA.

Holland, C.V., Taren, D.L., Crompton, D.W.T., Nesheim, M.C., Sanjur, D., Barbeau, I., Tucker, K., Tiffany, J., and Rivera, G. (1988) Intestinal helminthiases in relation to the socioeconomic environment of Panamanian children. *Social Science and Medicine*, 26, 209–213.

Holmes, J.C. (1961) Effects of concurrent infections on *Hymenolepis diminuta* (Cestoda) and *Moniliformis dubius* (Acanthocephala). I. General effects and comparison with crowding. *Journal of Parsitology*, 47, 209–216.

Holmes, J.C. (1962a) Effect of concurrent infections on *Hymenolepis diminuta* (Cestoda) and *Moniliformis dubius* (Acanthocephala). II. Effects on growth. *Journal of Parasitology*, 48, 87–96.

Holmes, J.C. (1962b) Effect of concurrent infections on *Hymenolepis diminuta* (Cestoda) and *Moniliformis dubius* (Acanthocephala). III. Effects in hamsters. *Journal of Parasitology*, 48, 97–100.

Holmes, J.C. (1973) Site selection by parasitic helminths: interspecific interactions, site segregation, and their importance to the development of helminth communities. *Canadian Journal of Zoology*, 51, 333–347.

Holmes, J.C. (1976) Host selection and its consequences. In *Ecological Aspects of Parasitology* (ed. C.R. Kennedy), North-Holland Publishing Company, Amsterdam, pp. 21–39.

Holmes, J.C. (1979) Parasite populations and host community structure. In *Host–Parasite Interfaces* (ed. B.B. Nickol), Academic Press, New York, pp. 27–46.

Holmes, J.C. (1983) Evolutionary relationships between parasitic helminths and their hosts. In *Coevolution* (eds D.J. Futuyma, and M. Slatkin), Sinauer Associates Inc., Sunderland, Massachusetts, pp. 161–185.

Holmes, J.C. (1987) The structure of helminth communities. *International Journal for Parasitology*, 17, 203–208.

Holmes, J.C. (1988) Progress in ecological parasitology – parasite communities. *Parazitologiya*, 22, 113–122 (in Russian).

Holmes, J.C. (1990) Helminth communities in marine fishes. In *Parasite Communities: Patterns and Processes* (eds G.W. Esch, A.O. Bush and J.M. Aho), Chapman and Hall, New York, pp. 101–130.

Holmes, J.C. and Bethel, W.M. (1972) Modification of intermediate host behaviour by parasites. In *Behavioural Aspects of Parasite Transmission* (eds E.U. Canning and C.A. Wright), *Journal of the Linnean Society of London*, Supplement 1, 123–149.

Holmes, J.C., Hobbs, R.P. and Leong, T.S. (1977) Populations in perspective: community organization and regulation of parasite populations. In *Regulation of Parasite Populations* (ed G.W. Esch), Academic Press, Inc., New York, pp. 209–245.

Holmes, J.C. and Price, P.W. (1986) Communities of Parasites. In *Community Ecology: Patterns and Processes* (eds D.J. Anderson and J. Kikkawa), Blackwell Scientific Publications, Oxford, pp. 187–213.

Hood, D.E. and Welch, H.E. (1980) A seasonal study of the parasites of the red-winged balckbird (*Agelaius phoeniceus* L.) in Manitoba and Arkansas. *Canadian Journal of Zoology*, **58**, 528–537.

Hopkins, C.A. (1980) Immunity and *Hymenolepis diminuta*. In *Biology of the Tapeworm* Hymenolepis diminuta (ed H.P. Arai), Academic Press, New York and London, pp. 551–637.

Howell, M.J. (1985) Gene exchange between hosts and parasites. *International Journal for Parasitology*, **15**, 597–600.

Hugghins, E.J. (1956) Ecological studies on a strigeid trematode at Oakwood Lakes, South Dakota. *Proceedings of the South Dakota Academy of Science*, **35**, 204–206.

Hughes, R.N. (1989) *A Functional Biology of Clonal Animals*, Chapman and Hall, London.

Hunter, G.W. and Hamilton, J.M. (1941) Studies on host parasite reactions to larval parasites. IV. The cyst of *Uvulifer ambloplitis* (Hughes). *Transactions of the American Microscopical Society*, **60**, 498–507.

Hurd, H. (1990) Physiological and behavioural interactions between parasites and invertebrate hosts. *Advances in Parasitology*, **29**, 271–318.

Hutchinson, G.E. (1957) Concluding remarks. *Cold Spring Harbor Symposia on Quantitative Biology*, **22**, 415–427.

Hutchinson, G.W., Lee, E.H. and Fernando, M.A. (1972) Effects of variations in temperature on infective larvae and their relationship to inhibited development of *Obeliscoides cuniculi* in rabbits. *Parasitology*, **65**, 333–342.

Hynes, H.B.N. and Nicholas, W.L. (1963) The importance of the acanthocephalan *Polymorphus minutus* as a parasite of domestic ducks in the United Kingdom. *Journal of Helminthology*, **37**, 185–198.

Ihering, H. von (1891) On the ancient relations between New Zealand and South America. *Transactions and Proceedings of the New Zealand Institute*, **24**, 431–445.

Ihering, H. von (1902) Die Helminthen als Hilfsmittel der zoogeographischen Forschung. *Zoologizer Anzeiger*, **26**, 42–51.

Inada, T. (1981) Studies on Merluccid fishes. *Bulletin of the Far Seas Fishery Research Laboratory*, **18**, 1–172.

Ito, A., Lightowlers, M.W., Rickard, M.D. and Mitchell, G.F. (1988) Failure of auto-infection with *Hymenolepis nana* in seven inbred strains of mice initially given beetle-derived cysticercoids. *International Journal for Parasitology*, **18**, 321–324.

Ito, A., Onitake, K. and Andreassen, J. (1988) Lumen phase specific cross immunity between *Hymenolepis microstoma* and *H. nana* in mice. *International Journal for Parasitology*, **18**, 1019–1027.

Ito, A., and Smyth, J.D. (1987) Adult cestodes. Immunology of the lumen-dwelling cestode infections. In *Immune Responses in Parasitic Infections: Immunology, Immunopathology, and Immunoprophylaxis*, Vol. 2, (ed. E.J.L. Soulsby), CRC Press, Boca Raton, Florida. pp. 115–163.

Jablonski, D., Flessa, K.W. and Valentine, J.W. (1985) Biogeography and paleobiology. *Paleobiology*, **11**, 75–90.

Jacobson, K.C. (1987) *Infracommunity Structure of Enteric Helminths in the Yellow-*

bellied Slider, Trachemys scripta scripta. Master's Thesis, Wake Forest University, Winston-Salem, North Carolina, USA.

James, H.A. and Ulmer, M.J. (1967) New amphibian host records for *Mesocestoides* sp. (Cestoda: Cyclophyllidea). *Journal of Parasitology,* **53,** 59.

Janovy, J. Jr. and Kutish, G.W. (1988) A model of encounters between host and parasite populations. *Journal of Theoretical Biology,* **134,** 391–401.

Janzen, D.H. (1980) When is it coevolution? *Evolution,* **34,** 611–612.

Jarroll, E.L. (1979) Population biology of *Bothriocephalus rarus* Thomas (1937) in the red-spotted newt, *Notophthalmus viridescens* Raf. *Parasitology,* **79,** 183–193.

Jarroll, E.L. (1980) Population dynamics of *Bothriocephalus rarus* (Cestoda) in *Notophthalmus viridescens. American Midland Naturalist,* **103,** 360–366.

Jennings, J.B. and Calow, P. (1975) The relationship between high fecundity and the evolution of entoparaisitism. *Oecologia,* **21,** 109–115.

Johnson, S.K. and Rogers, W.A. (1973) Distribution of the genus *Ergasilus* in several Gulf of Mexico drainage basins. Agricultural Experiment Station Auburn University, Bulletin No. 445, 74pp.

Johnston, S.J. (1912) On some trematode parasites of Australian frogs. *Proceedings of the Linnean Society of New South Wales,* **37,** 285–362.

Jones, A.W. and Tan, B. D. (1971) Effect of crowding upon growth and fecundity in the mouse bile duct tapeworm *Hymenolepis microstoma. Journal of Parasitology,* **57,** 88–93.

Kabata, Z. and Ho, J-S. (1981) The origin and dispersal of hake (Genus *Merluccius:* Pisces: Teleostei) as indicated by its copepod parasites. *Oceanography and Marine Biology Annual Review,* **19,** 381–404.

Kates, K.C. (1944) Some observations on experimental infections of pigs with the thorn headed worm, *Macracanthorhynchus hirudinaceus. American Journal of Veterinary Research,* **5,** 166–172.

Kearn, G.C. (1980) Light and gravity responses of the oncomiracidium of *Entobdella soleae* and their role in host location. *Parasitology,* **81,** 71–89.

Keller, A.E., Leathers, W.S. and Knox, J.C. (1937) The present status of hookworm infestation in North Carolina. *American Journal of Hygiene,* **26,** 437–454.

Kelley, G.W., Jr. (1955) The daily egg production of *Haemonchus contortus* (Nematoda) in a calf. *Journal of Parasitology,* **41,** 218–219.

Kennedy, C.R. (1968) Population biology of the cestode *Caryophyllaeus laticeps* (Pallas, 1781) in dace, *Leuciscus leuciscus* L., of the River Avon. *Journal of Parasitology,* **54,** 538–543.

Kennedy, C.R. (1969) Seasonal incidence and development of the cestode *Caryophyllaeus laticeps* (Pallas), in the River Avon. *Parasitology,* **59,** 783–794.

Kennedy, C.R. (1971) The effect of temperature upon the establishment and survival of the cestode *Caryophyllaeus laticeps* in orfe, *Leuciscus idus. Parasitology,* **63,** 59–66.

Kennedy, C.R. (1975) *Ecological Animal Parasitology,* Blackwell Scientific Publications, Oxford.

Kennedy, C.R. (1978a) An analysis of the metazoan parasitocoenoses of brown trout *Salmo trutta* from British lakes. *Journal of Fish Biology,* **13,** 255–263.

Kennedy, C.R. (1978b) The parasite fauna of resident char *Salvelinus alpinus* from Artic Islands, with special reference to Bear Island. *Journal of Fish Biology*, 13, 457–466.

Kennedy, C.R. (1983) General Ecology. In *Biology of the Eucestoda*, Vol. 1 (eds C. Arme and P.W. Pappas), Academic Press, London, pp. 27–80.

Kennedy, C.R. (1985) Site segregation by species of Acanthocephala in fish, with special reference to eels, *Anguilla anguilla*. *Parasitology*, 90, 375–390.

Kennedy, C.R. (1987) Long term stability in the population levels of the eye fluke *Tylodelphys podicipina* (Digenea: Diplostomatidae) in perch. *Journal of Fish Biology*, 31, 571–581.

Kennedy, C.R. (1990) Helminth communities in freshwater fish: structured communities or stochastic assemblages? In *Parasite Communities: Patterns and Processes* (eds G.W. Esch, A.O. Bush and J.M. Aho), Chapman & Hall, New York, pp. 131–156.

Kennedy, C.R. and Burrough, R.J. (1977) The population biology of two species of eye fluke, *Diplostomum gasterostei* and *Tylodelphys clavata*, in perch. *Journal of Fish Biology*, 11, 619–633.

Kennedy, C.R. and Burrough, R.J. (1978) Parasites of trout and perch in Malham Tarn. *Field Studies*, 4, 617–629.

Kennedy, C.R., Bush, A.O. and Aho, J.M. (1986) Patterns in helminth communities: why are fish and birds different? *Parasitology*, 93, 205–215.

Kennedy, C.R. and Rumpus, A. (1977) Long term changes in the size of the *Pomphorhynchus laevis* (Acanthocephala) population in the River Avon. *Journal of Fish Biology*, 10, 35–42.

Kennedy, C.R. and Walker, P.J. (1969) Evidence for an immune response in dace, *Leuciscus leuciscus*, to infections by the cestode *Caryophyllaeus laticeps*. *Journal of Parasitology*, 55, 579–582.

Keymer, A.E. (1980) The influence of *Hymenolepis diminuta* on the survival and fecundity of the intermediate host, *Tribolium confusum*. *Parasitology*, 81, 405–421.

Keymer, A.E. (1981) Population dynamics of *Hymenolepis diminuta* in the intermediate host. *Journal of Animal Ecology*, 50, 941–950.

Keymer, A.E. (1982) Density-dependent mechanisms in the regulation of intestinal helminth populations. *Parasitology*, 84, 573–587.

Keymer, A. and Pagel, M. (1990) In *Hookworm Infection: Current Status and New Directions* (eds G.A. Schad and K.S. Warren), Taylor and Francis Ltd., London.

Khalil, L.F. and Cable, R.M. (1969) Germinal development in *Philopthalmus megalurus*. *Zeitschrift für Parasitenkunde*, 31, 211–213.

Kisielewska, K. (1970) Ecological organization of helminth groupings in *Clethrionomys glareolus* (Schreb) (Rodentia). I. Structure and seasonal dynamics of helminth groupings in a host population in the Bialowieza National Park. *Acta Parasitologica Polonica*, 18, 121–147.

Kitaoka, M., Oku, Y., Okamoto, M. and Kamiya, M. (1990) Development and sexual maturation of *Taenia crassiceps* (Cestoda) in the golden hamster. *Journal of Parasitology*, 76, 399–402.

308   *References*

Klassen, G.H. and Beverley-Burton, M. (1987) Phylogenetic relationships of *Ligictaluridus* spp. (Monogenea: Ancyrocephalidae) and their ictalurid (Siluriformes) hosts: an hypothesis. *Canadian Journal of Zoology*, 54, 84–90.

Kluge, A.G. and Farris, J.S. (1969) Quantitative phyletics and the evolution of anurans. *Systematic Zoology*, 18, 1–32.

Kohler, P. (1985) The strategies of energy conservation in helminths. *Molecular and Biochemical Parasitology*, 17, 1–18.

Køie, M. (1979) On the morphology and life history of *Derogenes varicus* (Müller, 1784) Loos, 1901 (Trematoda, Hemiuridae). *Zeitschrift für Parasitenkunde*, 59, 67–78.

Kolzow, R.G. and Nollen, P.M. (1978) Effect of stressful conditions on the development and movement of reproductive cells in *Schistosoma japonicum*. *Journal of Parasitology*, 64, 994–997.

Kostitzin, V.A. (1934) *Symbiose, Parasitisme et Evolution*, Paris, Hermann.

Kostitzin, V.A. (1939) *Mathematical Biology*, London, George C. Harrap.

Kuris, A.M. (1974) Trophic interactions: similarity of parasitic castrators to parasitoids. *Quarterly Review of Biology*, 49, 129–148.

Kuris, A.M. (1990) Guild structure of larval trematodes in molluscan hosts: prevalence, dominance and significance of competition. In *Parasite Communities: Patterns and Processes* (eds G.W. Esch, A.O. Bush and J.M. Aho), Chapman and Hall, New York, pp. 69–100.

Kuris, A.M. and Blaustein, A.R. (1977) Ectoparasitic mites on rodents: Application of the island biogeography theory? *Science*, 195, 596–598.

Lang, B.F. (1984) The mitochondrial genome of the fission yeast *Schizosaccharomyces pombe*: highly homologous introns are inserted at the same position of the otherwise less conserved *coxl* genes in *Schizosacharomyces pombe* and *Aspergillus nidulans*. *EMBO Journal*, 3, 2129–2136.

Larsh, J.E., Jr. (1964). *Outline of Medical Parasitology*, McGraw-Hill, Inc., New York.

Larsh, J.E., Race. G.J. and Esch, G.W. (1965) A histopathologic study of mice infected with the larval stage of *Multiceps serialis*. *Journal of Parasitology*, 51, 45–52.

Larsh, J.E., Jr. and Weatherly, N.F. (1974) Studies on delayed (cellular) hypersensitivity in mice infected with *Trichinella spiralis*. IX. Delayed dermal sensitivity in artificially sensitized donors. *Journal of Parasitology*, 60, 93–98.

Lauckner, G. (1980) Diseases of Mollusca: Gastropoda. In *Diseases of Marine Animals*, Vol. I: General Aspects, Protozoa to Gastropoda (ed. O. Kinne), John Wiley and Sons, New York, pp. 311–424.

Lawton, J.H. and Strong, D.R. (1981) Community patterns and competition in folivorous insects. *American Naturalist*, 188, 317–338.

Lebedev, B.I. (1969) Basic regularities in the distribution of monogeneans and trematodes of marine fishes in the world ocean. *Zoologicheskii Zhurnal*, 48, 41–50 (In Russian).

Lemly, A.D. (1983) *Ecology of Uvulifer ambloplitis (Trematoda, Strigeida) in a Population of Bluegill Sunfish, Lepomis macrochirus (Centrarchidae)*. PhD dissertation, Wake Forest University, Winston-Salem, North Carolina, USA.

Lemly, A.D. and Esch, G.W. (1983) Differential viability of metacercariae of *Uvulifer ambloplitis* (Hughes 1927) in juvenile centrarchids. *Journal of Parasitology*, **69**, 746–749.

Lemly, A.D. and Esch, G.W. (1984a) Population biology of the trematode *Uvulifer ambloplitis* (Hughes, 1927) in the snail intermediate host, *Helisoma trivolvis*. *Journal of Parasitology*, **70**, 461–465.

Lemly, A.D. and Esch, G.W. (1984b) Population biology of the trematode *Uvulifer ambloplitis* (Hughes, 1927) in juvenile bluegill sunfish, *Lepomis macrochirus*, and largemouth bass, *Micropterus salmoides*. *Journal of Parasitology*, **70**, 466–474.

Lemly, A.D. and Esch, G.W. (1984c) Effects of the trematode *Uvulifer ambloplitis* on juvenile bluegill sunfish, *Lepomis macrochirus*: ecological implications. *Journal of Parasitology*, **70**, 475–492.

Lemly, A.D. and Esch G.W. (1985) Black-spot caused by *Uvulifer ambloplitis* (Trematoda) among juvenile centrarchids in the Piedmont area of North Carolina. *Proceedings of the Helminthological Society of Washington*, **52**, 30–35.

Lensky, R.E. and Levin, B.R. (1985) Constraints on the coevolution of bacteria and virulent phage: a model, some experiments, and predictions for natural communities. *American Naturalist*, **125**, 585–602.

Leong, T.S. and Holmes, J.C. (1981) Communities of metazoan parasites in open water fishes of Cold Lake, Alberta. *Journal of Fish Biology*, **18**, 693–713.

Leslie, J.F., Cain, G.D., Meffe, G.K. and Vrijenhoek, R. (1982) Enzyme polymorphism in *Ascaris suum* (Nematoda). *Journal of Parasitology*, **68**, 576–587.

Lester, R.J.G. (1977) An estimate of mortality in a population of *Perca flavescens* owing to the trematode *Diplostomun adamsi*. *Canadian Journal of Zoology*, **55**, 288–292.

Levin, S. (1983) Some approaches to modelling the coevolutionary interactions. In *Coevolution* (ed. M. Nitecki), University of Chicago Press, Chicago, pp. 21–65.

Levin, S. et al. (1982) Evolution of parasites and hosts (group report). In *Population Biology of Infectious Diseases* (eds R.M. Anderson and R.M. May), Springer Verlag, Berlin, pp. 213–243.

Levine, N.D. (1973) Protozoan parasites of domestic animals and man, 2nd ed., Burgess Publishing Co., Minneapolis.

Lewin, R. (1985) Fish to bacterium gene transfer. *Science*, **227**, 1020.

Lewin, R. (1986) Supply-side ecology. *Science*, **234**, 25–27

Lewert, R.M. and Lee, C.L. (1954a) Studies on the passage of helminth larvae through host tissues. I. Histochemical studies on the extracellular changes caused by penetrating larvae. *Journal of Infectious Diseases*, **95**, 13–35.

Lewert, R.M. and Lee, C.L. (1954b) Studies on the passage of helminth larvae through host tissues. II. Enzymatic activity of larvae in vivo and in vitro. *Journal of Infectious Diseases*, **95**, 36–51.

Lewontin, R.C. (1965) Selection for colonizing ability. In *Genetics of Colonizing Species* (eds H.E. Baker and G.L. Stebbins), Academic Press, New York, pp. 79–91.

Lewontin, R.C. (1974) The genetic basis of evolutionary change, Columbia University Press, New York.

Lie, K.J., Basch, P.F. and Umathevy, T. (1966) Studies on Echinostomatidae (Trematoda) in Malasya. XII. Antagonism between 2 spp. of echinostome trematodes in the same lymnaeid snail. *Journal of Parasitology*, 52, 454–457.

Lim, H.K. and Heynemann, D. (1972) Intramolluscan intertrematode antagonism: a review of factors influencing the host–parasite system and its possible role in biological control. *Advances in Parasitology*, 10, 191–268.

Lincicome, D.R. (1971) The goodness of parasitism: a new hypothesis. In *Aspects of the Biology of Symbiosis* (ed. T. Cheng), University Park Press, Baltimore, Maryland, pp. 139–227.

Loos-Frank, B. (1980) The common vole, *Microtus arvalis* Pall. as intermediate host for *Mesocestoides* (Cestoda) in Germany. *Zeitschrift für Parasitenkunde*, 63, 129–136.

Lotka, A.J. (1934) *Theorie Analytique des Associations Biologiques*, Hermann, Paris.

Lotz, J.M. and Font, W.F. (1985) Structure of enteric helminth communities in two populations of *Eptesicus fuscus* (Chiroptera). *Canadian Journal of Zoology*, 63, 2969–2978.

Love, M.S. and Moser, M. (1983) A check list of parasites of California, Oregon, and Washington Marine and Estuarine Fishes. NOAA Technical Report NMFS SSRF-777, U.S. Department of Commerce.

Lubbock, R. (1981) The clownfish/anemone symbiosis: a problem of cellular recognition. *Parasitology*, 82, 159–173.

Lydeard, C., Mulvey, M., Aho, J.M. and Kennedy, P.K. (1989) Genetic variability among natural populations of the liver fluke *Fascioloides magna* in white-tailed deer, *Odocoileus virginianus*. *Canadian Journal of Zoology*, 67, 2021–2025.

Lymbery, A.J. and Thompson, C.A. (1988) Electrophoretic analysis of genetic variation in *Echinococcus granulosus* from domestic hosts in Australia. *International Journal for Parasitology*, 18, 803–811.

Lymbery, A.J. and Thompson, C.A. (1989) Genetic differences between cysts of *Echinococcus granulosus* from the same host. *International Journal for Parasitology*, 19, 961–964.

MacArthur, R.H. (1972) *Geographical Ecology. Patterns in the Distribution of Species*, Harper and Row, New York.

MacArthur, R.H. and Wilson, E.O. (1967) *The Theory of Island Biogeography*, Princeton University Press, Princeton.

MacDonald, G. (1965) The dynamics of helminth infections with special reference to schistosomes. *Transactions of the Royal Society of Tropical Medicine and Hygiene*, 59, 489–506.

MacKenzie, K. (1983) Parasites as biological tags in fish populations studies. In *Advances in Applied Biology* (ed. R.H. Coaker), Academic Press, Cambridge, pp. 251–253.

MacKenzie, K. (1985) Relationships between the herring, *Clupea harengus* L., and its parasites. *Advances in Marine Biology*, 24, 263–319.

MacKenzie, K. (1987) Parasites as indicators of host populations. *International Journal for Parasitology*, 17, 345–352.

MacKintosh, J.H. (1981) Behaviour of the house mouse. *Symposium of the Zoological Society of London*, 47, 337–365.

MacPherson, C.N.L. and McManus, D.P. (1982) A comparative study of *Echinococcus granulosus* from human and animal hosts in Kenya using isoelectric focusing and isoenzyme analysis. *International Journal for Parasitology*, 12, 515–521.

Madhavi, R. and Anderson, R.M. (1985) Variability in the susceptibility of the fish host, *Poecilia reticulata*, to infection with *Gyrodactylus bullatarudis* (Monogenea). *Parasitology*, 91, 531–544.

Mann, C. (1991) Lynn Margulis: Science's unruly earth mother. *Science*, 252, 378–381.

Manter, H.W. (1934) Some digenetic trematodes from the deep-water fish at Tortugas, Florida, Carnegie Institute Publication No. 435, Papers from Tortugas Laboratory, 27, 257–345.

Manter, H.W. (1940) The geographical distribution of digenetic trematodes of marine fishes in the tropical American Pacific. *Allan Hancock Pacific Expeditions*, 2, 531–547.

Manter, H.W. (1955) The zoogeography of trematodes of marine fishes. *Experimental Parasitology*, 4, 62–86.

Manter, H.W. (1963) The zoogeographical affinities of trematodes of South American freshwater fishes. *Systematic Zoology*, 12, 45–70.

Marcogliese, D.J. and Esch, G.W. (1989) Alterations in seasonal dynamics of *Bothriocephalus acheilognathi* in a North Carolina cooling reservoir over a seven year period. *Journal of Parasitology*, 75, 378–382.

Marcogliese, D.J., Goater, T.M. and Esch, G.W. (1990) *Crepidostomum cooperi* (Allocreadidae) in the burrowing mayfly, *Hexagenia limbata* (Ephemeroptera) related to trophic status of a lake. *American Midland Naturalist*, 124, 309–317.

Margolis, L. (1956) Report on parasite studies of sockeye and pink salmon collected in 1955, with special reference to the utilization of parasites as a means of distinguishing between Asiatic and American stocks of salmon in high seas - a progress report on work being carried out as part of F.R.B.'s commitments to INPFC. Fisheries Research Board of Canada, Manuscript Reports of the Biological Station, No. 624.

Margolis, L. (1963) Parasites as indictors of the geographical origin of sockeye salmon, *Oncorhynchus nerka* (Walbaum) occurring in the North Pacific Ocean and adjacent seas. *Bulletin of the International North Pacific Fish Community*, 11, 101–156.

Margolis, L. (1965) Parasites as an auxiliary source of information about the biology of Pacific salmons (genus *Oncorhynchus*). *Journal of the Fisheries Research Board of Canada*, 22, 1387–1395.

Margolis, L. (1990) Trematodes as population markers for North Atlantic steelhead trout. *Bulletin de la Société Francaise de Parasitologie*, 8, Supplement 2, 735.

Margulis, L. (1981) *Symbiosis in Cell Evolution*, Freeman, San Francisco.

Martin, W.E. (1955) Seasonal infection of the snail *Cerithidea californica* Haldeman, with larval trematodes. In *Essays in the Natural Sciences in Honor of Captain Allan Hancock*, Allan Hancock Foundation, University of Southern California Press, Los Angeles, pp. 203–210.

Martin, W.E. (1972) An annotated key to the cercariae that develop in the snail *Cerithidea californica*. *Bulletin of the Southern California Academy of Sciences*, 71, 39–43.

May, R.M. (1977) Dynamical aspects of host–parasite associations: Crofton's model revisited. *Parasitology*, 75, 259–276.

May, R.M., and Anderson, R.M. (1979) Population biology of infectious diseases. II. *Nature*, London, **280**, 455–461.

May, R.M. and Anderson, R.M. (1983) Parasite–host coevolution. In *Coevolution* (ed. D.J. Futuyma and M. Slatkin), Sinauer Associates Inc., Sunderland, Massachusetts, pp. 186–206.

McGladdery, S.E. and Burt, M.D. (1985) Potential of parasites for use as biological indicators of migration, feeding and spawning behaviour of northwestern Atlantic herring (*Clupea harengus*). *Canadian Journal of Fisheries and Aquatic Research*, **42**, 1957–1968.

McGowan, J.A. (1971) Oceanic biogeography of the Pacific. In *The Micropaleontology of Oceans* (eds B.M. Funnel and W.R. Riedel), Cambridge University Press, Cambridge.

McKenna, P.B. (1973) The effect of storage on the infectivity and parasitic development of third-stage *Haemonchus contortus* larvae in sheep. *Research in Veterinary Science*, **14**, 312–316.

McManus, D.P. and Smyth, J.D. (1979) Isoelectric focusing of some enzymes from *Echinococcus granulosus* (horse and sheep strains) and *E. multilocularis*. *Transactions of the Royal Society of Tropical Medicine and Hygiene*, **73**, 259–265.

McManus, D.P. and Simpson, A.J.G. (1985) Identification of the *Echinococcus* (hydatid disease) organisms using cloned DNA markers. *Molecular and Biochemical Parasitology*, **17**, 171–178.

Metcalf, M. (1929) Parasites and the aid they give in problems of taxonomy, geographical distribution, and paleogeography. *Smithsonian Miscellanea Collection*, **81**, 1–36.

Metcalf, M. (1940) Further studies on the opalinid ciliate infusorians and their hosts. *Proceedings of the United States Museum*, **87**, 465–634.

Meyer, M.C. (1972) The pattern of circulation of *Diphyllobothrium sebago* (Cestoda: Pseudophyllidea) in a enzootic area. *Journal of Wildlife Diseases*, **8**, 215–220.

Meyerhoff, E. and Rothschild, M. (1940) A prolific trematode. *Nature*, London, **146**, 367.

Michaeli, D., Senyk, G., Maoz, A. and Fuchs, S. (1972) *Ascaris* cuticle collagen and mammalian collagens: Cell mediated and humoral immunity relationships. *Journal of Immunology*, **109**, 103–109.

Michel, J.F. (1974) Arrested development of nematodes and some related phenomena. *Advances in Parasitology*, **12**, 279–366.

Michel, J.F., Lancaster, M.B. and Hong, C. (1975a) Arrested development of *Ostertagia ostertagi* and *Cooperia oncophora*: effect of temperature at the free-living third stage. *Journal of Comparative Pathology*, **85**, 133–138.

Michel, J.F., Lancaster, M.B. and Hong, C. (1975b) Arrested development of *Obeliscoides cuniculi*: the effect of size of inoculum. *Journal of Comparative Pathology*, **85**, 307–315.

Minchella, D.J. and Lo Verde, P.T. (1983) Laboratory comparison of the relative success of *Biomphalaria glabrata* stocks which are susceptible and insusceptible to infection with *Schistosoma mansoni*. *Parasitology*, **86**, 335–344.

Mitchell, G.F. (1979) Responses to infection with metazoan and protozoan parasites in mice. *Advances in Immunology*, **28**, 451–511.

Mitter, C. and Brooks, D.R. (1983) Phylogenetic aspects of coevolution. In *Coevolution* (eds D.T. Futuyma and M. Slatkin), Sinauer Associates, Inc., Massachusetts, pp. 65–98.

Møller, A.P. (1990) Parasites and sexual selection: current status of the Hamilton and Zuk hypothesis. *Journal of Evolutionary Biology*, **3**, 319–328.

Montgomery, S.S.J. and Montogomery, W.I. (1990) Structure stability and species interactions in helminth communities of wood mice, *Apodemus sylvaticus*. *International Journal for Parasitology*, **20**, 225–242.

Moore, J. (1981) Asexual reproduction and environmental predictability in cestodes (Cyclophyllidea: Taeniidae). *Evolution*, **35**, 723–741.

Moore, J, (1983a) Responses of an avian predator and its isopod prey to an acanthocephalan parasite. *Ecology*, **64**, 1000–1015.

Moore, J. (1983b) Altered behaviour in cockroaches (*Periplaneta americana*) infected with an archiacanthocephalan, *Moniliformis moniliformis*. *Journal of Parasitology*, **69**, 1174–1176.

Moore, J. (1984a) Altered behavioral response in intermediate hosts – an acanthocephalan parasite strategy. *American Naturalist*, **123**, 572–577.

Moore, J. (1984b) Parasites that change the behavior of their hosts. *Scientific American*, **250** (5), 82–89.

Moore, J., Freehling, M., Horton, D. and Simberloff, D. (1987) Host age and sex in relation to intestinal helminths of bobwhite quail. *Journal of Parasitology*, **73**, 230–233.

Moore, J. and Goniteli, N.J. (1990) A phylogenetic perspective on the evolution of altered host behaviours: a critical look at the manipulation hypothesis. In *Parasitism and Host Behaviour* (eds C.J. Barnard and J. M. Behnke), Taylor and Francis Ltd., New York, pp. 193–233.

Moore, J. and Simberloff, D. (1990) Gastrointestinal helminth communities of bobwhite quail. *Ecology*, **71**, 344–359.

Mouahid, A. and Théron, A. (1987) *Schistosoma bovis*: variability of cercarial production as related to the snail hosts: *Bulinus truncatus*, *B. wrighti* and *Planorbarius metidjensis*. *International Journal for Parasitology*, **17**, 1431–1434.

Mueller, J.F. and Strano, A.J. (1974a) *Sparganum proliferum*, a sparganum infected with a virus? *Journal of Parasitology*, **60**, 15–19.

Mueller, J.F. and Strano, A.J. (1974b) The ubiquity of type-C viruses in sparagna of *Spirometra* spp. *Journal of Parasitology*, **60**, 398.

Munger, J.C., Karasov, W.H. and Chang, D. (1989) Host genetics as a cause of overdispersion of parasites among hosts: how general a phenomenon? *Journal of Parasitology*, **75**, 707–710.

Muzzall, P.M. (1991) Helminth infracommunities of the newt, *Notophthalmus viridescens*, from Turkey marsh, Michigan. *Journal of Parasitology*, **77**, 87–91.

Myers, A.A. and Giller, P.S. (1988) Process, pattern and scale in biogeography. In *Analytical Biogeography* (eds A.A. Myers and P.S. Giller), Chapman and Hall, New York, pp. 3–12.

Nadler, S.A. (1987) Genetic variability in endoparasitic helminths. *Parasitology Today*, **3**, 154–155.

Nadler, S.A. (1990) Molecular approaches to studying helminth population genetics and phylogeny. *International Journal for Parasitology*, **20**, 11–29.

Nawalinsky, T.A. and Schad, G.A. (1974) Arrested development in *Ancylostoma duodenale*: course of a self-induced infection in man. *American Journal of Tropical Medicine and Hygiene*, **23**, 895–898.

Neraasen, T.G. and Holmes, J.C. (1975) The circulation of cestodes among three species of geese resting on the Anderson River Delta, Canada. *Acta Parasitologica Polonica*, **23**, 277–289.

Nevo, E. (1978) Genetic variation in natural populations: patterns and theory. *Theoretical Population Biology*, **13**, 121–177.

Nevo, E., Beiles, A. and Ben-Shlomo, R. (1984) The evolutionary significance of genetic diversity: ecological, demographic and life history correlates. In *Evolutionary Dynamic of Genetic Diversity, Lecture Notes in Biomathematics*, Vol. 53 (ed. G.S. Mani), Springer, Berlin, pp. 13–213.

Noble, E.R. (1973) Parasites and fishes in a deep sea environment. *Advances in Marine Biology*, **11**, 121–195.

Noble, E.R., Noble, G.A., Schad, G.A. and MacInnes, A.J. (1989) *Parasitology. The Biology of Animal Parasites*, 6th ed., Lea and Febiger, Philadelphia.

Nollen, P.M. (1983) Patterns of sexual reproduction among parasitic platy-helminths. *Symposia of the British Society for Parasitology*, **20**, 99–120.

Olsen, O.W. (1986) *Animal Parasites. Their Life Cycles and Ecology*, Dover Publications, Inc., Mineola, New York.

Oya, H., Costello, L.C. and Smith, W.N. (1963) The comparative biochemistry of developing *Ascaris* eggs. II. Changes in cytochrome c oxidase activity during embryonation. *Journal of Cellular and Comparative Physiology*, **62**, 287–294.

Paperna, I. and Overstreet, R.M. (1981) Parasites and diseases of mullets (Mugilidae). In *Aquaculture of Grey Mullets* (ed. O.H. Oren), International Bilological Program 26, Cambridge University Press, pp. 411–494.

Pagel, M.D. and Harvey, P.H. (1988) Recent developments in the analysis of comparative data. *Quarterly Review of Biology*, **64**, 413–440.

Pappas, P.W. and Read, C.P. (1972a) Trypsin inactivation by intact *Hymenolepis diminuta*. *Journal of Parasitology*, **58**, 864–871.

Pappas, P.W. and Read, C.P. (1972b) Inactivation of α- and β-chymotrypsin by intact *Hymenolepis diminuta* (Cestoda). *Biological Bulletin of the Marine Biology Laboratory, Woods Hole*, **143**, 605–616.

Park, T. (1948) Experimental studies of competition. I. Competition between populations of the flour beetles, *Tribolium confusum* Duval and *Tribolium castaneum* Herbst. *Ecological Monographs*, **18**, 265–308.

Parry, G.D. (1981) The meaning of r- and K- selection. *Oecologia*, **48**, 260–264.

Pence, D.B. (1990) Helminth community of mammalian hosts: concepts at the infracommunity, component and compound community levels. In *Parasite Communities: Patterns and Processes* (eds G.W. Esch, A.O. Bush and J.M. Aho), Chapman and Hall, New York, pp. 233–260.

Pence, D.B. and Meinzer, W.P. (1979) Helminth parasitism in the coyote, *Canis latrans*, from the Rolling Plains of Texas. *International Journal for Parasitology*, **9**, 339–344.

Pennycuick, L. (1971) Frequency distributions of parasites in a population of three-spined sticklebacks, *Gasterosteus aculeatus* L., with particular reference to the negative binomial distribution. *Parasitology*, **63**, 389–406.

Pesigan, T.P., *et al.* (1958) Studies on *Schistosoma japonicum* infection in the Philippines. 1. General considerations and epidemiology. *Bulletin of the World Health Organization*, **18**, 345–455.

Pianka, E.R. (1970) On *r*- and K- selection. *American Naturalist*, **104**, 592–597.

Pianka, E.R. (1976) Natural selection and optimal reproductive tactics. *American Zoologist*, **16**, 775–784.

Pianka, E.R. (1983) *Evolutionary Ecology*, 3rd ed., Harper and Row Publishers, New York.

Pickard, G.L. (1975) *Descriptive Physical Oceanography*, 2nd ed., Pergamon Press, New York.

Platt, N.E. (1975) Infestation of cod (*Gadus morhua* L.) with larvae of codworm (*Terranova decipiens* Krabbe) and herringworm, *Anisakis* sp. (Nematoda: Ascaridata), in North Atlantic and Arctic waters. *Journal of Applied Ecology*, **12**, 437–450.

Platt, N.E. (1976) Codworm – a possible biological indicator of the degree of mixing of Greenland and Iceland cod stocks. *Journal du Conseil Internationel pour l'éxploration de la Mer*, **37**, 41–45.

Polyanski, Yu. I. (1961a) Ecology of parasites of marine fishes. In *Parasitology of Fishes* (eds V.A. Dogiel, G.K. Petrushevsky and Yu I. Polyansky), Oliver and Boyd, Edinburgh, pp. 48–83.

Polyansky, Yu. I. (1961b) Zoogeography of parasites of USSR marine fishes. In *Parasitology of Fishes* (eds V.A. Dogiel, G.K. Petrushevsky, and Yu I. Polyansky), Oliver and Boyd, Edinburgh, pp. 230–245.

Powell, J.R. (1975) Protein variation in natural populations of animals. *Evolutionary Biology*, **8**, 79–119.

Price, P.W. (1972) Parasitoids utilizing the same hosts: adaptive nature of differences in size and form. *Ecology*, **53**, 190–195.

Price, P.W. (1977) General concept on the evolutionary biology of parasites. *Evolution*, **31**, 405–420.

Price, P.W. (1980) *Evolutionary Biology of Parasites*, Princeton University Press, Princeton.

Price, P.W. (1984a) Communities of specialists: vacant niches in ecological and evolutionary time. In *Ecological Communities: Conceptual Issues and the Evidence* (eds D.R. Strong, D. Simberloff, L.G. Abele and A.B. Thistle), Princeton University Press, New Jersey, pp. 510–523.

Price, P.W. (1984b) Alternative paradigms in community ecology. In *A New Ecology: Novel Approaches to Interactive Systems* (eds P.W. Price, C.N. Slobodchikoff and W.S. Gaud), John Wiley and Sons, Inc., pp. 353–383.

Price, P.W. (1987) Evolution in parasite communities. *International Journal for Parasitology*, **17**, 209–214.

Price, P.W. and Clancy, K.M. (1983) Patterns in number of helminth parasite species in freshwater fishes. *Journal of Parasitology*, **69**, 449–454.

Price, P.W., Westoby, M., Rice, B., Atsatt. P.R., Fritz., R.S., Thompson, J.N. and Mobley, K. (1986) Parasite mediation in ecological interactions. *Annual Review of Ecology and Systematics*, **17**, 487–505.

Prudhoe, S. and Bray, R.A. (1973) Digenetic trematodes from fishes. BANZ Antarctic Research Expedition 1929-1931, *Reports*, Series B, **8**, 199–225.

Prudhoe, S. and Bray, R.A. (1982) *Platyhelminth Parasites of the Amphibia*, British Museum (Natural History), Oxford University Press.

Quinnell, R.J., Medley, G.F. and Keymer, A.E. (1990) The regulation of gastrointestinal helminth populations. *Philosophical Transactions of the Royal Society of London*, B, **330**, 191–201.

Rausch, R.L. (1985) Parasitology: retrospect and prospect. *Journal of Parasitology*, **71**, 139–151.

Read, A.F. (1988) Sexual selection and the role of parasites. *Trends in Ecology and Evolution*, **3**, 97–102.

Read, C.P. (1959) The role of carbohydrates in the biology of cestodes. VIII. Some conclusions and hypothesis. *Experimental Parasitology*, **8**, 365–382.

Richards, C.S. (1973) Susceptibility of adult *Biomphalaria glabrata* to *Schistosoma mansoni* infection. *American Journal of Tropical Medicine and Hygiene*, **22**, 749–756.

Richards, C.S. (1975a) Genetic factors in susceptibility of *Biomphalaria glabrata* for different strains of *Schistosoma mansoni*. *Parasitology*, **70**, 231–241.

Richards, C.S. (1975b) Genetic studies on variation in infectivity of *Schistosoma mansoni*. *Journal of Parasitology*, **61**, 233–236.

Richards, C.S. (1976a) Variations in infectivity for *Biomphalaria glabrata* on strains of *Schistosoma mansoni* from the same geographic area. *Bulletin of the World Health Organization*, **54**, 706–707.

Richards, C.S. (1976b) Genetic aspects of host–parasite relationships. *Symposia of the British Society of Parasitology*, **14**, 45–54.

Richards, C.S. (1984) Influence of snail age on genetic variation in susceptibility of *Biomphalaria glabrata* for infection with *Schistosoma mansoni*. *Malacologia*, **25**, 493–502.

Riggs, M.R. (1986) *Community Dynamics of the Asian Fish Tapeworm*, Bothriocephalus acheilognathi, *in a North Carolina Cooling Reservoir*. PhD dissertation, Wake Forest University, Winston-Salem, North Carolina, USA.

Riggs, M.R. and Esch, G.W. (1987) The suprapopulation dynamics of *Bothriocephalus acheilognathi* in a North Carolina cooling reservoir: abundance, dispersion and prevalence. *Journal of Parasitology*, **73**, 877–892.

Riggs, M.R., Lemly, A.D. and Esch, G.W. (1987) The growth, biomass and fecundity of *Bothriocephalus acheilognathi* in a North Carolina cooling reservoir. *Journal of Parasitology*, **73**, 893–900.

Roberts, L.S. (1966) Developmental physiology of cestodes. I. Host dietary carbohydrate and the 'crowding effect' in *Hymenolepis diminuta*. *Experimental Parasitology*, **18**, 305–310.

Roberts, L.S. (1980) Development of *Hymenolepis diminuta* in its definitive host. In *Biology of the Tapeworm* Hymenolepis diminuta (ed. H.P. Arai), Academic Press, New York, pp. 357–423.

Rohde, K. (1976a) Species diversity of parasites of the Great Barrier Reef. *Zeitschrift für Parasitenkunde*, **50**, 93–94.

Rohde, K. (1976b) Monogenean gill parasites of *Scomberomonus commersoni* Lacepede and other mackerel on the Australian East Coast. *Zeitschrift für Parasitenkunde*, **51**, 49–69.

Rohde, K. (1978a) Latitudinal differences in host-specificity of marine Monogenea and Digenea. *Marine Biology*, 47, 125–134.

Rohde, K. (1978b) Latitudinal gradients in species diversity and their causes. II. Marine parasitological evidence for a time hypothesis. *Biologisches Zentralblatt*, 97, 405–418.

Rohde, K. (1979) A critical evaluation of intrinsic and extrinsic factors responsible for niche restriction in parasites. *American Naturalist*, 114, 648–671.

Rohde, K. (1980a) Comparative studies on microhabitat utilization by ectoparasites of some marine fishes from the North Sea and Papua new Guinea. *Zoologisches Anzeiger*, 204, 27–63.

Rohde, K. (1980b) Diversity gradients of marine Monogenea in the Atlantic and Pacific Oceans. *Experientia*, 36, 1369–1371.

Rohde, K. (1981) Niche width of parasites in species-rich and species-poor communities. *Experientia*, 37, 359–361.

Rohde, K. (1982) *Ecology of Marine Parasites*, University of Queensland Press, St. Lucia, Australia.

Rohde, K. (1984) Ecology of marine parasites. *Helgolander Meeresuntersuchungen*, 37, 5–33.

Rohde, K. (1986) Differences in species diversity of Monogenea between the Pacific and Atlantic oceans. *Hydrobiologia*, 137, 21–28.

Rohde, K. (1989) Simple ecological systems, simple solutions to complex problems? *Evolutionary Theory*, 8, 305–350.

Rohde, K. (1990). Phylogeny of Platyhelminthes, with special reference to parasitic groups. *International Journal for Parasitology*, 20, 979–1007.

Rohde, K. (1991) Intra- and interspecific interactions in low density populations in resource-rich habitats. *Oikos*, 60, 91–104.

Rollinson, D. and Southgate, V.R. (1985) Schistosome and snail populations: genetic variability and parasite transmission. In *Ecology and Genetics of Host–Parasite Interactions* (eds D.Rollinson and R.M. Anderson), Academic Press, London, pp. 91–109.

Root, R.B. (1967) The niche exploitation pattern of the blue-green gnatcatcher. *Ecological Monographs*, 37, 317–350.

Rosenthal, G.A. (1983) Biochemical adaptations of the bruchiid beetle, *Caryedes brasiliensis* to l-canavanine, a higher plant allelochemical. *Journal of Chemical Ecology*, 9, 803–815.

Roughgarden, J. (1976) Resource partitioning among competing species – a coevolutionary approach. *Theoretical Population Biology*, 9, 388–424.

Roughgarden, J. (1979) *Theory of Population Genetics and Evolutionary Ecology: An Introduction*, Macmillan, New York.

Rummel, J.D. and Roughgarden, J. (1985) A theory of faunal buildup for competition communities. *Evolution*, 39, 1009–1033.

Russell, S.W., Baker, N.F. and Raizes, G.S. (1966) Experimental *Obeliscoides cuniculi* infections in rabbits: comparison with *Trichostrongylus* and *Ostertagia* infections in cattle and sheep. *Experimental Parasitology*, 19, 163–173.

Sackville Hamilton, N.R., Schmid, B. and Harper, J.L. (1987) Life-history concepts and the population biology of clonal organisms. *Proceedings of the Royal Society of London, B*, 232, 35–57.

Schad, G.A. (1963) Niche diversification in a parasitic species flock. *Nature, London,* **198**, 404–405.

Schad, G.A. (1977) The role of arrested development in the regulation of nematode populations. In *Regulation of Parasite Populations* (ed. G.W. Esch), Academic Press, Inc., New York, pp. 111–167.

Schad, G.A. and Anderson, R.M. (1985) Predisposition to hookworm infection in humans. *Science,* **228**, 1537–1539.

Schell, S.C. (1985) *Handbook of Trematodes of North America North of Mexico,* University Press of Idaho.

Schmidt, G.D. and Roberts, L.S. (1989) *Foundations of Parasitology,* 4th ed., Times Mirror/Mosby, College Publishing, St. Louis.

Schulte, F. and Poinar, G.O. (1991) On the geographical distribution and parasitism of *Rhabditis* (*Pelodera*) *orbitalis* (Nematoda: Rhabditidae). *Journal of the Helminthological Society of Washington,* **58**, 82–84.

Scott, M.E. (1982) Reproductive potential of *Gyrodactylus bullatarudis* (Monogenea) on guppies (*Poecilia reticulata*). *Parasitology,* **85**, 217–236.

Scott, M.E. (1985a) Experimental epidemiology of *Gyrodactylus bullatarudis* (Monogenea) on guppies (*Poecilia reticulata*): short- and long-term studies. In *Ecology and Genetics of Host–Parasite Interactions* (ed. D. Rollinson and R.M. Anderson), Academic Press, New York, pp. 21–38.

Scott, M.E. (1985b) Dynamics of challenge infections of *Gyrodactylus bullatarudis* (Monogenea) on guppies, *Poecilia reticulata* (Peters). *Journal of Fish Diseases,* **8**, 495–503.

Scott, M.E. (1987a) Regulation of mouse colony abundance by *Heligmosomoides polygyrus. Parasitology,* **95**, 111–124.

Scott, M.E. (1987b) Temporal changes in aggregation: a laboratory study. *Parasitology,* **94**, 583–595.

Scott, M.E. and Anderson, R.M. (1984) The population dynamics of *Gyrodactylus bullatarudis* (Monogenea) within laboratory populations of the fish host *Poecilia reticulata. Parasitology,* **89**, 159–194.

Scott, M.E. and Dobson, A. (1989) The role of parasites in regulating host abundance. *Parasitology Today,* **5**, 176–183.

Scott, M.E. and Nokes, D.J. (1984) Temperature-dependent reproduction and survival of *Gyrodactylus bullatarudis* (Monogenea) on guppies (*Poecilia reticulata*). *Parasitology,* **89**, 221–227.

Scott, M.E. and Robinson, M.A. (1984) Challenge infections of *Gyrodactylus bullatarudis* (Monogenea) on guppies, *Poecilia reticulata* (Peters), following treatment. *Journal of Fish Biology,* **24**, 581–586.

Selander, R.K. and Kaufman, D.W. (1973) Genic variability and strategies of adaptation in animals. *Proceedings of the National Academy of Science, USA,* **70**, 1875–1877.

Sey, O. (1991) *CRC Handbook of the Zoology of Amphistomes.* CRC Press, Inc., Boca Raton, Florida.

Sher, A.N., Smithers, S.R. and Mackenzie, P. (1975) Passive transfer of acquired resistance to *Schistosoma mansoni* in laboratory mice. *Parasitology,* **70**, 347–357.

Shostak A. and Esch, G.W. (1990a) Photocycle-dependent emergence by cercariae of *Halipegus occidualis* from *Helisoma anceps*, with special reference to cercarial emergence patterns as adaptations to transmission. *Journal of Parasitology*, **76**, 790–795.

Shostak A. and Esch, G.W. (1990b) Temperature effects on survival and excystment of cercariae of *Halipegus occidualis* (Trematoda). *International Journal for Parasitology*, **20**, 95–99.

Sibly, R.M. and Calow, P. (1985) Classification of habitats by selection pressures: a synthesis of life-cycle and r/K theory. In *Behavioral Ecology* (eds R.M. Sibly and R.H. Smith), Blackwell Scientific Publications, Oxford, pp. 75–90.

Sibly, R.M. and Calow, P. (1986) *Physiological Ecology of Animals: An Evolutionary Approach*, Blackwell Scientific Publications, London.

Simberloff, D. (1990) Free-living communities and alimentary tract helminths: hypothesis and pattern analyses. In *Parasite Communities: Patterns and Processes* (eds G.W. Esch, A.O. Bush and J.M. Aho), Chapman and Hall, New York, pp. 289–319.

Simmons, J.E. and Laurie, J.S. (1972) A study of *Gyrocotyle* in the San Juan Archipelago, Puget Sound, USA, with observations on the host, *Hydrolagus colliei* (Lay and Bennett). *International Journal for Parasitology*, **2**, 59–77.

Sindermann, C.J. (1957) Diseases of fishes of the western North Atlantic. V. Parasites as indicators of herring movements. *Maine Department of Sea and Shore Fisheries Research Bulletin*, No 27, 30 pp.

Smith, F.E. (1954) Quantitative aspects of population growth. In *Dynamics of Growth Processes* (ed. E.J. Boell), Princeton University Press, Princeton, New Jersey, pp. 277–294.

Smith, H.D. (1973) Observations on the cestode *Eubothrium salvelini* in juvenile sockeye salmon (*Oncorhynchus nerka*) at Babine Lake, British Columbia. *Journal of the Fisheries Research Board of Canada*, **30**, 947–964.

Smith-Trail, D.R. (1980) Behavioural interactions between parasites and hosts: host suicide and the evolution of complex life cycles. *American Naturalist*, **116**, 77–91.

Smyth, J.D. (1969) The *Physiology of Cestodes*, W.H. Freeman, San Francisco.

Smyth, J.D. (1973) Some interface phenomena in parasitic protozoa and platyhelminths. *Canadian Journal of Zoology*, **51**, 367–377.

Smyth, J.D. and Halton, D.W. (1983) The *Physiology of Trematodes*, 2nd ed., Cambridge University Press, Cambridge.

Smyth, J.D. and Smyth, M.M. (1964) Natural and experimental hosts of *Echinococcus granulosus* and *E. multilocularis*, with comments on the genetics of speciation in the genus *Echinococcus*. *Parasitology*, **54**, 493–514.

Smyth, J.D. and Smyth, M.M. (1969) Self insemination in *Echinococcus granulosus* in vivo. *Journal of Helminthology*, **43**, 383–388.

Sousa, W.P. (1990) Spatial scale and the processes structuring a guild of larval trematode parasites. In *Parasite Communities: Patterns and Processes* (eds G.W. Esch, A.O. Bush and J.M. Aho), Chapman and Hall, New York, pp. 41–67.

Sousa, W.P. (1991) Interspecific antagonism and species coexistence in a diverse guild of larval trematode parasites. *Ecological Monographs*.

Southgate, V.R. and Knowles, R.J. (1975) Observations on *Schistosoma bovis* (Sonsino, 1876). *Journal of Natural History*, 9, 273–314.

Southwood, T.R.E. (1961) The number of species of insect associated with various trees. *Journal of Animal Ecology*, 30, 1–8.

Southwood, T.R.E. (1977) Habitat, the templet for ecological strategies? *Journal of Animal Ecology*, 46, 337–365.

Southwood, T.R.E. (1988) Tactics, strategies and templets. *Oikos*, 52, 3–18.

Specht, D. and Voge, M. (1965) Asexual multiplication of *Mesocestoides* tetrathyridia in laboratory animals. *Journal of Parasitology*, 51, 268–272.

Stearns, S.C. (1976) Life-history tactics: a review of ideas. *Quarterly Review of Biology*, 51, 3–47.

Stearns, S.C. (1977) The evolution of life history traits: a critic of the theory and a review of the data. *Annual Review of Ecology and Systematics*, 8, 145–171.

Stock, T.M. and Holmes, J.C. (1987a) *Dioecocestus asper* (Cestoda: Dioecocestidae): an interference competitor in an enteric helminth community. *Journal of Parasitology*, 73, 1116–1123.

Stock, T.M. and Holmes, J.C. (1987b) Host specificity and exchange of intestinal helminth among four species of grebes (Podicipedidae). *Canadian Journal of Zoology*, 65, 669–676.

Stock, T.M. and Holmes, J.C. (1988) Functional relationships and microhabitat distribution of enteric helminth of grebes (Podicipedidae): the evidence for interactive communities. *Journal of Parasitology*, 74, 214–227.

Stoll, N.R. (1947) This wormy world. *Journal of Parasitology*, 33, 1–18.

Stoye, M. (1973) Untersuchungen über die Moglichkeit pranataler und galaktogener infektionen mit *Ancylostoma caninum* Ercolani, 1859 (Ancylostomidae beim Hund). *Zentralblatt Veterinaermedizin*, B, 20, 1-39.

Stunkard, H.W. (1955) Induced gametogenesis in a monogenetic trematode, *Polystoma stellai* Vigueras, 1955. *Journal of Parasitology*, 45, 389–394.

Sukhdeo, M.V.K. and Mettrick, D.F. (1986) The behaviour of juvenile *Fasciola hepatica*. *Journal of Parasitology*, 72, 492–497.

Sukhdeo, M.V.K. and Sukhdeo, S.C. (1989) Gastrointestinal hormones: environmental cues for *Fasciola hepatica*? *Parasitology*, 98, 239–243.

Suriano, D.M. and Sutton, C. (1980) Contribucion al conocimiento de la fauna parasitológica Argentina. VII. Digeneos de peces de la plataforma del mar argentino. *Revista del Museo de la Plata, Zoologiá*, 12, 261–271.

Svetovidov, A.N. (1948) *Fauna of USSR, Fishes, Gadiformes*. Published for the National Science Foundation, Washington, D.C., by the Israel Program of Scientific Translation, Jerusalem, 1962.

Szidat, L. (1955) La fauna the parásitos de *Merluccius hubbsi* como caracter auxiliar para la solución de problemas sistemáticos y zoogeográficos del genero *Merluccius*. *Comunicaciones del Instituto Nacional de Investigación en Ciencias Naturales*, Buenos Aires, 3, 1–54.

Szidat, L. (1961) Versuch einer Zoogeographie des SudAtlantik mit Hilfe von Leitparasiten der Meeresfische. *Parasitologische Schriftenreihe*, 13, 1–98.

Tallman, E.J., Corkum, K.C. and Tallman, D.A. (1985) The trematode fauna of two intercontinental migrants: *Tringa solitaria* and *Calidris melanotos* (Aves: Charadriiformes). *American Midland Naturalist*, 113, 374–383.

Tanner, J.T. (1987) *Guide to the Study of Animal Populations*, University of Tennessee Press, Knoxville.

Taylor, A.E.R. and Müller, R. (eds) (1976) *Genetic Aspects of Host–Parasite Relationships*. Symposia of the British Society of Parasitology, Vol. 14, Blackwell Scientific Publications, Oxford.

Taylor, L.R. and Taylor, R.A.J. (1977) Aggression, migration and population dynamics. *Nature*, London, 265, 415–421.

Terzakis, J.A. (1972) Virus-like particles and sporozoite budding. *Proceedings of the Helminthological Society of Washington* (special issue, Basic Research in Malaria), 39, 129–137.

Thomas, L.J. (1937) Environmental relation and life history of the tapeworm *Bothriocephalus rarus* Thomas. *Journal of Parasitology*, 23, 133–152.

Thompson, J.N. (1986a) Constraints on arms races in coevolution. *Trends in Ecology and Systematics*, 1, 105–107.

Thompson, J.N. (1986b) Patterns in coevolution. In *Coevolution and Systematics* (eds A.R. Stone and D.L. Hawksworth), The Systematics Association, Clarendon Press, Oxford, pp. 119–143.

Thompson, J.N. (1987) Symbiont-induced speciation. *Biological Journal of the Linnean Society*, 32, 385–393.

Threlfall, W. (1968) Studies on the helminth parasites of the American herring gull (*Larus argentatus* Pont.) in Newfoundland. *Canadian Journal of Zoology*, 46, 1119–1126.

Tielens, A.G.M., Van den Huevel, J.M. and Van den Bergh, S.G. (1984) The energy metabolism of *Fasciola hepatica* during its development in the final host. *Molecular and Biochemical Parasitology*, 13, 301–307.

Tinsley, R.C. and Owen, R.W. (1975) Studies on the biology of *Protopolystoma xenopodis* (Monogenoidea): the oncomiracidium and life cycle. *Parasitology*, 71, 445–463.

Turner, M.J. and Donelson, J.E. (1990) Cell biology of African trypanosomes. In *Modern Parasite Biology: Cellular, Immunological and Molecular Aspects* (ed. D.J. Wyler), W.H. Freeman and Company, New York, pp. 51–63.

Van Dobben, W.H. (1952) The food of the cormorant in the Netherlands. *Ardea*, 40: 1–63.

Van Valen, L. (1973) A new evolutionary law. *Evolutionary Theory*, 1, 1–30.

Villalba, C. and Fernández, J. (1986) Dos nuevas species de tremátodos parásitos de peces marinos en Chile. *Parasitologia al Dia*, 10, 45–51.

Voge, M. (1969) Systematics of cestodes – present and future. In *Problems in Systematics of Parasites* (ed. G. Schmidt), University Park Press, Baltimore, Maryland, pp. 49–73.

Wakelin, D. (1984) *Immunity to Parasites. How Animals Control Parasite Infections*, Edward Arnold, London.

Wakelin, D. (1987) The role of the immune response in helminth population regulation. *International Journal for Parasitology*, 17, 291–297.

Wallace, B.M. and Pence, D. (1986) Population dynamics of the helminth community from migrating blue-winged teal: loss of helminth without replacement on the wintering grounds. *Canadian Journal of Zoology*, 64, 1765–1773.

Wallis, G.P. and Miller, B.R. (1983) Electrophoretic analysis of the ticks *Ornithodoros* (*Pavlocskyella*) *erraticus* and *O.* (P.) *sonrai* (Acari: Argasidae). *Journal of Medical Entomology*, **20**, 570–571.

Ward, P.I. (1988) Sexual dichromatism and parasitism in British and Irish freshwater fish. *Animal Behavior*, **36**, 1210–1215.

Ward, P.I. (1989) Sexual showiness and parasitism in freshwater fish: combined data from several isolated water systems. *Oikos*, **55**, 428–429.

Waring, R.B., Brown, T.A., Ray, J.A., Scazzocchio, C. and Davies, R.W. (1984) Three variant introns of the same general class in the mitochondrial gene for cytochrome oxidase subunit-1 in *Aspergillus nidiulans*. *EMBO Journal*, **3**, 2121–2128.

Warner, R.E. (1968) The role of introduced diseases in the extinction of the Hawaiian endemic avifauna. *Condor*, **70**, 101–120.

Wassom, D.L., DeWitt, C.W. and Grundmann, A.W. (1974) Immunity to *Hymenolepis citelli* by *Peromyscus maniculatus*: genetic control and ecological implications. *Journal of Parasitology*, **60**, 47–52.

Wassom, D.L., Dick, T.A., Arnason, N., Strickland, D. and Grundmann, A.W. (1986) Host genetics: a key factor in regulating the distribution of parasites in natural host populations. *Journal of Parasitology*, **72**, 334–337.

Wassom, D.L., Guss, V.M. and Grundmann, A.W. (1973) Host resistance in a natural host–parasite system. Resistance to *Hymenolepis citelli* by *Peromyscus maniculatus*. *Journal of Parasitology*, **59**, 117–121.

Weekes, P.J. (1990) The nature and structure of an hyperapolytic cestode population. *Bulletin de la Société Française de Parasitologie*, **8**, 700.

Weinmann, C.J. (1966) Immunity mechanisms in cestode infections. In *Biology of Parasites* (ed. E.J.L. Soulsby), Academic Press, New York, pp. 301–320.

Whitfield, P.J. (ed) (1983) The reproductive biology of parasites. *Symposia of the British Society for Parasitology*, Vol. 20, Cambridge University Press, Cambridge.

Whitfield, P.J. and Evans, N.A. (1983) Parthenogenesis and asexual multiplication among parasitic platyhelminths. *Symposia of the British Society for Parasitology*, **20**, 121–160.

Wilbur, H.M. (1980) Complex life cycles. *Annual Review of Ecology and Systematics*, **11**, 67–93.

Wilbur, H.M., Tinkle, D.W. and Collins, J.P. (1974) Environmental certainty, trophic level, and resource availability in life history evolution. *American Naturalist*, **108**, 805–817.

Williams, G.C. (1975) *Sex and Evolution*, Princeton University Press, Princeton, New Jersey.

Williams, H.H., Colin, J.A. and Halvorsen, O. (1987) Biology of gyrocotylideans with emphasis on reproduction, population ecology and phylogeny. *Parasitology*, **95**, 173–207.

Williams, H.H. and McVicar, A. (1968) Sperm transfer in Tetraphyllidea (Platyhelminthes: Cestoda). *Nytt Magasin Zoologi*, **16**, 61–71.

Williams, J.A. and Esch, G.W. (1991) Infra- and component community dynamics in the pulmonate snail, *Helisoma anceps*, with special emphasis on the hemiurid trematode, *Halipegus occidualis*. *Journal of Parasitology*, 77, 246–253.

Williamson, M. (1988) Relationship of species number to area, distance and other variables. In *Analytical Biogeography* (eds A.A. Myers and P.S. Giller), Chapman and Hall, New York, pp. 91–115.

Wilson, D.S. (1980) *Natural Selection of Populations and Communities*, Benjamin Cummings, Menlo Park, California.

Wilson, E.O. (1969) The species equilibrium. *Brookhaven Symposium of Biology*, 22, 38–47.

Wisniewski, W.L. (1955) Zagadnienia biocenologiczne w parazytologyi wiad. *Parazytologiya*, 1, 7–41.

Wisniewski, W.L. (1958) Characterization of the parasitofauna of an eutrophic lake. *Acta Parasitologica Polonica*, 6, 1–64.

Wooten, R. (1973) The metazoan parasite fauna of fish from Hanningfield Reservoir, Essex in relation to features of the habitat and host populations. *Journal of Zoology*, London, 171, 323–331.

Wright, C.A. (1971) Review of 'Genetics of a molluscan vector of schistosomiasis' by C.S. Richards. *Tropical Disease Bulletin*, 68, 333–335.

Yamaguti, S. (1934) Studies of the helminth fauna of Japan. Part 4. Cestodes of fish. *Japanese Journal of Zoology*, 6, 1–112.

Zar, J.H. (1984) *Biostatistical Analysis*, 2nd ed., Prentice Hall, Inc., Englewood Cliffs, New Jersey.

Taxonomic host index

Engraulis 221
Eptesicus fuscus (=big brown bat) 164
Esox lucius (=Northern pike) 82

Gadiformes 205
Gadus morhua (=cod) 225, 227
Gambusia affinis (=mosquitofish) 59, 61–4, 87
Gammarus lacustris 173
Gammarus pseudolimneatus 36
Gammarus pulex 94–5
Gasterosteus aculeatus (=three-spined stickleback) 18, 153
Geomydoecus setzeri (=pocket gopher) 253
Glossina morsitans 10
Gobiomorphus cotidianus (=bullie) 124
Goniobasis 138

Helisoma anceps 6, 11, 31, 36, 131, 165, 167–9
Helisoma trivolvis 99–101
Hexagenia limbata (=mayfly) 50
Hyalella azteca (=amphipod) 173
Hydrolagus colliei 44
Hyla 191
Hyla septemtrionalis (=tree frog) 40

Ictalurus punctatus (=channel catfish) 61

Larus argentatus (=herring gull) 67
Lepomis macrochirus (=blnegill) 8, 9, 39, 100–5
Leptophleba marginata (=mayfly) 89
Lepus californicus (=jackrabbit) 125
Leuciscus idus (=orfe) 57
Leuciscus leuciscus (=dace) 56–7
Littorina littorea 131
Lota lota (=burbot) 82
Lymnaea rubiginosa 166

Megaceryle alcyon (=kingfisher) 99
Meriones unguiculatus (=Mongolian gerbil) 126
Merlucciidae 285
Merluccius (=hake) 192, 205, 227–30
Merluccius albidus 204

Merluccius angustimanus 204
Merluccius australis 204, 228–30
Merluccius bilinearis 204, 228
Merluccius capensis 204, 228
Merluccius gayi 204, 228–30
Merluccius hubbsi 204, 228–30
Merluccius merluccius 204, 228
Merluccius paradoxus 204
Merluccius polli 204, 228
Merluccius productus 204, 228
Merluccius senegalensis 204, 228
Mesocricetus auratus (=golden hamster) 125
Micromesistius merlangus 227
Micropterus dolomieiu (=smallmouth bass) 38
Micropterus salmoides (=largemouth bass) 38, 61, 100
Mugil cephalus (=striped mullet) 195, 205, 210–12, 223, 284
Mugilidae (=mullets) 36, 205, 210
Mus musculus (=house mouse) 124

Natica 202–3
Notophthalmus viridescens (=red-spotted newt) 85, 158, 262
Notropis lutrensis (=red shiner) 61, 64, 87
Numenius americanus (=long-billed curlew) 161, 173

Ochotona (=pika) 163
Odocoileus virginianus (=white-tailed deer) 178, 238, 260
Oncorhynchus 195, 226
Oncorhynchus kisutch (=coho salmon) 82
Oncorhynchus mykiss (=steelhead trout) 224
Oncorhynchus nerka (=sockeye salmon) 65, 193, 224
Orthogeomys bottae (=pocket gopher) 253
Orthogeomys cherriei (=pocket gopher) 253
Oryctolagus cuniculus (=rabbit) 70
Osmerus mordax (=smelt) 67

Taxonomic parasite index

Subject index